Problems on Algorithms

Habib Izadkhah

Problems on Algorithms

A Comprehensive Exercise Book for Students
in Software Engineering

 Springer

Habib Izadkhah
Faculty of Mathematics, Statistics,
and Computer Science
University of Tabriz
Tabriz, Iran

ISBN 978-3-031-17045-4   ISBN 978-3-031-17043-0 (eBook)
https://doi.org/10.1007/978-3-031-17043-0

This Springer imprint is published by the registered company Springer Nature Switzerland AG
The registered company address is: Gewerbestrasse 11, 6330 Cham, Switzerland

Preface

Learning is a mysterious process. No one can say what the precise rules of learning are. However, it is an agreed upon fact that the study of good examples plays a fundamental role in learning. In the short span of a semester, it is difficult to cover enough material to give students the confidence that they have mastered some portion of the subject. Consequently, it is well known that problem-solving helps one acquire routine skills in designing and analyzing of algorithms. The salient features of the theory are presented in class along with a few examples, and then the students are expected to teach themselves the finer aspects of the theory through worked examples. The book aims at presenting a lot of problems, aiming to improve the learning process of students.

With approximately 2500 problems, this supplement provides a collection of practical problems on the basic and advanced data structures, design, and analysis of algorithms. To make this book suitable for self-instruction, about one-third of the algorithms are supported by solutions, and some other are supported by hints and comments. This book is intended for students wishing to deepen their knowledge of algorithm design in an undergraduate or beginning graduate class on algorithms, for those teaching courses in this area, for use by practicing programmers who wish to hone and expand their skills, and as a self-study text for graduate students who are preparing for the qualifying examination on algorithms for a Ph.D. program in Computer Science or Computer Engineering. About all, it's a good source for exam problems for those who teach algorithms and data structure. The format of each chapter is just a little bit of instruction followed by lots of problems.

This book is intended to augment the problem sets found in any standard algorithm textbook. In this book, three levels of difficulty, simple or relatively simple, moderate level or slightly difficult. This book, also, emphasizes the creative aspects of algorithm design.

This book

- begins with four chapters on background material that most algorithm instructors would like their students to have mastered before setting foot in an algorithm class.

The introductory chapters include mathematical induction, complexity notations, recurrence relations, and basic algorithm analysis methods.

- provides many problems on basic and advanced data structures including basic data structures (arrays, stack, queue, and linked list), hash, tree, search, and sorting algorithms.
- provides many problems on algorithm design technique: divide and conquer, dynamic programming, greedy algorithms, graph algorithms, and backtracking algorithms.
- is rounded out with chapter on NP-completeness.

Tabriz, Iran   Habib Izadkhah

Contents

Chapter 1
Mathematical Induction

Abstract Mathematical induction is a technique for proving results or establishing statements for natural numbers. This chapter illustrates the method through a variety of examples and provides 50 exercises on mathematical induction. To this end, this chapter provides exercises on summations, inequalities, floors and ceilings, divisibility, postage stamps, Fibonacci numbers, binomial coefficients, and several other problems.

1.1 Lecture Notes

The principle of mathematical induction is used to prove that a given statement (formula, equality, inequality, and more) is true for all positive integer numbers greater than or equal to some integer N.

Let P(n) denotes the statement, where n is a positive integer. Proof consists of two steps:

Step 1: We first prove that the statement P(n) is true for the smallest possible value of the positive integer n.

Step 2: We assume that P(k) is true and prove that P(k+1) is also true.

Example 1 Using mathematical induction, prove that

$$1 + 2 + 3 + \cdots + n = n(n+1)/2$$

is true for all positive integers n.

Solution:
Let the statement P(n) be

$$1 + 2 + 3 + \cdots + n = n(n+1)/2$$

STEP 1: We first show that P(1) is true.
Left Side = 1
Right Side = 1(1 + 1) / 2 = 1
Both sides of the statement are equal hence P(1) is true.

STEP 2: We now assume that P(k) is true

$$1 + 2 + 3 + \cdots + k = k(k + 1)/2$$

and show that P(k + 1) is true by adding k + 1 to both sides of the above statement

$$1 + 2 + 3 + \cdots + k + (k + 1) = k(k + 1)/2 + (k + 1)$$

$$= (k + 1)(k/2 + 1)$$

$$= (k + 1)(k + 2)/2$$

The last statement may be written as

$$1 + 2 + 3 + \cdots + k + (k + 1) = (k + 1)(k + 2)/2$$

which is the statement P(k + 1).

Example 2 Using mathematical induction, prove that

$$1^2 + 2^2 + 3^2 + \cdots + n^2 = n(n + 1)(2n + 1)/6$$

is true for all positive integers n.

Solution:
Let the statement P(n) be

$$1^2 + 2^2 + 3^2 + \cdots + n^2 = n(n + 1)(2n + 1)/6$$

STEP 1: We first show that P(1) is true. Left Side = 1^2 = 1
Right Side = 1(1 + 1) (2×1 + 1)/ 6 = 1
Both sides of the statement are equal hence P(1) is true.

STEP 2: We now assume that P(k) is true

$$1^2 + 2^2 + 3^2 + \cdots + k^2 = k(k + 1)(2k + 1)/6$$

and show that P(k + 1) is true by adding $(k + 1)^2$ to both sides of the above statement

$$1^2 + 2^2 + 3^2 + \cdots + k^2 + (k + 1)^2 = k(k + 1)(2k + 1)/6 + (k + 1)^2$$

Set common denominator and factor k + 1 on the right side

$$= (k + 1)[k(2k + 1) + 6(k + 1)]/6$$

Expand k(2k + 1) + 6(k + 1)

$$= (k + 1)[2k^2 + 7k + 6]/6$$

Now factor $2k^2 + 7k + 6$.

$$= (k + 1)[(k + 2)(2k + 3)]/6$$

We have started from the statement P(k) and have shown that

$$1^2 + 2^2 + 3^2 + \cdots + k^2 + (k + 1)^2 = (k + 1)[(k + 2)(2k + 3)]/6$$

which is the statement P(k + 1).

Example 3 Using mathematical induction, prove that for any positive integer number n

$$n^3 + 2n \text{ is divisible by } 3$$

Solution:
Let the statement P(n) be

$$n^3 + 2n \text{ is divisible by } 3$$

STEP 1: We first show that P(1) is true. Let n = 1 and calculate $n^3 + 2n$

$$1^3 + 2(1) = 3$$

3 is divisible by 3. Hence P(1) is true.

STEP 2: We now assume that P(k) is true $k^3 + 2k$ is divisible by 3 is equivalent to $k^3 + 2k = 3M$, where M is a positive integer.
We now consider the algebraic expression $(k + 1)^3 + 2(k + 1)$; expand it and group like terms

$$(k + 1)^3 + 2(k + 1) = k^3 + 3k^2 + 5k + 3$$

$$= [k^3 + 2k] + [3k^2 + 3k + 3]$$

$$= 3M + 3[k^2 + k + 1] = 3[M + k^2 + k + 1]$$

Hence $(k + 1)^3 + 2(k + 1)$ is also divisible by 3 and therefore statement P(k + 1) is true.

Example 4 Show that $n! > 3^n$ for $n \geq 7$.

Solution:

For any $n \geq 7$, let P(n) be the statement that $n! > 3^n$.

Base Case. The statement P(7) says that $7! = 7 \times 6 \times 5 \times 4 \times 3 \times 2 \times 1 = 5040 >$ $3^7 = 2187$, which is true.

Inductive Step. Fix $k \geq 7$, and suppose that P(k) holds, that is, $k! > 3^k$.

It remains to show that P(k+1) holds, that is, that $(k+1)! > 3^{(k+1)}$.

$$(k+1)! = (k+1)k! > (k+1)3^k \geq (7+1)3^k = 8 \times 3^k > 3 \times 3^k = 3^{(k+1)}.$$

Therefore P(k+1) holds.

Thus by the principle of mathematical induction, for all $n \geq 1$, P(n) holds.

Example 5 Show that $3^n > n^2$.

Solution:

STEP 1: We first show that P(1) is true. Let n = 1 and calculate 3^1 and 1^2 and compare them
$3^1 = 3$ and $1^2 = 1$, so 3 is greater than 1 and hence P(1) is true.
Let us also show that P(2) is true.
$3^2 = 9$ and $2^2 = 4$, hence P(2) is also true.

STEP 2: We now assume that P(k) is true

$$3^k > k^2$$

Multiply both sides of the above inequality by 3

$$3 \times 3^k > 3 \times k^2$$

The left side is equal to 3^{k+1}. For $k > 2$, we can write

$$k^2 > 2k \ \ and \ \ k^2 > 1$$

We now combine the above inequalities by adding the left hand sides and the right hand sides of the two inequalities

$$2k^2 > 2k + 1$$

We now add k^2 to both sides of the above inequality to obtain the inequality

$$3k^2 > k^2 + 2k + 1$$

Factor the right side we can write

$$3 \times k^2 > (k + 1)^2$$

If $3 \times 3^k > 3 \times k^2$ and $3 \times k^2 > (k + 1)^2$ then

$$3 \times 3^k > (k + 1)^2$$

Rewrite the left side as 3^{k+1}

$$3^{k+1} > (k + 1)^2$$

which proves that P(k + 1) is true.

1.2 Exercises

1.2.1 Summations

Prove by induction on $n \geq 0$ that

1. $1 + 3 + 5 + \cdots + (2n - 1) = n^2$.

2. $\sum_{i=1}^{n} i = n(n + 1)/2$.

3. $\sum_{i=1}^{n} i^2 = n(n + 1)(2n + 1)/6$.

4. $\sum_{i=1}^{n} i^3 = n^2(n + 1)^2/4$.

5. $\sum_{i=1}^{n} i(i + 1) = n(n + 1)(n + 2)/3$.

6. $\sum_{i=1}^{n} i(i + 1)(i + 2) = n(n + 1)(n + 2)(n + 3)/4$.

7. $\sum_{i=1}^{n} i.i! = (n + 1)! - 1$.

8. $\sum_{i=1}^{n} 1/2^i = 1 - 1/2^n$.

9. $\sum_{i=1}^{n} i2^i = (n - 1)2^{n+1} + 2$.

10. Prove by induction on $n \geq 1$ that $\sum_{i=1}^{n} \frac{1}{i(i+1)} = \frac{n}{n+1}$.

11. Prove by induction on $n \geq 1$ that for every $a \neq 1$

$$\sum_{i=0}^{n} a^i = \frac{a^{n+1} - 1}{a - 1}$$

1.2.2   Inequalities

12. Prove by induction on $n \geq 3$ that if $2n + 1 \leq 2^n$.

13. Prove by induction on $n \geq 1$ that if $x > -1$, $(1 + x)^n \geq 1 + nx$.

14. Prove by induction on $n \geq 7$ that $3^n < n!$.

15. Prove by induction on $n \geq 5$ that $2^n > n^2$.

16. Prove by induction on $k \geq 1$ that $\sum_{i=1}^{n} i^k \leq n^k(n + 1)/2$.

17. Prove by induction on $n \geq 0$ that

$$\sum_{i=1}^{n} \frac{1}{i^2} > 1 - \frac{1}{n}$$

1.2.3   Floors and Ceilings

Suppose $x \in \mathbb{R}^+$. The floor of x, denoted $\lfloor x \rfloor$, returns the largest integer less than or equal to x. The ceiling of x, denoted $\lceil x \rceil$, returns the smallest integer greater than or equal to x.

18. Prove by induction on $n \geq 0$ that

$$\left\lfloor \frac{n}{2} \right\rfloor = \begin{cases} n/2 & \text{if } n \text{ is even} \\ (n - 1)/2 & \text{if } n \text{ is odd} \end{cases}$$

19. Prove by induction on $n \geq 0$ that

$$\left\lceil \frac{n}{2} \right\rceil = \begin{cases} n/2 & \text{if } n \text{ is even} \\ (n + 1)/2 & \text{if } n \text{ is odd} \end{cases}$$

20. Prove by induction on $n \geq 1$ that for all $m \in \mathbb{N}$

$$\lceil \frac{n}{m} \rceil = \lfloor \frac{n+m-1}{m} \rfloor.$$

1.2.4 Divisibility

21. Prove by induction on $n \geq 1$ that $7^n - 1$ is divisible by 6.

22. Prove by induction that $n^3 + 2n$ is divisible by 3 for every non-negative integer n.

23. Prove by induction that $n^5 - n$ is divisible by 5 for every non-negative integer n.

24. Prove by induction that $5^{n+1} + 2 \times 3^n + 1$ is divisible by 8 for every non-negative integer n.

25. Prove that an integer number is divisible by 3 if and only if the sum of its digits is divisible by 3.

26. Prove that an integer number is divisible by 9 if and only if the sum of its digits is divisible by 9.

27. Prove that the sum of the cubes of three successive natural numbers is divisible by 9.

1.2.5 Postage Stamps

28. Prove that any integer postage of 7 cents or more can be formed using only 3-cent and 5-cent stamps.

29. Prove that any integer postage of 34 cents or more can be formed using only 5-cent and 9-cent stamps.

30. Prove that any integer postage of 5 cents or more can be formed using only 2-cent and 7-cent stamps.

31. Prove that any integer postage of 59 cents or more can be formed using only 7-cent and 11-cent stamps.

32. Find the smallest value of k such that any integer postage greater than k cents can be formed by using only 4-cent and 9-cent stamps.

33. Find the smallest value of k such that any integer postage greater than k cents can be formed by using only 6-cent and 11-cent stamps.

1.2.6   Fibonacci Numbers

The Fibonacci sequence F_n for $n \geq 0$ is defined recursively as follows:

$$F_0 = 0, \ F_1 = 1, \text{and for } n \geq 2, \ F_n = F_{n-1} + F_{n-2}.$$

34. Prove by induction on n that $\sum_{i=0}^{n} F_i = F_{n+2} - 1$.

35. Prove by induction that $F_{n+k} = F_k F_{n+1} + F_{k-1} F_n$.

36. Prove by induction on $n \geq 1$ that $\sum_{i=1}^{n} F_i^2 = F_n F_{n+1}$.

37. Prove by induction on $n \geq 1$ that $F_{n+1} F_{n+2} = F_n F_{n+3} + (-1)^n$.

38. Prove by induction on $n \geq 2$ that $F_{n-1} F_{n+1} = F_n^2 + (-1)^n$.

1.2.7   Binomial Coefficients

The binomial coefficient

$$\binom{n}{r},$$

For $n, r \in N, r \leq n$ is defined as the number of possible ways to choose r objects from n without replacement, that is,

$$\binom{n}{r} = \frac{n!}{r!(n-r)!}$$

39. Prove by induction on n that

$$\sum_{m=0}^{n} \binom{n}{m} = 2^n$$

40. Prove by induction on $n \geq 1$ that for all $1 \leq m \leq n$,

$$\binom{n}{m} \leq n^m$$

41. Prove by induction that for all $n \in N$

$$\sum_{i=0}^{n} i \cdot \binom{n}{i} = n \cdot 2^{n-1}$$

1.2.8 Miscellaneous

42. Using induction, show that n straight lines separate the plane into $(n^2 + n + 2)/2$ regions. Suppose no two lines are parallel and no three lines have a common point.

43. Show that n straight lines in the plane, all passing through a single point, separate the plane into $2n$ regions.

44. If a tree has n vertices, it has n-1 edges. Prove this by induction.

45. A complete binary tree with n levels has $2^n - 1$ vertices. Prove this by induction.

46. Prove by induction that if there exists an n-node tree in which all the nonleaf nodes have k children, then $n \bmod k = 1$.

47. Prove by induction that for every n such that $n \bmod k = 1$, there exists an n-node tree in which all of the nonleaf nodes have k children.

48. Between any pair of vertices in a tree there is a unique path. Prove this by induction.

49. Prove that any set of regions defined by n lines in the plane can be colored with two colors so that no two regions that have a common edge have the same color.

50. One of the interesting problems of algorithm design is the problem of tiling or to tile the given floor/board (Fig. 1.1) using $1 \times m$ tiles.
   The purpose is tiling this plot of land using tiles with the following shapes (L shaped tile, where a L shaped tile is a 2×2 square with one cell of size 1×1 missing.), as shown in Fig. 1.2.
   In such a way that one of the checkered earth houses is not covered. It can be assumed that this house will be used to build a garden or a small pond. Note that the size of the sides of the small squares of the tiles is the same as the checkered plate

Fig. 1.1 A sample of
floor/board

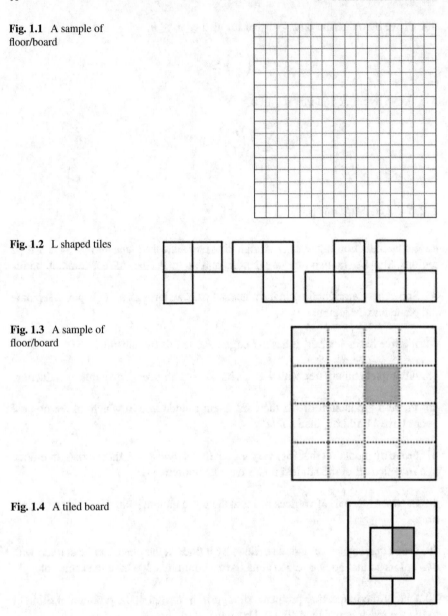

Fig. 1.2 L shaped tiles

Fig. 1.3 A sample of
floor/board

Fig. 1.4 A tiled board

above one meter. We also do not have the right to break these tiles into smaller pieces.
For example, suppose in Fig. 1.3 we want the specified house not to be covered:

This floor can be tiled as Fig. 1.4.

Given a n by n board where n is of form 2^k where $k \geq 1$ (basically n is a power of
2 with minimum value as 2). The board has one missing cell (of size 1×1). Prove
by induction that this board can be tiled by the tiles shown in Fig. 1.2.

1.3   Solutions

1.

STEP 1: We first show that P(1) is true.
Left Side = $2 \times 1 - 1 = 1$
Right Side = $1^2 = 1$
Both sides of the statement are equal hence P(1) is true.
STEP 2: We now assume that P(k) is true

$$1 + 3 + 5 + \cdots + (2k - 1) = k^2$$

and show that P(k + 1) is true by adding $(2(k + 1) - 1)$ to both sides of the above statement

$$1 + 3 + 5 + \cdots + (2k - 1) + (2(k + 1) - 1) = k^2 + (2(k + 1) - 1)$$

$$= k^2 + (2k + 2 - 1)$$

$$= k^2 + 2k + 1$$

$$= (k + 1)^2$$

The last statement may be written as

$$1 + 3 + 5 + \cdots + (2k - 1) + (2(k + 1) - 1) = (k + 1)^2$$

Which is the statement P(k + 1).

2.

STEP 1: We first show that P(1) is true.

$$P(1) = 1 = \frac{1 \times (1+1)}{2}$$

Both sides of the statement are equal hence P(1) is true.

STEP 2: We now assume that P(k) is true

$$P(k) = 1 + 2 + \cdots + k = \frac{k \times (k + 1)}{2}$$

and show that P(k + 1) is true by adding k + 1 to both sides of the above statement

$$P(k+1) = 1 + 2 + \cdots + k + (k+1) = p(k) + (k+1) =$$

$$\frac{k \times (k+1)}{2} + k + 1 = \frac{k^2 + k}{2} + \frac{2k + 2}{2} = \frac{k^2 + 3k + 2}{2} = \frac{(k+1) \times (k+2)}{2}$$

Since $P(k) \Rightarrow P(k+1)$, then $P(n)$ is true for all positive numbers.

3.

STEP 1: We first show that P(1) is true.

$$\frac{n(n+1)(2n+1)}{6} = \frac{1(2)(3)}{6} = 1 = 1^2 = \sum_{i=1}^{1} i^2$$

Both sides of the statement are equal hence P(1) is true.
STEP 2: We now assume that P(k) is true.

$$\sum_{i=1}^{k} i^2 = \frac{k(k+1)(2k+1)}{6}$$

and show that P(k + 1) is true by adding k + 1 to both sides of the above statement

$$\frac{(k+1)((k+1)+1)(2(k+1)+1)}{6} = \frac{(k+1)(k+2)(2k+3)}{6}$$

$$\sum_{i=1}^{n} i^2 + (k+1)^2 = \frac{k(k+1)(2k+1)}{6} + (k+1)^2$$

$$\sum_{i=1}^{k+1} i^2 = \frac{k(k+1)(2k+1)}{6} + \frac{6(k+1)^2}{6}$$

$$= \frac{k(k+1)(2k+1) + 6(k+1)^2}{6}$$

$$= \frac{k(k+1)(2k+1) + 6(k+1)(k+1)}{6}$$

$$= \frac{(k+1)[k(2k+1) + 6(k+1)]}{6}$$

$$= \frac{(k+1)[2k^2 + k + 6k + 6]}{6}$$

$$= \frac{(k+1)\,[(k+2)\,(2k+3)]}{6}$$

$$= \frac{(k+1)\,((k+1)+1)\,(2\,(k+1)+1)}{6} = \sum_{i=1}^{k+1} i^2$$

So P(k + 1) is true.

4.

STEP 1: We first show that P(1) is true.

$$1^3 = 1^2(1+1)^2/4$$

$$\Longrightarrow 1 = 1$$

Both sides of the statement are equal hence P(1) is true.
STEP 2: We now assume that P(k) is true

$$1^3 + 2^3 + 3^3 + \cdots + k^3 = k^2(k+1)^2/4$$

and show that P(k + 1) is true:

$$1^3 + 2^3 + 3^3 + \cdots + k^3 + (k+1)^3 = (k+1)^2(k+2)^2/4$$

We place the expression $1^3 + 2^3 + 3^3 + \cdots + k^3 = k^2(k+1)^2/4$ in the statement above.

$$k^2(k+1)^2/4 + (k+1)^3 = (k+1)^2(k+2)^2/4$$

$$\Longrightarrow k^2(k+1)^2 + 4\,(k+1)^3 = (k+1)^2(k+2)^2$$

Now we take a factor of $(k+1)^2$ and delete it from both sides.

$$k^2 + 4\,(k+1)^1 = (k+2)^2$$

$$\Longrightarrow k^2 + 4k + 4 = k^2 + 4k + 4$$

As a result, the sides to the relationship are equal and the problem is true.

5.

STEP 1: We first show that P(1) is true.

$$\sum_{i=1}^{n} i(i+1) = n(n+1)(n+2)/3$$
$$p(1) \Rightarrow i(i+1) = (1+1)(1+2)/3$$
$$1(1+1)(1+2)/3$$
$$1(1+1) = 6/3$$
$$2 = 2$$

STEP 2: We now assume that P(k) is true.

$$P(k) \Rightarrow \sum_{i=1}^{k} i(i+1) = k(k+1)(k+2)/3$$
$$P(k+1) \Rightarrow \sum_{i=1}^{k+1} i(i+1) = (k+1)(k+2)(k+3)/3$$
$$\sum_{i=1}^{k+1} i(i+1) = \sum_{i=1}^{k} i(i+1) + (k+1)(k+2)$$
$$\Rightarrow k(k+1)\frac{(k+2)}{3} + (k+1)(k+2)$$
$$\Rightarrow (k+1)(k+2)(\frac{k}{3}+1)$$
$$\Rightarrow (k+1)(k+2)(\frac{k+3}{3}) = \frac{(k+1)(k+2)(k+3)}{3}$$

6.

$$\sum_{i=1}^{n} i\,(i+1)\,(i+2) = \frac{n\,(n+1)\,(n+2)\,(n+3)}{4}$$

STEP 1: We first show that P(1) is true.

$$1 \times 2 \times 3 + 2 \times 3 \times 4 + \cdots + n\,(n+1)\,(n+2) = \frac{n\,(n+1)\,(n+2)\,(n+3)}{4}$$

$$n = 1 \Rightarrow \frac{(1 \times 2 \times 3 \times 4)}{4} = 6 \text{ (on the right equation)} = 6 \text{ (on the left equation)}$$

Both sides of the statement are equal hence P(1) is true.
STEP 2: We now assume that P(k) is true:

$$1 \times 2 \times 3 + 2 \times 3 \times 4 + \cdots + k\,(k+1)\,(k+2) = \frac{k\,(k+1)\,(k+2)\,(k+3)}{4}$$

and show that P(k + 1) is true:

$$1 \times 2 \times 3 + 2 \times 3 \times 4 + \cdots + k\,(k+1)\,(k+2) + k + 1\,(k+2)\,(k+3) =$$

$$\frac{k\,(k+1)\,(k+2)\,(k+3)}{4} + k + 1\,(k+2)\,(k+3)$$

Finally:

$$\frac{k\,(k+1)\,(k+2)\,(k+3) + 4\,(k+1)\,(k+2)\,(k+3)}{4} = \frac{(k+1)\,(k+2)\,(k+3)\,(k+4)}{4}$$

7.

STEP 1: We first show that P(1) is true.

$$\sum_{i=1}^{n} i(i!) = (n+1)! - 1 \xrightarrow{n=1} \sum_{i=1}^{1} i(i!) = (1+1)! - 1$$

Both sides of the statement are equal, hence P(1) is true.
STEP 2: We now assume that P(k) is true

$$p(k) = \sum_{i=1}^{k} i(i!) = (k+1)! - 1$$

and show that P(k + 1) is true by adding k + 1 to both sides of the above statement

$$p(k+1) = \sum_{i=1}^{k+1} i(i!) = (k+2)! - 1$$

$$\sum_{i=1}^{k+1} i(i!) = \sum_{i=1}^{k} i(i!) + (k+1)(k+1)!$$
$$= (k+1) - 1 + (k+1)(k+1)!$$
$$= (k+2)(k+1)! - 1$$
$$= (k+2)! - 1$$

Since $P(k) \Rightarrow P(k+1)$, then $P(n)$ is true for all positive numbers.

8.

STEP 1: For P(1) is true.
STEP 2: We now assume that P(k) is true

$$\sum_{i=1}^{k} \frac{1}{2^i} = 1 - \frac{1}{2^k} \rightarrow \frac{1}{2^k} = \frac{1}{2^k}$$

and show that P(k + 1) is true by adding k + 1 to both sides of the above statement

$$\sum_{i=1}^{k+1} \frac{1}{2^i} = 1 - 1/2^{k+1}$$

$$\sum_{i=1}^{k+1} \frac{1}{2^i} = \frac{1}{2} + \frac{1}{4} + \frac{1}{8} + \cdots + 1/2^{k+1}$$

$$= \frac{1}{2} + \frac{1}{2}(\frac{1}{2} + \frac{1}{4} + \frac{1}{8} + \cdots + \frac{1}{2^k})$$

$$= \frac{1}{2} + \frac{1}{2}\sum_{i=1}^{k} 1/2^i \rightarrow -\frac{1}{2} + \frac{1}{4} \cdot \left(1 - \frac{1}{2^k}\right) = 1 - 1/2^{k+1}$$

10.

STEP 1: We first show that P(1) is true.

$$P\,(1): \ \sum_{i=1}^{1} \frac{1}{i\,(i+1)} = \frac{1}{1(1+1)} = \frac{1}{2} = \frac{n}{n+1} = \frac{1}{1+1} = \frac{1}{2}$$

Both sides of the statement are equal hence P(1) is true.
STEP 2: We now assume that P(k) is true

$$P\,(k) = \sum_{i=1}^{k} \frac{1}{i\,(i+1)} = \frac{k}{k+1}$$

and show that P(k + 1) is true by adding k + 1 to both sides of the above
statement

$$P\,(k+1): \ \sum_{i=1}^{k+1} \frac{1}{i\,(i+1)} = \frac{k+1}{k+2}$$

$$\sum_{i=1}^{k+1} \frac{1}{i\,(i+1)} = \sum_{i=1}^{k} \frac{1}{i\,(i+1)} + \frac{1}{(k+1)\,(k+2)} = \frac{k}{k+1} + \frac{1}{(k+1)\,(k+2)}$$

$$= \frac{k\,(k+2)+1}{(k+1)(k+2)} = \frac{k^2 + 2k + 1}{(k+1)(k+2)} = \frac{(k+1)^2}{(k+1)(k+2)} = \frac{k+1}{k+2}.$$

11.

We now assume that P(k) is true

$$a^0 + a^1 + a^2 + \cdots + a^k = \frac{a^{k+1} - 1}{a-1}$$

and show that P(k + 1) is true.

$$a^0 + a^1 + a^2 + \cdots + a^k + a^{k+1} = \frac{a^{(k+1)+1} - 1}{a-1}.$$

We have:

$$\underbrace{a^0 + a^1 + a^2 + \cdots + a^k}_{\frac{a^{k+1}-1}{a-1}} + a^{k+1} = \frac{a^{k+1}-1}{a-1} + a^{k+1} = \frac{a^{k+1}-1}{a-1} + \frac{a^{k+2}-a^{k+1}}{a-1}$$

$$= \frac{a^{k+1} - 1 + a^{k+2} - a^{k+1}}{a-1} = \frac{a^{k+2}-1}{a-1} = \frac{a^{(k+1)+1}-1}{a-1}$$

In this case, we conclude that this expression is also true for n = k + 1.

13.

$$(1+x)^n \geq 1 + nx \; ; \; x > -1, \; n \geq 1$$

STEP 1: We first show that P(1) is true.

$$n = 1 \rightarrow 1 + x \geq 1 + x$$

Both sides of the statement are equal hence P(1) is true.
STEP 2: We now assume that P(k) is true

$$(1+x)^k \geq 1 + kx$$

and show that P(k + 1) is true by adding k + 1 to both sides of the above statement

$$(1+x)^{k+1} \geq 1 + (k+1)x$$

$$(1+x)^k \geq 1 + kx \rightarrow (1+x)(1+x)^k \geq (1+kx)(1+x) \rightarrow$$

$$\rightarrow (1+x)^{k+1} \geq 1 + kx + x + kx^2$$

$$(1+x)^{k+1} \geq kx^2 + x(k+1) + 1 \geq (k+1)x + 1$$

14.

First we check for n = 7

$$3^7 < 7!$$

$$\Longrightarrow 2187 < 5040$$

So that's true. Now suppose $n = k$, P(k), and $k \geq 7$

$$3^k < k!$$

and show that P(k + 1) is true.

$$3^{k+1} < (k+1)!$$

$$\implies 3^k \times 3 < (k+1) \times k!$$

Since we know $3 < k + 1$ because $k \geq 7$ then:

$$\implies 3^k < k!$$

So for k + 1, it is proved, in a nutshell:

$$(k+1)! = (k+1)\,k! > (k+1)\,3^k \geq (7+1)\,3^k = 8 \times 3^k > 3 \times 3^k = 3^{k+1}$$

15.

STEP 1: We first show that P(5), i.e., $n = 5$, is true.

$$32 > 25$$

hence P(5) is true.
STEP 2: We now assume that P(k) is true

$$2^k > k^2$$

and show that P(k + 1) is true

$$2^{k+1} > (k+1)^2$$

We know that $2^{k+1} = 2^1 \times 2^k$. So, we have:

$$2 \times 2^k = 2^k + 2^k \overset{2^k > k^2}{\implies} 2^k + 2^k > k^2 + k^2 = k^2 + k \times k$$

k is at least 5 and this means:

$$2^k + 2^k > k^2 + 4k = k^2 + 2k + 2k \overset{k>4}{\longrightarrow} 2^k + 2^k > k^2 + 2k + 8 > k^2 + 2k + 1$$

We know that $(k+1)^2 = k^2 + 2k + 1$, so it is proved.

$$2^k + 2^k = 2^{k+1} > (k+1)^2 = k^2 + 2k + 1$$

16.

$$n = 1 \implies 1^2 \le 1^2(1+1)/2$$

Both sides of the statement are equal hence P(1) is true.

$$n = m \implies 1^2 + 2^2 + 3^2 + \cdots + m^2 \le m^2(m+1)/2$$

$$n = m+1 \implies \frac{m^2(m+1)}{2} + (m+1)^2 \le \frac{(m+1)^2((m+2)-2)}{2}$$

$$m^2(m+1) \le m(m^2 + 2m + 1)$$

$$m^2 + m \le m^2 + 2m + 1$$

And obviously this inequality is also true.

17.

STEP 1: We first show that P(1) is true.
Left Side:

$$\int_{i=1}^{n} \frac{1}{i^2} \xrightarrow{n=1} \int_{i=1}^{1} \frac{1}{i^2} = \frac{1}{1^2} = 1$$

Right Side:

$$1 - \frac{1}{n} = 1 - \frac{1}{1} = 0$$

As result, we have:

$$1 > 0$$

Hence P(1) is true.
STEP 2: We now assume that P(k) is true by adding k + 1 to both sides of the above statement

$$\int_{i=1}^{k} \frac{1}{i^2} > 1 - \frac{1}{k} = \frac{k-1}{k} \tag{1.1}$$

and show that P(k + 1) is true:

$$\int_{i=1}^{k+1} \frac{1}{i^2} > 1 - \frac{1}{k+1} = \frac{k}{k+1}$$

$$\sum_{i=1}^{k+1} \frac{1}{i^2} = \sum_{i=1}^{k} \frac{1}{i^2} + \frac{1}{(k+1)^2}$$

Regarding 1.1, we know that $\sum_{i=1}^{k} \frac{1}{i^2} > \frac{k-1}{k}$. So given that the expression $\frac{1}{(k+1)^2}$ is a positive value then we have:

$$\sum_{i=1}^{k+1} \frac{1}{i^2} > \frac{k-1}{k}$$

Then, we compare $\frac{k-1}{k}$ and $\frac{k}{k+1}$

$$\frac{k}{k+1} > \frac{k-1}{k}$$

According to the above relations we conclude that the statement

$$\int_{i=1}^{k+1} \frac{1}{i^2} > \frac{k}{k+1}$$

is true.

18.

STEP 1: We must first show the correctness of P(1) as an odd number and P(2) as an even number:
Left Side:

$$P(1) \rightarrow \left\lfloor \frac{1}{2} \right\rfloor = 0$$

Right Side:

$$P(1) \rightarrow \frac{1-1}{2} = 0$$

Left Side:

$$P(2) \rightarrow \left\lfloor \frac{2}{2} \right\rfloor = 1$$

Right Side:

$$P(2) \rightarrow \frac{2}{2} = 1$$

Both P(1) and P(2) are true.
STEP 2: We now assume that P(k) is true

$$\left\lfloor \frac{k}{2} \right\rfloor = \begin{cases} \frac{k}{2}, & \text{if } k \text{ is even} \\ \frac{k-1}{2}, & \text{if } k \text{ is odd} \end{cases}$$

and show that P(k + 1) is true by adding k + 1 to both sides of the above statement

$$\left\lfloor \frac{k+1}{2} \right\rfloor = \begin{cases} \frac{k+1}{2}, & if\ k+1\ is\ even\ k+1 = 2M\ ,\ M \in N \\ \frac{k}{2}, & if\ k+1\ is\ odd\ k+1 = 2M - 1\ ,\ M \in N \end{cases}$$

If $k+1$ is even:

$$K + 1 = 2M\ ,\ M = \frac{k+1}{2} \quad \left\lfloor \frac{2M}{2} \right\rfloor = \lfloor M \rfloor = M = \frac{k+1}{2}$$

If $k+1$ is odd:

$$K + 1 = 2M - 1\ ,\ M = \frac{k}{2} \quad \left\lfloor \frac{2M-1}{2} \right\rfloor = \left\lfloor M - \frac{1}{2} \right\rfloor = M - \left\lfloor \frac{1}{2} \right\rfloor = M = \frac{k}{2}$$

Therefore, the statement P(k+1) is true.

20.

STEP 1: We first show that P(1) is true.
For $n = 1$ and $m = 2$, we have:

$$\left\lceil \frac{1}{2} \right\rceil = \left\lfloor \frac{1+2-1}{2} \right\rfloor \implies 1 = 1$$

or for $n = 1$, we can write:

$$\left\lceil \frac{1}{m} \right\rceil = \left\lfloor \frac{1+m-1}{m} \right\rfloor \implies \left\lceil \frac{1}{m} \right\rceil = 1$$

hence P(1) is true.
STEP 2: We now assume that P(k) is true

$$\left\lceil \frac{k}{m} \right\rceil = \left\lfloor \frac{k+m-1}{m} \right\rfloor$$

and show that P(k + 1) is true.

$$\left\lceil \frac{k+1}{m} \right\rceil = \left\lfloor \frac{k+1+m-1}{m} \right\rfloor \implies \left\lceil \frac{k}{m} + \frac{1}{m} \right\rceil = \left\lfloor \frac{k}{m} + 1 \right\rfloor \implies \left\lceil \frac{k}{m} + \frac{1}{m} \right\rceil = \left\lfloor \frac{k}{m} \right\rfloor + 1$$

Because $m \in N$, $1 = \left\lceil \frac{1}{m} \right\rceil$, so we have

$$\implies \left\lceil \frac{k}{m} \right\rceil + 1 = \left\lfloor \frac{k}{m} \right\rfloor + 1$$

So for k + 1, it is proved.

21.

STEP 1: We first show that P(1) is true.

$$P(1) = 7^1 - 1 = 7 - 1 = 6$$

6 is divisible by 6, so P(1) is true.
STEP 2: Assume that P(k) is true, i.e., $7^k - 1$ is divisible by 6. We have

$$7^k - 1 = 6M , \ MZ^+$$

Now we need to show the correctness of P(k + 1), i.e., $7^{k+1} - 1$ is divisible by 6:

$$7^{k+1} - 1 = \ 7^k 7 - 1$$

$$= \ 7^k(6+1) - 1$$

$$= 67^k + 7^k - 1$$

$$\xrightarrow{7^k - 1 = 6M} = 67^k + 6M$$

$$= 6(7^k + M)$$

Hence $7^{k+1} - 1$ is divisible by 6 and hence P(k + 1) is correct.

22.

$$n = 0 \quad \rightarrow \quad 0^3 + 2 \times 0 \quad \rightarrow True \ disible \ by \ 3$$

$$n = k \quad \rightarrow \quad k^3 + 2k$$

$$n = k+1 \ \rightarrow \quad (k+1)^3 + 2(k+1) \quad \rightarrow \quad (k^3 + 3k^2 + 3k + 1) + (2k+2)$$

$$\rightarrow \quad (k^3 + 2k) + (3k^2 + 3k + 3)$$

$$= (k^3 + 2k) + 3(k^2 + k + 1)$$

$$(k^3 + 2k) \ is \ divisble \ by \ 3 \ and \ 3(k^2 + k + 1) \ is \ divisble \ by \ 3 \ too$$

$$\rightarrow \ (k^3 + 2k) + 3(k^2 + k + 1) \ is \ divisble \ by \ 3 \quad \rightarrow \quad n^3 + 2n \ for \ n \geq 0 \ is \ divisble \ by \ 3$$

23.

STEP 1: We first show that P(1) is true.

$$(1)^5 - 1 = 0$$

Because zero is divisible by any non-zero number. So 5 is also divisible and P(1) is true.

STEP 2: We now assume that P(k) is true

$$k^5 - k = 5n$$

and show that P(k + 1) is true by adding k + 1 to both sides of the above statement

$$(k + 1)^5 - (k + 1) = \left(k^5 + 5k^4 + 10k^3 + 10k^2 + 5k + 1\right) - (k + 1)$$

$$= k^5 + 5k^4 + 10k^3 + 10k^2 + 5k - k$$

$$= (k^5 - k) + 5k^4 + 10k^3 + 10k^2 + 5k$$

$$= 5n + 5(k^4 + 2k^3 + 2k^2 + k)$$

Therefore $(k + 1)^5 - (k + 1)$ is divisible by 5 and P(k + 1) is true.

24.

STEP 1: We first show that P(1), i.e., $n = 1$, is true.

$$25 + 6 + 1 = 32$$

32 is divisible by 8.

STEP 2: We now assume that P(k) is true

$$8 \mid 5^{k+1} + 2\left(3^k\right) + 1$$

and show that P(k + 1) is true.

$$8 \mid 5^{k+2} + 2\left(3^{k+1}\right) + 1$$

$$5^{k+2} + 2\left(3^{k+1}\right) + 1$$

$$= 5(5^{k+1}) + 6\left(3^k\right) + 1$$

$$= 5(5^{k+1}) + 10\left(3^k\right) + 5 - 4\left(3^k\right) - 4$$

$$= 5\left(5^{k+1} + 2\left(3^k\right) + 1\right) - 4\left(3^k - 1\right)$$

We know:

$$8\mid 5^{k+1} + 2\left(3^k\right) + 1 \implies 8\mid 5\left(5^{k+1} + 2\left(3^k\right) + 1\right)$$

Because $\left(3^k - 1\right)$ for every k is a member of even natural numbers:

$$8\mid 4\left(3^k - 1\right)$$

$$\implies 8\mid 5\left(5^{k+1} + 2\left(3^k\right) + 1\right) - 4\left(3^k - 1\right)$$

$$\implies 8\mid 5^{k+2} + 2\left(3^{k+1}\right) + 1$$

So for k + 1, it is proved.

34.

STEP 1: We first show that P(1) is true.

$$n = 1 \qquad f_3 - 1 = 2 - 1 = 1$$

Both sides of the statement are equal hence P(1) is true.
STEP 2: We now assume that P(k) is true

$$n = k \qquad f_0 + f_1 + f_2 + \cdots + f_k = f_{k+2} - 1$$

and show that P(k + 1) is true by adding k + 1 to both sides of the above statement

$$f_{k+2} - 1 + f_{k+1} = f_{k+3} - 1$$

$$f_{k+1} + f_{k+2} = f_{k+3}$$

And this phrase always stands. Because we know that in the Fibonacci sequence the value of one sentence is equal to the sum of the previous two sentences.

36.
STEP 1: We first show that P(1) is true.

$$n = 1 \rightarrow f_1^2 = f_1 f_2 \rightarrow 1 = 1$$

Both sides of the statement are equal hence P(1) is true.
STEP 2: We now assume that P(k) is true

$$\sum_{i=1}^{n} f_i^2 = f_n f_{n+1}$$

and show that P(k + 1) is true by adding k + 1 to both sides of the above statement

$$\sum_{i=1}^{n+1} f_i^2 = f_{n+1} f_{n+2}$$

$$\sum_{i=1}^{n} f_i^2 + f_{n+1}^2 = f_n f_{n+1} + f_{n+1}^2 \rightarrow \sum_{i=1}^{n+1} f_i^2 = f_{n+1}(f_n + f_{n+1}) = f_{n+1} f_{n+2}$$

38.

$$f_{n-1} \times f_{n+1} = f_n^2 + (-1)^n.$$

STEP 1: We first show that P(2), $n = 2$, is true.

$$f_{2-1} \times f_{2+1} = f_2^2 + (-1)^2 \Rightarrow 1 \times 2 = 1^2 + 1$$

STEP 2: We now assume that P(k) is true

$$f_{k-1} \times f_{k+1} = f_k^2 + (-1)^k$$

and show that P(k + 1) is true.

$$f_{k-1} \times f_{k+1} - (-1)^k = f_k^2 \Rightarrow f_{k-1} \times f_{k+1} + (-1)^{k+1} = f_k^2$$

$$\xRightarrow{f_{k-1}=(f_{k+1}-f_k), f_{k+1}=(f_{k+2}-f_k)} (f_{k+1} - f_k) \times (f_{k+2} - f_k) + (-1)^{k+1} = f_k^2$$

$$\rightarrow f_{k+1} \times f_{k+2} - f_k \times f_{k+1} - f_k \times f_{k+2} + \boxed{f_k^2} + (-1)^{k+1} = \boxed{f_k^2}$$

$$\xRightarrow{f_{k+2}=(f_{k+1}+f_k)} f_{k+1} \times (f_{k+1} + f_k) - f_k \times f_{k+1} - f_k \times f_{k+2} + (-1)^{k+1} = 0$$

$$\rightarrow f_{k+1}^2 + \boxed{f_k \times f_{k+1}} - \boxed{f_k \times f_{k+1}} - f_k \times f_{k+2} + (-1)^{k+1} = 0$$

Now we move $-f_k \times f_{k+2}$ to the right and we have:

$$f_{k+1}^2 + (-1)^{k+1} = f_k \times f_{k+2}$$

so it is proved.

40.

STEP 1: We first show that P(1) is true.

$$\binom{n}{m} \leq n^m \quad ==> \quad \binom{1}{m} \leq 1^m \quad ==> \quad 1 \leq 1$$

Both sides of the statement are equal hence P(1) is true.
STEP 2: We now assume that P(k) is true.

$$n = k \quad ==> \quad \binom{k}{m} \leq k^m$$

and show that P(k + 1) is true by adding k + 1 to both sides of the above statement

$$n = k + 1 \quad ==> \quad \binom{k+1}{m} \leq (k+1)^m$$

$$\frac{\binom{k}{m}}{\binom{k+1}{m}} \leq \frac{k^m}{(k+1)^m} \quad ==> \quad \frac{\frac{k!}{(k-m)!m!}}{\frac{(k+1)k!}{(k+1-m)(k-m)!m!}} \leq \frac{k^m}{(k+1)^m}$$

$$\frac{k+1-m}{k+1} \leq \left(\frac{k}{k+1}\right)^m$$

41.

STEP 1: We first show that P(1) is true.
Left Side:

$$\int_{i=0}^{1} 0.\binom{1}{0} + 1.\binom{1}{1} = 0 \times 1 + 1 \times 1 = 1$$

Right Side:

$$1.2^{1-1} = 1 \times 2^0 = 1 \times 1 = 1$$

Both sides of the statement are equal hence P(1) is true.
STEP 2: We now assume that P(k) is true

$$\int_{i=0}^{K} i.\binom{K}{i} = K.2^{K-1}$$

and show that P(k + 1) is true:

$$\int_{i=0}^{K+1} i.\binom{K+1}{i} = (K+1).2^K$$

$$\int_{i=0}^{K+1} i. \binom{K+1}{i} = \int_{i=0}^{K+1} i. \frac{(k+1)!}{(k+1-i)!i!}$$

$$= \sum_{i=0}^{K+1} i. \frac{(k+1)k!}{(k-(i-1))!i.(i-1)!}$$

$$= \sum_{i=0}^{K+1} \frac{(k+1)k!}{(k-(i-1))!(i-1)!}$$

$$= \sum_{i=0}^{K+1} \frac{(k+1)k!}{(k-(i-1))!(i-1)!}$$

$$= (k+1) \sum_{i=0}^{K+1} \frac{k!}{(k-(i-1))!(i-1)!}$$

$$= (k+1) \sum_{i=0}^{K+1} \binom{K}{i-1}$$

Because of the coefficient i in the expression

$$\int_{i=0}^{K+1} i. \binom{K+1}{i}$$

At i = 0 it becomes zero and zero has no effect on our equation. We can consider i from 1 to solve the question.

$$= (k+1) \sum_{i=1}^{K+1} \binom{K}{i-1}$$

$$= (k+1) \sum_{i-1=1-1}^{K+1-1} \binom{K}{i-1}$$

$$= (k+1) \sum_{i-1=0}^{K} \binom{K}{i-1}$$

We know:

$$\sum_{m=0}^{n} \binom{n}{m} = 2^n$$

hence

$$(k+1) \sum_{i-1=0}^{K} \binom{K}{i-1} = (k+1).2^k$$

hence P(k + 1) is correct.

Chapter 2
Growth of Functions

Abstract Having come up with your new and innovative algorithm, how do you measure its efficiency? Certainly, we would prefer to design an algorithm which to be as efficient as possible, therefore we will require some method to prove that it will really operate as well as we had expected. Also, some algorithms are more efficient than others. But what do we mean by efficient? Do we mean CPU/memory/disk usage, and so on? And how do we measure the efficiency of an algorithm? Generally speaking, complexity (asymptotic analysis or asymptotic notation) is used to describe the efficiency of an algorithm. In asymptotic analysis, we don't measure the actual running time, on the contrary, we prefer to evaluate the efficiency of an algorithm in terms of input size. We calculate, how the amount of time taken by an algorithm increases with the number of items of input. This chapter discusses useful theorems involving asymptotic notations and then illustrates the complexity analysis of the algorithms through a variety of examples. This chapter presents a comprehensive set of exercises related to the complexity analysis of the algorithms (275 exercises).

2.1 Lecture Notes

Having come up with your new and innovative algorithm, how do you measure its efficiency?

Certainly, we would prefer to design an algorithm which to be as efficient as possible, therefore we will require some method to prove that it will really operate as well as we had expected. Also, some algorithms are more efficient than others. But what do we mean by efficient? Do we mean CPU/memory/disk usage, and so on? And how do we measure the efficiency of an algorithm?

To mix performance (the amount of CPU/memory/disk usage) with complexity (how well the algorithm scales) is one of the most common misconceptions that occurs when analyzing the efficiency of an algorithm.

© The Author(s), under exclusive license to Springer Nature Switzerland AG 2022
H. Izadkhah, *Problems on Algorithms*,
https://doi.org/10.1007/978-3-031-17043-0_2

Determining the running time of an algorithm, i.e. CPU time, is not a particularly good indication of efficiency. There exist many aspects of the running time that can be influenced greatly by the performance of the underlying hardware on which the code will run, the compiler used to generate the machine code, in addition to the quality of the code. It is determined, therefore, how a designed algorithm acts as the size of the input increases. When n (input size) goes to infinity, for example, what effect would that have on running time?

Generally speaking, complexity (asymptotic analysis or asymptotic notation) is used to describe the efficiency of an algorithm. In asymptotic analysis, we don't measure the actual running time, on the contrary, we prefer to evaluate the efficiency of an algorithm in terms of input size. We calculate, how does the amount of time taken by an algorithm increases with the number of items of input. There exist two types of efficiency: time efficiency and space efficiency. Other measures could include transmission speed, temporary disk usage, long-term disk usage, power consumption, total cost of ownership, response time to external stimuli, etc. Time efficiency, also called time complexity, indicates the amount of time that an algorithm takes to run. Space efficiency, also called space complexity, describes how much working storage needed by the algorithm in addition to the space required for its input and output. Today, with the advancement of storage hardware and increased storage capacity, there is little to worry about the value of space required by an algorithm. However, the problem of time has not been reduced quite to the same extent. Furthermore, the analysis experience has shown that for most algorithms, we can obtain considerable improvement in speed than in space.

The time efficiency depends on several aspects, including:

- The quality of the code;
- The nature of the processor; and
- The input data.

In the analysis of time efficiency, the first two are ignored and focus particularly on the third. That is, we will assume that any algorithms being compared have similar code quality and are being executed on the same processor. Hence, the time efficiency will be assumed to depend only on the input data. Also, because of this, we will not measure the time efficiency in any particular time units. Order of growth associates input size to the running time of the algorithm, indeed, the order of growth only specifies how the running time scales as the input grows. An algorithm designer is only interested in how an algorithm is performed on large inputs, because even slow algorithms end up quickly on small inputs.

Let us assume that a sample computer can do 10,000 operations per second. We designed several algorithms that require $\log n$, n, n^2, n^3, n^4, n^6, 2^n, and $n!$ operations to perform a given task on n input items, Table 2.1 shows the behavior of the designed algorithms on different input items. Tables 2.2 and 2.3 show the explosive growth of 2^n and $n!$ in more details, respectively.

As shown in Table 2.1, with increasing n, the run-time in algorithms with complexity $logn$, n, $nlogn$, and n^2 has not increased so much. For example, if we increase

Table 2.1 Time required to process n input items on different algorithms

	n			
	10	50	100	1,000
$\log n$	0.0003 s	0.0006 s	0.0007 s	0.0010 s
n	0.0010 s	0.0050 s	0.0100 s	0.1000 s
$n \log n$	0.0033 s	0.0282 s	0.0664 s	0.9966 s
n^2	0.0100 s	0.2500 s	1.0000 s	100.00 s
n^3	0.1000 s	12.500 s	100.00 s	1.1574 days
n^4	1.0000 s	10.427 min	2.7778 h	3.1710 years
n^6	1.6667 min	18.102 days	3.1710 years	3171.0 centuries
2^n	0.1024 s	35.702 centuries	4×10^{16} centuries	1×10^{166} centuries
$n!$	362.88 s	1×10^{51} centuries	3×10^{144} centuries	1×10^{2554} centuries

Table 2.2 Time required to process n input items using a 2^n algorithm

n						
15	20	25	30	35	40	45
3.28 s	1.75 min	55.9 min	1.24 days	39.8 days	3.48 years	1.12 centuries

Table 2.3 Time required to process n input items using a $n!$ algorithm

n						
11	12	13	14	15	16	17
1.11 h	13.3 h	7.20 days	101 days	4.15 years	66.3 years	11.3 centuries

the size of n from 10 to 100, the *logn* runtime from 0.0003 to 0.0007 s has increased. But for algorithms with complexity 2^n and $n!$, run-time has greatly increased. For 2^n, the runtime has increased from 0.1024 s to 4×10^{16} centuries!

In general, the purpose of analyzing algorithms is to study their growth rate. By analyzing algorithms, one can choose among the various available algorithms for a problem, an algorithm with the lowest growth rate.

2.1.1 Orders of Growth

Order of growth in algorithm means how the time for computation increases when you increase the input size. Order of growth provides only a raw description of the behavior of an algorithm. To analyze the efficiency of an algorithm in terms of orders of growth, five asymptotic notations including O (big-oh), Ω (big-omega), Θ (big-theta), o (little-oh), and ω (little-omega) are employed. These notations are used to symbolically express the asymptotic behavior of a given function. In the following discussion, we present, first, these notations informally, and then, after several examples, formal descriptions are presented. We also assume that $f(n)$ and $g(n)$ are positive functions from positive numbers to positive numbers. We are interested to characterize an algorithm's efficiency in terms of running time, hence, $f(n)$ will be an algorithm's running time, and $g(n)$ will be the growth rate of $f(n)$. To calculate $f(n)$, we need to measure the number of elementary steps that the algorithm takes. In other words, the runtime for an algorithm can be measured by counting the number of steps required to solve the problem.

O-Notation

$O(g(n))$ characterizes the set of all functions with a lower or same order of growth as $g(n)$ (when n tends towards a particular value or infinity). It only provides an asymptotic upper bound on the growth rate of the function. For example, the following relationships are all true:

$$1000n + 5000 \in O(nlogn), \quad n^2 \in O(n^2), \quad \frac{1}{10}n(n-1) \in O(n^2)$$

The first function ($1000n + 5000$) is linear, i.e. n, and hence have a lower order of growth than $nlogn$, while the last two functions are quadratic and hence has the same order of growth as n^2. On the other hand,

$$n^2 \notin O(n), \quad \frac{n^3}{1000} \notin O(n^2), \quad n^5 + n^2 + 100000 \notin O(n^3)$$

Note that the functions n^2 and $\frac{n^3}{1000}$ are quadratic and cubic and hence have a higher order of growth than n and n^2, respectively, and the last function has the fifth-degree polynomial hence has a higher order of growth than n^3.

Also, we have:

$$O(n^3) = \{n^3, n^2logn, n^2, \ldots\}$$

The second asymptotic notation, $\Omega(g(n))$, characterizes the set of all functions with a higher or same order of growth as $g(n)$ (when n tends towards a particular value or infinity). For example,

$$n^4 \in \Omega(n^3), \quad \frac{1}{1000}n(n-1) \in \Omega(n^2), \quad n^5 + n^2 + 100000 \notin \Omega(2^n)$$

Finally, $\Theta(g(n))$ denotes the set of all functions that have the same order of growth as $g(n)$ (when n tends towards a particular value or infinity). For example:

$$\Theta(n^2) = \{2n^2, 2n^2 + \sqrt{n}, n^2 + n, n^2, n^2 + sinn, n^2 + logn, \ldots\}$$

Fig. 2.1 Big-oh notation:
$f(n) \in O(g(n))$

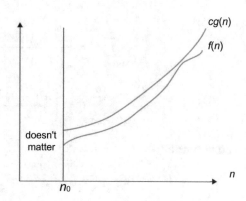

After this informal explanations, in the following, we give the formal definitions.

Definition 2.1 A function $f(n)$ has order $O(g(n))$, denoted $f(n) \in O(g(n))$, if and only if for all sufficiently large values of n, the value of $f(n)$ is bounded above at most a positive constant multiple of $g(n)$, i.e., if there exists a positive constant c and a nonnegative integer n_0, such that for all $n \geq n_0$ satisfying

$$f(n) \leq cg(n)$$

The definition is depicted in Fig. 2.1.

Theorem 2.1 *If* $f(n) = a_m n^m + a_{m-1} n^{m-1} + a_{m-2} n^{m-2} + \cdots + a_1 m_0 + a_0$, *then* $f(n) = O(n^m)$?

Proof We can write:

$$\begin{aligned}
f(n) &\leq |f(n)| \\
&= \left| a_m n^m + a_{m-1} n^{m-1} + \cdots + a_1 n + a_0 \right| \\
&\leq \left| a_m n^m \right| + \left| a_{m-1} n^{m-1} \right| + \cdots + |a_1 n| + |a_0| \\
&\leq n^m \sum_{i=0}^{m} |a_i|
\end{aligned}$$

Therefore, for $C = \sum_{i=0}^{m} |a_i|$, we have $f(n) \in O(n^m)$.
According to Theorem 2.1, we will have:

$$\sum_{i=1}^{n} i^0 = n \in O(n)$$

$$\sum_{i=1}^{n} i^1 = \frac{n(n+1)}{2} \in O(n^2)$$

$$\sum_{i=1}^{n} i^2 = \frac{n(n+1)(2n+1)}{6} \in O(n^3)$$

$$\sum_{i=1}^{n} i^3 = \left[\frac{n(n+1)}{2}\right]^2 \in O(n^4)$$

and in general:

$$\sum_{i=1}^{n} i^k = 1^k + 2^k + \cdots + n^k \in O(n^{k+1})$$

The following relationships are hold:

1. If the program have k instructions with the growth order $O(n^{m_1})$, $O(n^{m_2})$, ..., $O(n^{m_k})$, then the program's growth order is $O(n^m)$ in which $m = Maximum$ $\{m_1, m_2, \ldots, m_k\}$.
2. $O(f(n))O(g(n)) = O(f(n)g(n))$
3. $f(n) = O(f(n))$
4. $O(cf(n)) = O(f(n))$, $c > 0$
5. $O(f(n)g(n)) = f(n)O(g(n))$
6. $O(f(n)) + O(g(n)) = O(f(n) + g(n))$.

Example 2.1 Show that if $t(n) = (4n^3 + 5n^2 + 7n)(8\log n)$ then $t(n) = O(n^3 \log n)$.

Suppose $t_1(n) = (4n^3 + 5n^2 + 7n)$ and $t_2(n) = (8\log n)$ then we have: $t_1(n) = O(n^3)$ and $t_2(n) = O(\log n)$, hence:

$$t(n) = t_1(n)t_2(n) = O(n^3)O(\log n) = O(n^3 \log n)$$

Example 2.2 We want to calculate the asymptotic behavior:

$$f(n) = \left\lfloor \frac{n}{k} \right\rfloor + \frac{1}{2}k^2 + \frac{5}{2}k - 3$$

where

$$k = \lfloor \sqrt[3]{n} \rfloor$$

For solving, we have:

$$k = \lfloor \sqrt[3]{n} \rfloor \Longrightarrow k = \sqrt[3]{n}\left(1 + O\left(\frac{1}{\sqrt[3]{n}}\right)\right) = n^{\frac{1}{3}}(1 + O(n^{\frac{1}{3}}))$$

$$k^2 = n^{\frac{2}{3}}(1 + O(n^{-\frac{1}{3}}))^2 = n^{\frac{2}{3}}(1 + (O(n^{-\frac{1}{3}}))^2 + 2O(n^{-\frac{1}{3}}))$$

$$k^2 = n^{\frac{2}{3}} + n^{\frac{2}{3}}O(n^{\frac{2}{3}}) + 2n^{\frac{2}{3}}O(n^{-\frac{1}{3}})$$

$$k^2 = n^{\frac{2}{3}} + O(n^{\frac{1}{3}}) + O(1) = n^{\frac{2}{3}} + O(n^{\frac{1}{3}})$$

$$\left\lfloor \frac{n}{k} \right\rfloor = n^{\frac{2}{3}}(1 + O(n^{-\frac{1}{3}})) + O(1) = n^{\frac{2}{3}} + O(n^{\frac{1}{3}})$$

$$f(n) = n^{\frac{2}{3}} + O(n^{\frac{1}{3}}) + \frac{1}{2}(n^{\frac{2}{3}} + O(n^{\frac{1}{3}})) + \frac{5}{2}n^{\frac{1}{3}}(1 + O(n^{-\frac{1}{3}}))$$

$$f(n) = \frac{2}{3}n^{\frac{2}{3}} + \frac{3}{2}O(n^{\frac{1}{3}}) + \frac{5}{2}n^{\frac{1}{3}} + \frac{5}{2}O(1) - 3$$

$$f(n) = \frac{3}{2}n^{\frac{2}{3}} + O(n^{\frac{1}{3}})$$

Ω-Notation

Definition 2.2 A function $f(n)$ is said to be in $\Omega(g(n))$, denoted $f(n) \in \Omega(g(n))$, if and only if for all sufficiently large values of n, the value of $f(n)$ is bounded below at most a positive constant multiple of $g(n)$, i.e., if there exists a positive constant c and a nonnegative integer n_0, such that for all $n \geq n_0$ satisfying

$$f(n) \geq cg(n)$$

below is an example of the formal proof that $5n^2 \in \Omega(n)$:

$$5n^2 \geq n, \quad \text{for all } n \geq 0,$$

for $c = 1$ and $n_0 = 0$.

For other examples, we have:

$$\Omega(n^2) = \{n^2 logn, n^2 logloglogn, n^2, n^3, 100n^2 - 10000n, 100n^2 + 10000n, \ldots\}$$

The Ω definition is illustrated in Fig. 2.2.

Fig. 2.2 Big-omega
notation: $f(n) \in \Omega(g(n))$

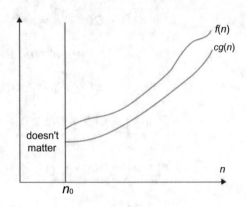

Θ-*Notation*

Definition 2.3 A function $f(n)$ is said to be in $\Theta(g(n))$, denoted $t(n) \in \Theta(g(n))$, if and only if for all sufficiently large values of n, the value of $f(n)$ is bounded both above and below at most a positive constant multiple of $g(n)$, i.e., if there exists a positive constants c_1 and c_2 and a nonnegative integer n_0, such that for all $n \geq n_0$ satisfying

$$c_2 g(n) \leq f(n) \leq c_1 g(n)$$

The definition is illustrated in Fig. 2.3.

Theorem 2.2 *For any two functions $f(n)$ and $g(n)$, we have $f(n) = \Theta(g(n))$ if and only if $f(n) = O(g(n))$ and $f(n) = \Omega(g(n))$.*

Example 2.3 Prove $(\log n!) \in \Theta(n \log n)$?

Fig. 2.3 Big-theta notation:
$f(n) \in \Theta(g(n))$

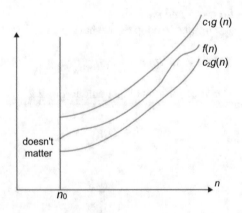

First solution: Using Stirling's formula $(n! \approx \sqrt{2\pi n} \left(\frac{n}{e}\right)^n)$

$$n! = \left(\frac{n}{e}\right)^n \sqrt{2\pi n} \Rightarrow \log n! = \log\left(\frac{n}{e}\right)^n + \log\sqrt{2\pi n}$$

$$\Rightarrow \log n! = n \log\frac{n}{e} + \log\sqrt{2\pi n}$$

$$\Rightarrow \log n! = n \log n - \log e^n + \log\sqrt{2\pi n}$$

$$\Rightarrow \log n! = n \log n + \log\frac{\sqrt{2\pi n}}{e^n}$$

$$\Rightarrow \log n! \in \Theta(n \log n)$$

Second solution: Here we have to prove that $\log n! \in O(n \log n)$ and $\log n \in \Omega(n \log n)$

$$\log n! = \log n + \log n - 1 + \cdots + \log 1 \leq \underbrace{\log n + \log n + \cdots + \log n}_{n} \leq n \log n$$

$$\Rightarrow \log n! \in \Theta(n \log n)$$

$$\log n! = \log n + \log n - 1 + \cdots + \log 1 \geq \frac{n}{2} \log\frac{n}{2}$$

$$\Rightarrow \log n! \geq \frac{n}{2} \log n - \frac{n}{2} \log 2 \geq \frac{1}{4} n \log n \Rightarrow \log n! \in \Omega(n \log n)$$

$$\Rightarrow \log n! \in \Theta(n \log n)$$

Third solution: Note the following:

$$(n!)^2 = (1 \times 2 \times \cdots \times (n-1)n)^2$$

$$\Rightarrow (n!)^2 = (1 \times 2 \times \cdots \times (n-1)n)(n(n-1) \times \cdots 1) = \prod_{x=1}^{n} x(n - x + 1)$$

$$\begin{cases} y = -x^2 + (n+1)x \\ x = \frac{n+1}{2} \Rightarrow y_{max} = \frac{(n+1)^2}{4} \\ x = 1 \Rightarrow y = n, \ x = n \Rightarrow y = n \Rightarrow y_{min} = n \end{cases}$$

$$\prod_{x=1}^{n} n \le (n!)^2 \le \prod_{x=1}^{n} \frac{(n+1)^2}{4} \Rightarrow n^n \le (n!)^2 \le \left(\frac{n+1}{2}\right)^{2n}$$

$$\Rightarrow n \log n \le 2 \log n! \le 2n \log \frac{n+1}{2} \le 2n \log n \Rightarrow \frac{n}{2} \log n \le \log n! \le n \log n$$

$$\Rightarrow \log n! \in \Theta(n \log n)$$

Let function $g(n, m)$ denote a function with two parameters n and m that can go to infinity independently at different rates. $O(g(n, m))$ is defined as follows:

$$O(g(n, m)) = \{f(n, m) : \text{ there exist positive constants } c, n_0, \text{ and } m_0$$

$$\text{such that } 0 \le f(n, m) \le cg(n, m) \text{ for all } n \ge n_0 \text{ and } m \ge m_0\}.$$

o-Notation

The notation $o(n)$, pronounced "Little-O of n" and is defined as follow:

$$o(g(n)) = \{f(n) : \text{ for all constants } c > 0, \text{ there exists a constant}$$

$$n_0 > 0 \text{ such that } 0 \le f(n) < cg(n) \text{ for all } n \ge n_0\}.$$

Little-o notation is related to Big-O notation, which both describe upper bounds, although Little-o is the stronger statement. Therefore, Big-O ($f(n) \in O(g(n))$) means that $f(n)$ asymptotic growth is no faster than $g(n)$, whereas $f(n) \in o(g(n))$ means that $f(n)$ asymptotic growth is strictly slower than $g(n)$. See the following examples:

$$n^{2.9999} = o(n^3), \quad \frac{n^3}{\log n} = o(n^3), \quad n^3 \notin o(n^3), \quad \frac{n^3}{1000} \notin o(n^3).$$

Example 2.4 Does $\log^2(\log n) = o(\log n)$?

All polylogarithmic functions grow more slowly than all positive polynomial functions. It means that for constants $a, b > 0$, we have

$$\log^b n = o(n^a)$$

Substitute $\log n$ for n, 2 for b, and 1 for a, gives $\log^2(\log n) = o(\log n)$.

ω-Notation

The notation $\omega(n)$, pronounced "Little-omega of n" and is defined as follow:

$f(n) \in \omega(g(n)) = \{$for any real constant $c > 0$, there exists a constant

$n_0 \geq 1$ such that $f(n) > cg(n)$ for every $n \geq n_0\}$.

Little-ω notation is related to Ω notation, which both describe lower bounds, although Little-ω is the stronger statement. Therefore, $\Omega(f(n) \in \Omega(g(n)))$ means that $g(n)$ asymptotic growth is no faster than $f(n)$, whereas $f(n) \in \omega(g(n))$ means that $g(n)$ asymptotic growth is strictly slower than $f(n)$. See the following examples:

$$n^{3.0001} = \omega(n^3), \quad n^3 logn = \omega(n^3), \quad n^3 \notin \omega(n^3).$$

In this regard, we have the following relationships:

$$\omega(g(n)) \subset \Omega(g(n) - \Theta(g(n))), \quad \omega(g(n)) \subset \Omega(g(n) - O(g(n))).$$

Figure 2.4 should help you visualize the relationships between these notations. Informally, to summarizes these notations, we have:

- $f(n) = O(g(n)$, if eventually f grows slower than some multiple of g,
- $f(n) = o(g(n)$, if eventually f grows slower than any multiple of g,
- $f(n) = \Omega(g(n)$, if eventually f grows faster than some multiple of g,
- $f(n) = \omega(g(n)$, if eventually f grows faster than any multiple of g,
- $f(n) = \Theta(g(n)$, if eventually f grows grows at the same rate as g,

The asymptotic notations have far more similarities than differences. Table 2.4 summarizes the key features of these four definitions.

Fig. 2.4 Relationships between asymptotic notations

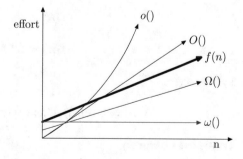

Table 2.4 Key features of these four definitions

Definition	$? c > 0$	$? n_0 \geq 1$	$f(n) \ ? c. g(n)$
$O(n)$	$\exists$	$\exists$	$\leq$
$o(n)$	$\forall$	$\exists$	$<$
$\Omega(n)$	$\exists$	$\exists$	$\geq$
$\omega(n)$	$\forall$	$\exists$	$>$

Comparison of Functions

Numerous of the relational properties of real numbers, such as transitivity, reflexivity, symmetry and transpose symmetry, hold for asymptotic notations as well. Let $f(n)$ and $g(n)$ be functions that map positive integers to positive real numbers. We have

Transitivity:

$$f(n) = \Theta(g(n)) \ \text{and} \ g(n) = \Theta(h(n)) \Rightarrow f(n) = \Theta(h(n)),$$

$$f(n) = O(g(n)) \ \text{and} \ g(n) = O(h(n)) \Rightarrow f(n) = O(h(n)),$$

$$f(n) = \Omega(g(n)) \ \text{and} \ g(n) = \Omega(h(n)) \Rightarrow f(n) = \Omega(h(n)),$$

$$f(n) = o(g(n)) \ \text{and} \ g(n) = o(h(n)) \Rightarrow f(n) = o(h(n)),$$

$$f(n) = \omega(g(n)) \ \text{and} \ g(n) = \omega(h(n)) \Rightarrow f(n) = \omega(h(n)).$$

Reflexivity:

$$f(n) = \Theta(f(n)),$$

$$f(n) = O(f(n)),$$

$$f(n) = \Omega(f(n)).$$

Symmetry:

$$f(n) = \Theta(g(n)) \ \text{if and only if} \ g(n) = \Theta(f(n)).$$

Transpose symmetry:

$$f(n) = O(g(n)) \ \text{if and only if} \ g(n) = \Omega(f(n)),$$

$$f(n) = o(g(n)) \ \text{if and only if} \ g(n) = \omega(f(n)).$$

Table 2.5 Common growth functions

Class	Name	Comments
$O(1)$	Constant	Pronounced "order 1" and denoting a function that runs in constant time; in other words, performance isn't affected by the size of the problem
$O(\log n)$	Logarithmic	Pronounced "order log N" and denoting a function that runs in logarithmic time
$O(\log \log n)$	Double logarithmic	–
$O((\log n)^c),\ c > 1$	Polylogarithmic	Pronounced "order $(\log n)^c$" and denoting a function that runs in logarithmic time; Note that $O(\log n)$ is exactly the same as $O(\log(n^c))$. The logarithms with different bases are equal and vary only by a constant factor so that the big-oh notation ignores that
$O(n)$	Linear	–
$O(n \log n)$	Linearithmic, or loglinear, or quasilinear	Pronounced "order N log N" and denoting a function that runs in time proportional to the size of the problem and the logarithmic time e.g. several divide-and-conquer algorithms
$O(n^2)$	Quadratic	Pronounced "order N squared" and denoting a function that runs in quadratic time; typically, describes the efficiency of algorithms with two nested loops e.g. primary sorting algorithms and some operations on matrices
$O(n^3)$	Cubic	Typically, characterizes efficiency of algorithms with three nested loops
$O(n^c)$	Polynomial	Typically, characterizes efficiency of algorithms with c embedded loops
$O(2^n)$	Exponential	Typical for algorithms that generate all subsets of an n-element set
$O(c^n),\ c > 1$	Exponential	Note that $O(n^c)$ and $O(c^n)$ are very different. The latter grows much, much faster, no matter how big the constant c is. A function that grows faster than any power of n is called superpolynomial. One that grows slower than an exponential function of the form c n is called subexponential
$O(n!)$	Factorial	Pronounced "order N factorial" and denoting a function that runs in factorial time; typical for algorithms that generate all permutations of an n-element set

Table 2.5 shows a number of common orders of growth in increasing order, so that most the time efficiencies of a large number of algorithms fall into only these orders of growth.

2.1.2 Useful Theorems Involving the Asymptotic Notations

Using the formal description of the asymptotic notations, their general characteristics can be proved. The following theorem, in particular, is useful in analyzing algorithms that has two consecutively parts.

Theorem 2.3 If $f_1(n) \in O(g_1(n))$ and $f_2(n) \in O(g_2(n))$, then

$$f_1(n) + f_2(n) \in O(max\{g_1(n), g_2(n)\}).$$

Proof Since $f_1(n) \in O(g_1(n))$, there exist some positive constant c_1 and some non-negative integer n_1 such that

$$f_1(n) \le c_1 g_1(n) \text{ for all } n \ge n_1.$$

Similarly, since $f_2(n) \in O(g_2(n))$,

$$f_2(n) \le c_2 g_2(n) \text{ for all } n \ge n_2.$$

Let us denote $c_3 = max\{c_1, c_2\}$ and consider $n \ge max\{n_1, n_2\}$ so that we can use both inequalities. Adding them yields the following:

$$f_1(n) + f_2(n) \le c_1 g_1(n) + c_2 g_2(n)$$

$$\le c_3 g_1(n) + c_3 g_2(n) = c_3[g_1(n) + g_2(n)]$$

$$\le c_3 2 \max \{g_1(n), g_2(n)\}.$$

Hence, $f_1(n) + f_2(n) \in O(max\{g_1(n), g_2(n)\})$, with the constants c and n_0 required by the O definition being $2c_3 = 2max\{c_1, c_2\}$ and $max \{n_1, n_2\}$, respectively.

Application of Theorem 2.3. Suppose that the program P consists of two subprograms $P1$ which takes time $f(n)$ and $P2$ which takes time $g(n)$, and $P2$ is executed after $P1$. In this case, the complexity of time P, according to this theory, is equal to $max\{f(n), g(n)\}$. Intuitively, the time complexity of an algorithm is equal to the time complexity of a part of the algorithm that has the highest execution time. The theorem also applies to more than two subprograms.

Theorem 2.4 If $f_1(n) \in O(g_1(n))$ and $f_2(n) \in O(g_2(n))$, then

$$f_1(n).f_2(n) \in O(g_1(n), g_2(n)).$$

Proof Since $f_1(n) \in O(g_1(n))$, there exist some positive constant c_1 and some non-negative integer n_1 such that

$$f_1(n) \leq c_1 g_1(n) \text{ for all } n \geq n_1.$$

Similarly, since $f_2(n) \in O(g_2(n))$,

$$f_2(n) \leq c_2 g_2(n) \text{ for all } n \geq n_2.$$

Let us denote $c = c_1.c_2$ and consider $n_0 = max\{n_1, n_2\}$ so for all $n \geq n_0$ we have:

$$f_1(n).f_2(n) \leq c_1 g_1(n).c_2 g_2(n)$$

$$\leq c(g_1(n).g_2(n))$$

Application of Theorem 2.4 This theory is used to analyze loops. If there is a loop whose statements have the complexity of time $g(n)$ and this loop is repeated $f(n)$ times, then the time complexity of the whole loop is equal to $f(n).g(n)$.

Lemma 2.1 *Determining that a function* f(n) *is polynomially bounded is equal to showing that*

$$log(f(n)) = O(logn)$$

Proof This is true because if *f(n)* be polynomially bounded, then there exist constants c, k, n_0 such that for all $n \geq n_0$, $f(n) \leq cn^k$. Hence, $log(f(n)) \leq kclog\,n$, which, since c and k are constants, means that $log(f(n)) = O(log\,n)$.

Example 2.5 Does $\lceil log\ n \rceil!$ polynomially bounded?

To proof, we use the following two facts:

1. $log(n!) = \Theta(n\,log\,n)$ (see Example 2.3),
2. $\lceil log\ n \rceil = \Theta(log\,n)$, because

- $\lceil log\ n \rceil \geq log\,n$,
- $\lceil log\ n \rceil < log\,n + 1 \leq 2log\,n$ for all $n \geq 2$.

$$log\,(\lceil log\ n \rceil!) = \Theta(\lceil log\ n \rceil log\lceil log\ n \rceil)$$
$$= \Theta(log\,n\,loglog\,n) = \omega(log\,n).$$

Therefore, $log\lceil log\ n \rceil! \notin O(log\,n)$, and so $\lceil log\ n \rceil!$ is not polynomially bounded.

Example 2.6 Does $\lceil loglog\ n \rceil!$ polynomially bounded?

$$
\begin{aligned}
\log\lceil \log\log n \rceil! &= \Theta(\lceil \log\log n \rceil \log\lceil \log\log n \rceil) \\
&= \Theta(\log\log n\ \log\log\log n) \\
&= o((\log\log n)^2) \\
&= o(\log^2\log n) \\
&= o((\log n)^2)\ \text{(see Example 2.4)}.
\end{aligned}
$$

Therefore, $\log\lceil \log\log n \rceil! = O(\log n)$, and so $\lceil \log\log n \rceil!$ is polynomially bounded.

2.1.3 Applying Limits for Analyzing Orders of Growth

Though the formal descriptions of addressed asymptotic notations are necessary for proving the general characteristics of asymptotic functions, they are seldom utilized for comparing the orders of growth of two particular functions. A very practical way to compare the asymptotic behavior of two functions is based on computing the limit of the ratio of those functions. Let $f(n)$ and $g(n)$ denote the two positive functions, three cases may occur:

$$
\lim_{n\to\infty} \frac{f(n)}{g(n)} = \begin{cases} 0 & \text{indicates that } f(n) \text{ has a smaller order of growth than } g(n), \\ c & \text{indicates that } f(n) \text{ has the same order of growth as } g(n), \\ \infty & \text{indicates that } f(n) \text{ has a larger order of growth than } g(n). \end{cases}
$$

Note that:

- the first case means $f(n) \in o(g(n))$,
- the first two cases mean that $f(n) \in O(g(n))$,
- the second case means that $f(n) \in \Theta(g(n))$,
- the last two cases mean that $f(n) \in \Omega(g(n))$,
- the last case means $f(n) \in \omega(g(n))$.

This technique is often more helpful than the one based on the formal descriptions because it can use the advantage of the powerful calculus methods developed for computing limits, such as L'Hopital's rule

$$
\lim_{n\to\infty} \frac{f(n)}{g(n)} = \lim_{n\to\infty} \frac{f'(n)}{g'(n)}
$$

In the following, we give six examples of using the L'Hopital's rule to compare orders of growth of two functions.

Example 2.7 Compare the orders of growth of $f(n) = a^n$ and $g(n) = b^n$, where $b > a > 1$.

$$\lim_{n\to\infty} \frac{a^n}{b^n} = \lim_{n\to\infty} \left(\frac{a}{b}\right)^n = 0.$$

This means that $f(n) \in o(g(n))$.

Example 2.8 Compare the orders of growth of $f(n) = log_a n$ and $g(n) = log_b n$.

$$\lim_{n\to\infty} \frac{log_a n}{log_b n} = \lim_{n\to\infty} \frac{(1/n)(log_a e)}{(1/n)(log_b e)} = \frac{log_a e}{log_b e} > 0.$$

or

$$\lim_{n\to\infty} \frac{log_a n}{log_b n} = \lim_{n\to\infty} \frac{1/(n\ln a)}{1/(n\ln b)} = \frac{\ln b}{\ln a} > 0.$$

This means that $f(n) \in \Theta(g(n))$.

Example 2.9 Compare the orders of growth of $f(n) = \frac{1}{10}n(n-1)$ and $g(n) = n^2$.

$$\lim_{n\to\infty} \frac{f'(n)}{g'(n)} = \lim_{n\to\infty} \frac{\frac{1}{10}n(n-1)}{n^2} = \frac{1}{10}\lim_{n\to\infty} \frac{(n^2-n)}{n^2} = \frac{1}{10}\lim_{n\to\infty}\left(1-\frac{1}{n}\right) = \frac{1}{10}.$$

This means that all three $f(n) \in O(g(n))$, $f(n) \in \Theta(g(n))$ and $f(n) \in \Omega(g(n))$ are correct.

Example 2.10 Compare the orders of growth of $f(n) = log_2 n$ and $g(n) = \sqrt{n}$.

$$\lim_{n\to\infty} \frac{(log_2 n)'}{(\sqrt{n})'} = \lim_{n\to\infty} \frac{(log_2 e)\frac{1}{n}}{\frac{1}{2\sqrt{n}}} = 2\,log_2 e \lim_{n\to\infty} \frac{\sqrt{n}}{n} = 2\,log_2 e \lim_{n\to\infty} \frac{1}{\sqrt{n}} = 0.$$

Since the limit is equal to zero, then $log_2 n$ grows slower than any multiple of $\sqrt{n}$. It means that $f(n)$ has a smaller order of growth than $g(n)$, thus $f(n)=o(g(n))$.

Example 2.11 Compare the orders of growth of $n!$ and 2^n. Taking advantage of Stirling's formula,

Stirling's formula:

$$n! \approx \sqrt{2\pi n}\left(\frac{n}{e}\right)^n \quad \text{for large values of } n.$$

we get

$$\lim_{n\to\infty} \frac{n!}{2^n} = \lim_{n\to\infty} \frac{\sqrt{2\pi n}\left(\frac{n}{e}\right)^n}{2^n} = \lim_{n\to\infty} \sqrt{2\pi n}\frac{n^n}{2^n e^n} = \lim_{n\to\infty} \sqrt{2\pi n}\left(\frac{n}{2e}\right)^n = \infty.$$

The following relation can be helpful in comparing the order of growth of some functions together.

$$\frac{f(n)}{g(n)} = 2^{log\ \frac{f(n)}{g(n)}} = 2^{log\ f(n) - log\ g(n)}$$

thus, we have:

$$\lim_{n \to \infty} (log\ f(n) - log\ g(n)) = -\infty \implies f(n) \in o(g(n)).$$

$$\lim_{n \to \infty} (log\ f(n) - log\ g(n)) = c > 0 \implies f(n) \in \Theta(g(n)).$$

$$\lim_{n \to \infty} (log\ f(n) - log\ g(n)) = +\infty \implies f(n) \in \omega(g(n)).$$

Example 2.12 Compare the orders of growth of $f(n) = 2^n$ and $g(n) = 2^{2n}$.

$$\lim_{n \to \infty} (log\ 2^n - log\ 2^{2n}) = \lim_{n \to \infty} (n - 2n) = \lim_{n \to \infty} (-n) = -\infty \implies f(n) \in o(g(n)).$$

2.1.4   Iterated Function

Let $f(n)$ be a function over the reals, then, the iterates of f are: $f(n), f(f(n)), f(f(f(n)))$, ... For positive integers i, we use the notation $f^{(i)}(n)$ to denote an elegant way to represent repeated operations for function $f(n)$. That is, function $f(n)$ iteratively applied i times to an initial value of n. Iterative function is defined as follow:

$$f^{(i)}(n) = \begin{cases} n & i = 0 \\ f(f^{(i-1)}(n)) & i > 0 \end{cases}$$

For example, if $f(n) = 10n$, then $f^{(i)}(n) = 10^i n$. The notation $log^*\ n$ (pronounced "log star of n") is used to express the iterated logarithm, which is defined as follows.

$$log^*\ n = min\{i = 0 : log^{(i)}n \le 1\}.$$

It is important to note that there is difference between $log^{(i)}n$ and $log^i n$. In $log^{(i)}n$, the logarithm function applied i times in succession to an initial value of n, whereas, in $log^i n$, the logarithm function raised to the ith power. The iterated logarithm is a very slowly growing function, for example, see the following iterated logarithms:

$log^*2 = 1,$
$log^*4 = 2,$

$log^* 16 = 3$,

$log^* 65536 = 4$,

$log^* (2^{65536}) = 5$.

Note that, in the most cases, the input of an algorithm is much less than 2^{65536}, therefore, we rarely encounter an input size n such that $log^* n > 5$.

2.2 Exercises

2.2.1 Size of Problem

1. Suppose a computer can perform one billion (10^9) basic operations per second, or in other words, each basic operation takes one nanosecond (10^{-9}). Fill in the table below for algorithms with different time complexities and input sizes.

input size/time complexity	log n	n	n^2	2^n
10				
10^2				
10^2				
10^3				

2. In the table below, in each row, a function $f(n)$ is written, which shows the execution time of an algorithm in microseconds. For each time t that appears in the columns and each function $f(n)$, specify the largest size of the problem that the algorithm can solve in time t.

	1 s	1 min	1 h	1 day	1 month	1 year	1 century
log n							
$\sqrt{n}$							
n							
n log n							
n^2							
n^3							
2^n							
n!							

3. An algorithm with time complexity $nlogn$ runs on a computer for 1 s. How long will the same algorithm run on another computer at 100 times faster?

4. A program with input size 10 and time complexity n^2 runs on a computer in 1 millisecond. How long does the same problem with input size 100 run on the same computer?

5. The execution time of a program for input size 50 is equal to 10 s. If the time complexity of this program is n^2, the execution time of this program for input size 100 is approximately closer to which of the following times?

  a. 10 s
  b. 20 s
  c. 40 s
  d. 100 s

6. The execution time of a program for input size 1000 is 30 s. If its time complexity is n^2, what is the largest input size to solve it in two minutes?

  a. 2000
  b. 4000
  c. 6000
  d. 60000

7. The execution time of a program for input sizes 100 and 1000 is 6 s and 10 min, respectively. Which of the following is most likely the complexity of this program?

  a. constant
  b. n
  c. n^2
  d. n^3

2.2.2   True or False?

8. $n^2 \log n = O(n^3)$.

9. $n! = O(n^n)$.

10. $n^4 = O(n^3)$.

11. $10n^3 + 1000 = O(n^3)$.

12. $n \log n = O(n\sqrt{n})$.

13. $\sqrt{n} = O(\log n)$.

14. $\log n = O(\sqrt{n})$.

15. $n^4 + n^3 = O(n^2(10 + n^2))$.

16. $n^3(1 + 2\sqrt{n}) = O(n^3)$.

17. $n(1+2\sqrt{n}) = O(n \log n)$.

18. $10n^2+\sqrt{n} = O(n^2)$.

19. $10n^2+5\sqrt{n} = O(1000n + \sqrt{n} + n\sqrt{n})$.

20. $2\log n+\sqrt{n} = O(n)$.

21. $\sqrt{n}\log n = O(n)$.

22. $1/n = O(\log n)$.

23. $\log n = O(1/n)$.

24. $\log n = O(n^{-1/2})$.

25. $n + 100\sqrt{n} = O(\sqrt{n}\log n)$.

26. $2^{n+1} = O(2^n)$.

27. $2^{2n} = O(2^n)$.

2.2.3 Rank the Functions

List the following functions according to their asymptotic order from the lowest to the highest:

28. $O(\log n)$, $O(\sqrt{n})$, $O(n!)$, $O(a^n)$, $O(n)$

29. n^{1000}, $n!$, $(1.005)^n$

30. $O(1 + \varepsilon)^n$, $O(n \log n)$, $O\left(\frac{n^2}{\log n}\right)$

31. $(\log n)^{\log n}$, $\frac{n^2}{\log n}$, $2^{\sqrt{2\log n}}$, $n\sqrt{n}$, $\sqrt{\log(n!)}$

32. $(n - 10)!$, $5\log(10n + 1)^{10}$, 2^{2n}, $0.0001n^4 + 2n^3 + 1$, ln^2n, $\sqrt[3]{n}$

33. Compare the following functions according to their order of growth.

$$f(n) = (\log n)^{\log n}, \quad g(n) = 4^{\log n}, \quad h(n) = \log^2 n$$

34. List the following functions according to their order of growth from the lowest to the highest:

$$\log(\log^* n) \qquad 2^{\log^* n} \qquad (\sqrt{2})^{\log n} \qquad n^2 \qquad n! \qquad \log n!$$

$$\left(\tfrac{3}{2}\right)^n \qquad n^3 \qquad \log^2 n \qquad \log n! \qquad 2^{2^n} \qquad n^{1/\log n}$$

$$\ln\ln n \qquad \log^* n \qquad n 2^n \qquad n^{\log\log n} \qquad \ln n \qquad 1$$

$$2^{\log n} \qquad (\log n)^{\log n} \qquad e^n \qquad 4^{\log n} \qquad (n+1)! \qquad \sqrt{\log n}$$

$$\log^*(\log n) \qquad 2^{\sqrt{2\log n}} \qquad n \qquad 2^n \qquad n\log n \qquad 2^{2^{n+1}}$$

35. Consider the following eighteen functions:

$$\begin{array}{ccc}
\ln^2 n & 10000 & \log(\log^2 n) \\
n\log n & n- n^4 + 10n^5 & 2n/\log n \\
n^2 & n^4 & \log n \\
n^{1/3} + \log n & (\log n)^3 & n! \\
\sqrt{n} & 10n^2 + \log n & 2^n \\
(2/5)^n & (3/5)^n & n
\end{array}$$

Group these functions so that $f(n)$ and $g(n)$ are in the same group if and only if $f(n) = O(g(n))$ and $g(n) = O(f(n))$), and list the groups in increasing order.

36. Draw a line from each of the five functions in the center to the best big-Ω value on the left, and the best big-O value on the right.

$\Omega(1/n)$		$O(1/n)$
$\Omega(1)$		$O(1)$
$\Omega(\log\log n)$		$O(\log\log n)$
$\Omega(\log^2 n)$		$O(\log^2 n)$
$\Omega(\log n)$		$O(\log n)$
$\Omega(\sqrt[3]{n})$		$O(\sqrt[3]{n})$
$\Omega(n/\log n)$	$1/(\log n)$	$O(n/\log n)$
$\Omega(n)$	$4n^3 - 2n^2 + 10$	$O(n)$
$\Omega(n^{1.00001})$	$(n^2 + n)/(\log^2 n + \log n)$	$O(n^{1.00001})$
$\Omega(n^2/\log n)$	$2^{\log^2 n}$	$O(n^2/\log^2 n)$
$\Omega(n^2/\log^2 n)$	4^n	$O(n^2/\log n)$
$\Omega(n^2)$		$O(n^2)$
$\Omega(n^{3/2})$		$O(n^{3/2})$
$\Omega(3^n)$		$O(3^n)$
$\Omega(6^n)$		$O(6^n)$
$\Omega(n^n)$		$O(n^n)$
$\Omega(n^{n^2})$		$O(n^{n^2})$

37. Considering the following functions, which choice is the correct?

g_6	g_5	g_4	g_3	g_2	g_1
$2\sqrt{2logn}$	n	2^n	$nlogn$	n^{2+logn}	$log^*(n^n)$

a. $g_1 < g_2 < g_3 < g_4 < g_5 < g_6$
b. $g_6 < g_5 < g_3 < g_1 < g_2 < g_4$
c. $g_1 < g_6 < g_5 < g_3 < g_2 < g_4$
d. $g_6 < g_1 < g_5 < g_3 < g_2 < g_4$

38. Which of the following is the correct?

a. $n^3 \log n = O(n^{3+\varepsilon})\ \ 0 < \varepsilon < 0.1$
b. $\sqrt{n} = O(\log n)$
c. $n^{1+\varepsilon} = O(\log n)\ \ 0 < \varepsilon < 0.1$
d. $n^2 = O\left(\frac{n^2}{\log n}\right)$

39. Considering the following functions, which choice is the correct?

g_6	g_5	g_4	g_3	g_2	g_1
$2^{log^* n}$	$(\sqrt{2})^{logn}$	n^2	$n!$	$logn!$	$2^{2^{n+1}}$

a. $g_6 < g_5 < g_4 < g_3 < g_2 < g_1$
b. $g_5 < g_4 < g_6 < g_2 < g_3 < g_1$
c. $g_5 < g_6 < g_2 < g_4 < g_3 < g_1$
d. $g_6 < g_4 < g_5 < g_2 < g_3 < g_1$

40. Considering the following functions, which of the following relationships shows the order of growth of these functions?

$$f_1(n) = n^2 \log^2 n$$

$$f_2(n) = n^2 \log^3 \sqrt{n}$$

$$f_3(n) = 4^{(logn+\lg logn)}$$

a. $f_1 = f_3 < f_2$
b. $f_2 < f_1 < f_3$
c. $f_1 < f_2 < f_3$
d. $f_1 = f_2 = f_3$

41. Which of the following is the correct?

a. $n^{\frac{1}{10}} \in \Omega(\log n)$
b. $\log(n!) \in O(\log n)$
c. $8n^2 - 3n - 4 \in O(n \log n)$
d. $1^n + 2^n + \cdots + n^n \in O(n^n)$

2.2.4   *Prove Using the Definition of Notation*

For each of the following pairs of functions, find $c \in \mathbb{R}^+$ such that $f(n) \leq cg(n)$ for all $n > 1$.

42. $f(n) = 2n^2 + 4n$, $g(n) = n^3$.

43. $f(n) = 5n^2 + n + 1$, $g(n) = 2n^2$.

44. $f(n) = 4n + 5\log n$, $g(n) = n\log n$.

45. $f(n) = 4 \times 2^n + n^3$, $g(n) = n2^n$.

46. $f(n) = 7n + 3$, $g(n) = (n^2 - 6n)/2$.

47. $f(n) = 2\lfloor \sqrt{n} \rfloor - 5$, $g(n) = n - \lceil \sqrt{n} \rceil$.

Prove by the Big-oh definition.

48. $n^2 + 10\,n \in O(n^2)$.

49. $5n^2 \in O(n^2)$.

50. $n(n-1)/2 \in O(n^2)$.

51. $n \in O(n^2)$.

52. $n! + 7n^5 \in O(n^n)$.

53. $\frac{12n^5}{\log n} + 7n^4 \in O(n^5)$.

54. $n = O(2^n)$.

55. $100000n + 1 = O(2^n)$.

56. $n^2 = O(2^n)$.

57. $2^n = O(n!)$.

58. $(n + 7)^2 = O(n^2)$.

59. $3n\lfloor \log n \rfloor = O(n^2)$.

60. $log(\lceil loglogn \rceil!) = o(n)$.

Prove by the Ω definition.

61. $5n^2 \in \Omega(n^2)$.

62. $n(n-1)/2 \in \Omega(n^2)$.

63. $n^2 + 10n \in \Omega(n^2)$.

64. $n^3 \in \Omega(n^2)$.

65. $2n^3 + 7n^2 \in \Omega(n^2)$.

66. $2n^3 + 7n^2 \in \Omega(n^3)$.

67. $n! = \Omega(2^n)$.

68. $log(\lceil logn \rceil!) = \omega(n)$.

Prove by the Θ definition.

69. $0.5n^2 + 3n \in \Theta(n^2)$.

70. $60n^2 + 5n + 1 = \Theta(n^2)$.

71. Does $2n + 3logn = \Theta(n)$? Prove your answer.

72. $\log n! = \Theta(n \log n)$.

73. $\sum_{i=1}^{n} i \log i = \Theta(n^2 \log n)$.

74. $7n^2 - 6n + 2 \in \Theta(n^2)$.

75. $n^3 + n^2 Log\, n \in \Theta(n^3)$.

76. $7n^2 2^n + 5n^2 Log\, n \in \Theta(n^2 2^n)$.

77. $2n^{2^n} + 72^n \in \Theta(n^{2^n})$.

78. $\sum_{i=0}^{n} i^3 \in \Theta(n^4)$.

79. $n^4 + 700n^2 \in \Theta(n^4)$.

Without using the notations definition, show whether the following statements are true or not.

80. $n^2 + 10n \in O(n^2)$.

81. $n^3 + 10n \in O(n^2)$.

82. $5n^2 \in O(n^2)$.

83. $n(n-1)/2 \in O(n^2)$.

84. $n \in O(n^2)$.

85. $n! + 7n^5 \in O(n^n)$.

86. $\frac{12n^5}{\log n} + 7n^4 \in O(n^5)$.

87. $n = O(2^n)$.

88. $100000n + 1 = O(2^n)$.

89. $n^2 = O(2^n)$.

90. $2^n = O(n!)$.

91. $5n^2 \in \Omega(n^2)$.

92. $n(n-1)/2 \in \Omega(n^2)$.

93. $n^2 + 10n \in \Omega(n^2)$.

94. $n^3 \in \Omega(n^2)$.

95. $2n^3 + 7n^2 \in \Omega(n^2)$.

96. $2n^3 + 7n^2 \in \Omega(n^3)$.

97. $n! = \Omega(2^n)$.

98. $0.5n^2 + 3n \in \Theta(n^2)$.

99. $60n^2 + 5n + 1 = \Theta(n^2)$.

100. $2n + 3\log n = \Theta(n)$.

101. $\log n! = \Theta(n \log n)$.

102. $\sum_{i=1}^{n} i \log i = \Theta(n^2 \log n)$.

103. $7n^2 - 6n + 2 \in \Theta(n^2)$.

104. $n^3 + n^2 \, Log \, n \in \Theta(n^3)$.

105. $7n^2 2^n + 5n^2 \, Log \, n \in \Theta(n^2 2^n)$.

106. $2n^{2^n} + 72^n \in \Theta(n^{2^n})$.

107. $\sum_{i=0}^{n} i^3 \in \Theta(n^4)$.

108. $n^4 + 700n^2 \in \Theta(n^4)$.

Let $f(n) \in$ *notation* $g(n)$. Show through computing the limit of the ratio of two functions that (Questions 109–128):

109. $n^2 + 10n \in O(n^2)$.

110. $5n^2 \in O(n^2)$.

111. $n(n-1)/2 \in O(n^2)$.

112. $n(n-1)/2 \in \Omega(n^2)$.

113. $n^3 \in \Omega(n^2)$.

114. $0.5n^2 + 3n \in \Theta(n^2)$.

115. $7n^2 - 6n + 2 \in \Theta(n^2)$.

116. $n^3 + n^2 \, Log \, n \in \Theta(n^3)$.

117. $7n^2 2^n + 5n^2 \, Log \, n \in \Theta(n^2 2^n)$.

118. $2n^{2^n} + 72^n \in \Theta(n^{2^n})$.

119. $n^4 + 700n^2 \in \Theta(n^4)$.

120. $\frac{12n^5}{log n} + 7n^4 \in O(n^5)$.

121. $2n^3 + 7n^2 \in \Omega(n^2)$.

122. $2n^3 + 7n^2 \in \Omega(n^3)$.

123. $(n + 1)^2 = O(n^2)$.

124. $3n\lfloor \log n \rfloor = O(n^2)$.

125. $n = O(2^n)$.

126. $100000n + 1 = O(2^n)$.

127. $n^2 = O(2^n)$.

128. $2^n = O(n!)$.

Let the running time, $T(n)$, for a number of algorithms is given below. Prove the correctness of the following phrases.

129. $T(n) = 6n + 4 \in \Omega(n)$.

130. $T(n) = 3n + 2 \notin \Omega(n^2)$.

131. $T(n) = 5^n + n^2 \in \Omega(2^n)$.

132. For $a > 1$ and $b > 1$, prove the following relation is hold.

$$\log_a^n = \Theta(\log_b^n)$$

133. Prove or disprove?

$$n^{1.001} + n \log n = \Theta(n^{1.001})$$

134. Prove that running time $T(n) = n^3 + 20n + 1$ is $O(n^3)$.

135. Prove that running time $T(n) = n^3 + 20n + 1$ is not $O(n^2)$.

136. Prove that running time $T(n) = n^3 + 20n$ is $\Omega(n^2)$.

2.2.5 Find Notations

Let $f(n) = N(g(n))$, where "N" indicates one of O, Ω, and Θ notations. Find for each of the following pairs of functions a "N" notation.

137. $f(n) = 5n^2 + 2n + 10, \quad g(n) = n$

138. $f(n) = n^3 + 100n^2 + 40, \quad g(n) = 10000n + 20$

139. $f(n) = 6\frac{n^3}{\log n}, \quad g(n) = n^3$

140. $f(n) = 2\sqrt{n} + 10, \quad g(n) = \log(n + 10)$

141. $f(n) = n^{2^n} + 102^n, \quad g(n) = n^{2^n}$

142. $f(n) = n^{1.001} + n\log n, \quad g(n) = n^{1.001}$

143. $f(n) = n\sqrt{n} + n, \quad g(n) = n^2 - 10n$

144. $f(n) = n\sqrt{n}, \quad g(n) = 2n\log(n^4 + 4)$

145. $f(n) = (n^{10} + 1000)/(1 + 2^{-n}), \quad g(n) = n^{10} + 3$

146. $f(n) = 0.00001 \times 2^n - n^4, \quad g(n) = 10000n^4 + n^2$

147. $f(n) = 4 \times 2^n + n^2, \quad g(n) = 2^n$

Compare the following pairs of functions. In each case, say whether $f = o(g)$, $f = \omega(g)$, or $f = \Theta(g)$, and prove your claim based on formal definition or limit computation (Questions 148–153).

148. $f(n) = 2n + \log n, \quad g(n) = 1000n + (\log n)^{10}$

149. $f(n) = \log n, \quad g(n) = \log\log(n^2)$

150. $f(n) = n^3/\log n, \quad g(n) = n(\log n)^3$

151. $f(n) = (\log n)^{10^6}, \quad g(n) = n^{10^{-6}}$

152. $f(n) = n\log n^2, \quad g(n) = (\log n)^{\log n}$

153. $f(n) = n^3 3^n, \quad g(n) = 4^n$

Find a theta notation for the following functions.

154. $\frac{(n+1)(n+3)}{n+2}$

155. $2\log n + 4n + 3n\log n$

156. $2 + 4 + 6 + \cdots + 2n$

157. $\frac{(n^2 + \log n)(n+3)}{n + n^2}$

158. $2 + 4 + 8 + 16 + \cdots + 2^n$

159. $\sqrt{10n^2 + 7n + 3}$

160. $2^{n+1} + 3^{n-1}$

161. $(n^2 + 1)^{10}$

162. $2n\log(n + 2)^2 + (n + 2)^2 \log \frac{n}{2}$

2.2.6 Property of Notations

163. Which of the following is correct?

   a. $if\ f(n) = O(g(n))\ and\ f(n) = \Theta(g(n))\ then\ f(n) = \Omega(g(n))$
   b. $if\ f(n) = \Omega(g(n))\ and\ f(n) = \Theta(g(n))\ then\ f(n) = O(g(n))$
   c. $if\ f(n) = \Omega(g(n))\ and\ f(n) = O(g(n))\ then\ f(n) = \Theta(g(n))$
   d. $if\ f(n) = O(g(n))\ and\ g(n) = \Omega(f(n))\ then\ f(n) = \Theta(g(n))$

164. Suppose f and g are two arbitrary functions as $f, g : N \to R^+$, as well as suppose that $\lim_{n \to \infty} \frac{f(n)}{g(n)} = +\infty$, which option is correct?

   a. $f(n) \in O(g(n)), g(n) \notin \Omega(f(n))$
   b. $g(n) \in O(f(n)), f(n) \in O(g(n))$
   c. $f(n) \in \Omega(g(n)), f(n) \notin \Theta(g(n))$
   d. $f(n) \in \Theta(g(n)), g(n) \notin \Omega(f(n))$

Which of the propositions of Questions 165–172 is true and which is false? Provide an encounter if the proposition is false.

165. If $f(n) = \Theta(h(n))$ and $g(n) = \Theta(h(n))$ Then $f(n) + g(n) = \Theta(h(n))$.

166. If $f(n) = \Theta(g(n))$ Then $cf(n) = \Theta(g(n))$ for each $c \neq 0$.

167. If $f(n) = \Theta(g(n))$ Then $2^{f(n)} = \Theta(2^{g(n)})$.

168. If $f(n) = \Theta(g(n))$ Then $\log f(n) = \Theta(\log g(n))$. (It is assumed that for all $n \geq 1$, $f(n) \geq 1$, $g(n) \geq 1$).

169. If $f(n) = O(g(n))$ Then $g(n) = O(f(n))$.

170. If $f(n) = O(g(n))$ Then $g(n) = \Omega(f(n))$.

171. If $f(n) + g(n) = \Theta(h(n))$ where in $h(n) = max\{f(n), g(n)\}$.

172. If $f(n) + g(n) = \Theta(h(n))$ where in $h(n) = min\{f(n), g(n)\}$.

173. Find functions f, g, h, and t that apply in the following relationships.

$$f(n) = \Theta(g(n)), \qquad h(n) = \Theta(t(n)),$$

$$f(n) - h(n) \neq \Theta(g(n) - t(n))$$

174. Find functions f and g that apply in the following relationships.

$$f(n) \neq O(g(n)), \qquad g(n) \neq O(f(n))$$

175. Find functions f and g that apply in the following relationships.

$$0 < f(n) < f(n+1), \qquad 0 < g(n) < g(n+1)$$

$$f(n) \neq O(g(n)), \qquad g(n) \neq O(f(n))$$

176. Suppose $f(n) = \Theta(g(n))$. Prove that $h(n) = O(f(n))$ iff $h(n) = O(g(n))$.

Suppose $T_1(n) \in \Omega(f(n))$ and $T_2(n) \in \Omega(g(n))$, Prove or disprove?

177. $T_1(n) + T_2(n) \in \Theta(max\{f(n), g(n)\})$

178. $T_1(n) * T_2(n) \in \Omega(max\{f(n), g(n)\})$

179. $T_1(n) + T_2(n) \in \Theta(max\{f(n), g(n)\})$

180. $T_1(n) * T_2(n) \in \Theta(f(n)) * (g(n))$

181. Suppose $f, g : N \to R \geq 0$, prove?

(a) $f(n) + g(n) = O(max\{f(n), g(n)\})$

(b) $f(n) + g(n) = \Theta(max\{f(n), g(n)\})$

(c) $f(n) + g(n) = \Omega(max\{f(n), g(n)\})$

182. Let $T_1(n) = O(F(n))$ and $T_2(n) = O(F(n))$, which choice is the correct?

a. $T_1(n) + T_2(n) = O(F(n))$
b. $T_1(n)/T_2(n) = O(1)$
c. $T_1(n) * T_2(n) = O(F(n))$
d. $T_1(n) = O(T_2(n))$

183. Show if $f(n) = a_m n^m + a_{m-1} n^{m-1} + a_{m-2} n^{m-2} + \cdots + a_1 m_0 + a_0$, then $f(n) = \Omega(n^m)$?

184. Show if $f(n) = a_m n^m + a_{m-1} n^{m-1} + a_{m-2} n^{m-2} + \cdots + a_1 m_0 + a_0$, then $f(n) = \Theta(n^m)$?

Prove the following statements.

185. $f_1(n) = O(g_1(n))$ and $f_2(n) = O(g_2(n)) \Rightarrow f_1(n) + f_2(n) = O(g_1(n) + g_2(n))$.

186. $f_1(n) = \Omega(g_1(n))$ and $f_2(n) = \Omega(g_2(n)) \Rightarrow f_1(n) + f_2(n) = \Omega(g_1(n) + g_2(n))$.

187. $f_1(n) = O(g_1(n))$ and $f_2(n) = O(g_2(n)) \Rightarrow f_1(n) + f_2(n) = O(\max\{g_1(n), g_2(n)\})$.

188. $f_1(n) = \Omega(g_1(n))$ and $f_2(n) = \Omega(g_2(n)) \Rightarrow f_1(n) + f_2(n) = \Omega(\min\{g_1(n), g_2(n)\})$.

189. $f_1(n) = O(g_1(n))$ and $f_2(n) = O(g_2(n)) \Rightarrow f_1(n) \times f_2(n) = O(g_1(n) \times g_2(n))$.

190. $f_1(n) = \Omega(g_1(n))$ and $f_2(n) = \Omega(g_2(n)) \Rightarrow f_1(n) \times f_2(n) = \Omega(g_1(n) \times g_2(n))$.

191. Suppose that $f_1(n) = \Theta(g_1(n))$ and $f_2(n) = \Theta(g_2(n))$. Justify whether the following statements are true.

(a) $f_1(n) + f_2(n) = \Theta(g_1(n) + g_2(n))$
(b) $f_1(n) + f_2(n) = \Theta(\max\{g_1(n), g_2(n)\})$
(c) $f_1(n) + f_2(n) = \Theta(\min\{g_1(n), g_2(n)\})$

192. Prove or disprove: for all functions $f(n)$ and $g(n)$, either $f(n) = O(g(n))$ or $g(n) = O(f(n))$.

193. Prove or disprove: If $f(n) > 0$ and $g(n) > 0$ for all n, then $O(f(n) + g(n)) = f(n) + O(g(n))$.

194. Prove or disprove?

if $f(n) = \Theta(h(n))$ **and** $g(n) = \Theta(h(n))$ **then** $f(n) + g(n) = \Theta(h(n))$

2.2.7 More Exercises

195. Show

$$2^{\sqrt{2\log n}} = \Theta(n^{\sqrt{\frac{2}{\log n}}})$$

196. Prove that $cn^k + d = O(2^n)$ for all $c, d, k \in \mathbb{R}^+$.

197. Multiply $2\log n + 10 + O(1/n)$ by $n^2 + O(\sqrt{n})$ and simplify your answer as much as possible.

198. For each of the following functions, indicate how much the function's value will change if its argument is increased fourfold.

(a) $\log_2 n$
(b) $\sqrt{n}$
(c) n^2
(d) n^3
(e) 2^n

199. How many of the following relationships are true?

$$3^{2^n} = \Omega(2^{3^n}), \quad n! = O(n^n), \quad n^{2+\frac{1}{\sqrt{n}}} = o(n^2)$$

a. 0
b. 1
c. 2
d. 3

200. P is a problem whose complexity is $\Omega(n^2)$ and also $O(n^3)$. Algorithm A solves P. Which of the following choices can be complexity of A?

a. $O(n^4)$
b. $o(n^2)$
c. $\Theta(n^3)$
d. $\Theta(n^2)$

201. Prove or disprove:

$$O\left(\left(\frac{n^2}{\log \log n}\right)^{1/2}\right) = O(\lfloor\sqrt{n}\rfloor)$$

202. Prove or disprove: $2^{(1+O(1/n))^2} = 2+O(1/n)$.

In questions 203–207, find two functions $f(n)$ and $g(n)$ that satisfy the following relationships. If no such f and g exist, write "None."

203. $f(n) = o(g(n))$ and $f(n) \neq \Theta(g(n))$.

204. $f(n) = \Theta(g(n))$ and $f(n) = o(g(n))$.

205. $f(n) = \Theta(g(n))$ and $f(n) \neq O(g(n))$.

206. $f(n) = \Omega(g(n))$ and $f(n) \neq O(g(n))$.

207. $f(n) = \Omega(g(n))$ and $f(n) \neq o(g(n))$.

208. Prove $1 + 2 + 3 + \cdots + n = \Theta(n^2)$?

209. Prove $1^k + 2^k + 3^k + \cdots + n^k = \Theta(n^{k+1})$?

210. Prove $1 + \frac{1}{2} + \frac{1}{3} + \cdots + \frac{1}{n} = \Theta(\log n)$?

211. Prove $o(f(n)) \subseteq O(f(n))\backslash\Omega(f(n))$?

212. Prove $o(f(n)) \subseteq O(f(n))\backslash\Theta(f(n))$?

213. For any real constants a and b, where $b > 0$, prove:

$$(n + b)^b = \Theta(n^b)$$

214. Explain why the statement, "The running time of algorithm A is at least $O(n^3)$," is meaningless.

215. Compare the following orders of growth.

$$f(n) = n, \quad g(n) = n^{1+\sin n}.$$

216. Let $f(n)$ and $g(n)$ denote the order of growth two algorithms A and B, respectively; and also we have $g(n) = o(f(n)logn)$. Which of the following is more accurate?

a. For all n, A is faster than B.
b. For all n, B is faster than A.
c. For all $n > c$, A is faster than B.
d. For all $n > c$, B is faster than A.

217. Two functions $f(n)$ and $g(n)$ are given, both of which are nonnegative for n. Determine whether each of the following statements is "always correct" or "sometimes correct"?

(a) $f(n) + g(n) = \Theta(max f(n), g(n))$
(b) If $f(n) = \Omega(n^2)$ and $g(n) = \Omega(n)$ then $f(g(n)) = \Omega(n^3)$
(c) $f(g(n)) = \Theta(g(f(n)))$

218. Suppose $f(n) = O(g(n))$. Which of the following for each function with positive values such as $f(n)$ and $g(n)$ is correct?

a. $g(n) = \Theta(f(n))$
b. $2^{g(n)} = O(2^{f(n)})$
c. $2^{g(n)} = \Omega(2^{f(n)})$
d. $log(g(n)) = O(log(f(n)))(for\ log(g(n) \le 1)$

219. The runtime of the two algorithms A and B are at worst no greater than $150\ n\ log\ n$ and n^2, respectively. Which of the following is correct?

a. For large n, program B is better than A.
b. For large n, program A is better than B.
c. Perhaps program A is easier to implement.
d. For some n, program B is faster than program A.

Find the order of growth of the following sums (Questions 220–230):

220. $\sum_{i=3}^{n+1} 1$

221. $\sum_{i=3}^{n+1} i$

222. $\sum_{i=0}^{n-1} i(i + 1)$

223. $\sum_{i=1}^{n} 3^{i+1}$

224. $\sum_{i=1}^{n} \sum_{j=1}^{n} ij$

225. $\sum_{i=1}^{n} \frac{1}{i(i+1)}$

226. $\sum_{i=0}^{n-1} (i^2 + 1)^2$

227. $\sum_{i=2}^{n-1} \log i^2$

228. $\sum_{i=1}^{n} (i+1)2^{i-1}$

229. $\sum_{i=1}^{n} (n-i)$

230. $\sum_{i=0}^{n-1} \sum_{j=0}^{i-1} (i+j)$

Explain which of the following statements are true (Questions 231–236).

231. The running time of algorithm A is at least $O(n^2)$.

232. The running time of algorithm A is at best $O(n^2)$.

233. The running time of algorithm A is at worst $O(n^2)$.

234. The running time of algorithm A is at average $\Theta(n^2)$.

235. The running time of algorithm A is at least $\Omega(n^2)$.

236. The running time of algorithm A is at best $\Omega(n^2)$.

237. Does the function $\lceil \sqrt{n} \rceil!$ polynomially bounded?

238. Compare the following functions in terms of asymptotically?

$$log(log^2 n), \quad log^2(log n)$$

239. Compare the following functions in terms of asymptotically?

$$log(log^* n), \quad log^*(log n)$$

Let

$$t(n) = \sum_{i=0}^{d} a_i n^i$$

where $a > 0$, prove the following properties.

240. If $k \geq d$, then $t(n) = O(n^k)$.

241. If $k \leq d$, then $t(n) = \Omega(n^k)$.

242. If $k = d$, then $t(n) = \Theta(n^k)$.

243. If $k > d$, then $t(n) = o(n^k)$.

244. If $k < d$, then $t(n) = \omega(n^k)$.

245. Compare the growth order of the following functions.

$$(\ln(n))^n, \qquad n^{\ln(n)}$$

246. Compare the growth order of the following functions.

$$(\ln(n))!, \qquad n^{\ln \ln \ln(n)}$$

247. Compare the growth order of the following functions.

$$e^{c\sqrt{n}}, \qquad e^{\sqrt{n}}$$

248. Show that $\lceil log(n + 1) \rceil = \lfloor logn \rfloor + 1$.

249. In the following relations, an error occurred. Find this mistake.

$$n = O(n), \ 2n = O(n), \ 3n = O(n), \ \ldots$$

$$\sum_{k=1}^{n} kn = \sum_{k=1}^{n} O(n) = O(n^2)$$

250. Simplify the following statement.

$$(\ln(n) + O(n^{-1}))(n + O(n^{\frac{1}{2}}))$$

251. Show that for large n we have:

$$1 + \frac{2}{n} O(n^{-2}) = \left(1 + \frac{2}{n}\right)(1 + O(n^{-2}))$$

252. Generalize the asymptotic functions to two-variable functions.

253. Suppose that $f(n) = n^{log \ n}$ and $g(n) = (logn)^n$. Which of the following phrases is true?

$$f(n) = O(g(n))$$

$$f(n) = \Omega(g(n))$$

254. Suppose that $f(n) = n^{\log \log \log n}$ and $g(n) = (\log n)!$. Which of the following phrases is true?

$$f(n) = O(g(n))$$

$$f(n) = \Omega(g(n))$$

255. Suppose that $f(n) = (n!)!$ and $g(n) = ((n-1)!)!(n-1)!^{n!}$. Which of the following phrases is true?

$$f(n) = O(g(n))$$

$$f(n) = \Omega(g(n))$$

256. The approximate value of the factorial n can be represented as follows:

$$n! = \sqrt{2\pi n}\left(\frac{n}{e}\right)^n \left(1 + \Theta\left(\frac{1}{n}\right)\right)$$

which is called sterling number. Using this approximate value, check the correctness of the following relationships:

$$n! = o(n^n)$$

$$n! = \omega(2^n)$$

$$\log(n!) = \Theta(n \log n)$$

$$\sqrt{2\pi n}\left(\frac{n}{e}\right)^n \leq n! \leq \sqrt{2\pi n}\left(\frac{n}{e}\right)^{n+\frac{1}{12n}}$$

257. Let us assume in the table below $k \geq 1$, $\epsilon > 0$ and $c > 1$. For each row, check the following relationships. Then, write in the house corresponding to each correct relationship, the letter T, and with each false letter F. If there is no relationship, write "None".

$$f(n) = \Omega(g(n)), \quad f(n) = o(g(n)), \quad f(n) = \Theta(g(n)),$$

$$f(n) = O(g(n)), \quad f(n) = \omega(g(n)).$$

$f(n)$	$g(n)$	O	o	Ω	ω	Θ
$\log^k n$	n^ε					
n^k	c^n					
$\sqrt{n}$	$n^{\sin(n)}$					
2^n	$n^{n/2}$					
$n^{\log c}$	$c^{\log n}$					
$\log(n!)$	$\log(n^n)$					

In the table below, 30 functions $g_1(n)$, $g_2(n)$, ..., $g_{30}(n)$ are written from left to right respectively.

$\log(\log^* n)$	$(2/3)^n$	n^2	$n!$	$(n+1)!$
$\log^*(\log n)$	n^3	$\log(n!)$	$(\log n)!$	$\ln(n)$
$\ln\ln(n)$	$\log^* n$	$n^{\log\log n}$	1	2^{2^n}
$2^{\log n}$	$(\sqrt{2})^{\log n}$	$4^{\log n}$	$\sqrt{\log n}$	e^n
$2\sqrt{2\log n}$	$\log n^2$	2^n	$2^{2^{n+1}}$	n^{2^n}
$(\log n)^{\log n}$	$2^{\log^* n}$	n	$n\log n$	$n^{\log^{-1} n}$

Answer questions 258–260.

258. List these functions as follows:

$$g_1(n) = \Omega(g_2(n))$$

$$g_2(n) = \Omega(g_3(n))$$

$$g_3(n) = \Omega(g_4(n))$$

$$\vdots$$

$$g_{29}(n) = \Omega(g_{30}(n))$$

259. We define the relation R on this set of functions as follows:

$$g_i(n)\,R\,g_j(n) \Leftrightarrow g_i(n) = \Theta(g_i(n))$$

Show that R is an equivalence relation and find the equivalence classes R.

260. Find the function $f(n)$ for all i = 1, 2, ..., 30 values, so that:

$$f(n) \neq O(g_i(n))$$

$$f(n) \neq \Omega(g_i(n))$$

261. True or False? For each of the following statements, briefly explain your answer.

(a) If I prove that an algorithm takes $O(n^2)$ worst-case time, it is possible that it takes $O(n)$ on some inputs.
(b) If I prove that an algorithm takes $O(n^2)$ worst-case time, it is possible that it takes $O(n)$ on all inputs.
(c) If I prove that an algorithm takes $\Theta(n^2)$ worst-case time, it is possible that it takes $O(n)$ on some inputs.
(d) If I prove that an algorithm takes $\Theta(n^2)$ worst-case time, it is possible that it takes $O(n)$ on all inputs.

Assume f(n) and g(n) are two positive functions. Does the following phrases are true? Justify your answer.

262. $f(n) = O(g(n)) \Rightarrow g(n) = O(f(n))$

263. $f(n) + g(n) = \Theta(\min(f(n),\ g(n)))$

264. $f(n) = o(g(n)) \Rightarrow 2^{f(n)} = o(2^{g(n)})$

265. $f(n) = O(g(n)) \Rightarrow 2^{f(n)} = O(2^{g(n)})$

266. $f(n) = O(g(n)) \Rightarrow logf(n) = O(log\ g(n))$

267. $f(n) = O(g(n)) \Rightarrow f(n)/h(n) = O(g(n)/h(n))$

268. $f(n) = O((f(n))^2)$

269. $f(n) = O(g(n)) \Rightarrow g(n) = \Omega(f(n))$

270. $f(n) = \Theta\left(f\left(\frac{n}{2}\right)\right)$

271. $f(n) + o(f(n)) = \Theta(f(n))$

272. $f(n) = O(g(n)) \Rightarrow f(n)^k = O(g(n)^k)$

273.

(a) Prove the equality $a^{log_b c} = c^{log_b a}$.
(b) Why is $n^{log_2 3}$ better than $3^{log_2 n}$?

274. Iterated logarithm is defined as follows:

$$log^{(i)}n = \begin{cases} n, & i = 0 \\ log(lg^{(i-1)}) & i \geq 1,\ log^{(i-1)}n > 0 \\ undefined & i < 0 \end{cases}$$

we define:

$$log^*n = \min \left\{ i \geq 0 \,\middle|\, log^{(i)}\, n \leq 1 \right\}$$

For example, $log^*2^{256} = 5$, because $log^*\, 2^{256} < 1$.

Using this definition, get the following iterated logarithms.

$$log^*2, \quad log^*4, \quad log^*16, \quad log^*65536, \quad log^*2^{65536}$$

275. Assume that $f(n)$ is a positive function, c is a constant and i is a positive integer. We define the functions $f^{(i)}$ and f_c^* as follows:

$$f^{(i)}(n) = \begin{cases} f(f^{(i-1)}(n)), & i \geq 1 \\ n, & i = 0 \end{cases}$$

$$f_c^*(n) = \min \left\{ i \geq 0 \,\middle|\, f_{(n)}^{(i)} \leq c \right\}$$

For example, if $f(n) = \frac{n}{2}$, we have $f_2^* = \frac{n}{2^k} \leq 2$. Because:

$$f(n) = \frac{n}{2},\, f(f(n)) = f\left(\frac{n}{2}\right) = \frac{n}{4}$$

$$f(f(f(n))) = f\left(f\left(\frac{n}{2}\right)\right) = f\left(\frac{n}{4}\right) = \frac{n}{8}$$

So

$$f^{(k)}(n) = \frac{n}{2^k} \Rightarrow f_2^*(n) = \frac{n}{2^k} \leq 2$$

and for $\frac{n}{k^2} \leq 2$, we must have $n \leq 2^{(k+1)}$ and or $k \geq log\, n - 1$. For each of the following functions, find $f_c^*(n)$ and write in the table.

$f(n)$	c	$f_c^*(n)$
$log\, n$	1	
$n - 1$	0	
$n/2$	1	
$\sqrt{n}$	2	
$\sqrt{n}$	1	
$n^{1/3}$	2	
$\frac{n}{log\, n}$	2	

2.3   Solutions

The following notations and relations can be helpful in solving the growth of functions problems.

$$a^{-1} = \frac{1}{a}, \qquad (a^m)^n = a^{mn}, \qquad a^m a^n = a^{m+n}.$$

Some notations:

$log\ n = log_2\ n$ (binary logarithm),

$ln\ n = log_e\ n$ (natural logarithm),

$log^k\ n = (log\ n)^k$ (exponentiation),

$log\ log\ n = log(log\ n)$ (composition).

Useful relations for logarithm function, where $a > 0$, $b > 0$, $c > 0$, and n, are real and logarithm bases are not 1:

$a = b^{log_b\ a}$,

$log_c\ (ab) = log_c\ a + log_c\ b$,

$log_b\ a^n = n log_b\ a$,

$log_b\ a = \frac{log_c\ a}{log_c\ b}$,

$log_b\ (1/a) = -log_b\ a$,

$log_b\ a = \frac{1}{log_a b}$,

$a^{log_b\ c} = c^{log_b a}$.

For each $b > 1$ and $x > 0$; $log_b n \leq O(n^x)$.

For each $r > 1$ and $d > 0$; $n^d \leq O(r^n)$.

For all $x > 0$, $log_2 x \leq x$.

For example:

1. $(log\ n)^{log\ n} = n^{log log\ n}$ because $a^{log_b c} = c^{log_b a}$.

2. $4^{log\ n} = n^2$ because $a^{log_b c} = c^{log_b a}$.

3. $2^{log\ n} = n$ because $a^{log_b c} = c^{log_b a}$.

4. $2 = n^{\frac{1}{log\ n}}$, because by applying $log_b\ a = \frac{1}{log_a b}$ to the power $\frac{1}{log\ n}$ and then using $a^{log_b\ c} = c^{log_b a}$.

5. $2^{\sqrt{2 log\ n}} = n^{\sqrt{\frac{2}{log n}}}$.

6. $(\sqrt{2})^{log\ n} = \sqrt{n}$, because $(\sqrt{2})^{log\ n} = 2^{\frac{1}{2} log n} = 2^{log \sqrt{n}} = \sqrt{n}$.

7. $log^*(log\ n) = (log^* n) - 1$.

Solutions

8. True
9. True
10. False
11. True
12. True
13. False
14. True
15. True
16. False
17. False
18. True
19. False
20. True
21. True
22. True
23. False
24. False
25. False
26. True
27. False

28.

$$O(log\ n) < O(\sqrt{n}) < O(n) < O(a^n) < O(n!)$$

29.
$$n^{1000} < (1.005)^n < n!$$

30.
$$O(n\log n) < O\left(\frac{n^2}{\log n}\right) < O(1+\varepsilon)^n$$

31.
$$\sqrt{\log(n!)} < 2^{\sqrt{2\log n}} < n\sqrt{n} < \frac{n^2}{\log n} < (\log n)^{\log n}$$

32.
$$\ln^2 n < 5\log(10n+1)^{10} < \sqrt[3]{n} < 0.0001n^4 < 2^{2n} < (n-10)!$$

33.
$$\left.\begin{array}{l} f(n) = 4^{\log n} = n^{\log 4} = n^{\log_2^4} = n^2 \\ g(n) = (\log n)^{\log n} = n^{\log\log n} \\ h(n) = \log^2 n \end{array}\right\} \Rightarrow h(n) < f(n) < g(n)$$

34.
$$1 = n^{\frac{1}{\log n}} < \log(\log^* n) < \log^*(\log n) \simeq \log^* n < 2^{\log^* n} < \ln\ln n < \sqrt{\log n} < \ln n$$

$$< \log^2 n < 2^{\sqrt{2\log n}} < (\sqrt{2})^{\log n} < n = 2^{\log n} < \log n!$$

$$= n\log n < n^2 = 4^{\log n} < n^3 < \log n! < (\log n)^{\log n} = n^{\log\log n} < \left(\frac{3}{2}\right)^n < e^n$$

$$< 2^n < n2^n < n! < (n+1)! < 2^{2^n} < 2^{2^{n+1}}$$

35.
$$\left(\frac{2}{5}\right)^n < \left(\frac{3}{5}\right)^n < 1000 < \log(\log n^2) < \log n < \ln^2 n < (\log n)^3 < \sqrt[3]{n} + \log n < \sqrt{n}$$

$$< \frac{2n}{\log n} < n < n\log n < n^2 = 10n^2 + \log n < n^4 < n - n^4 + 10n^5 < 2^n < n!$$

38.
$$n^3 \log n = O(n^{3+\varepsilon}) \quad 0 < \varepsilon < 0.1$$

39. 3

40.

$$f_1(n) = n^2(\log n)^2 \; f_2(n) = n^2(\log\sqrt{n})^3 = \frac{1}{8}n^2(\log n)^3$$

$$f_3(n) = 4^{(\log n)+(\log\log n)} = 4^{\log n} \times 4^{\log\log n} = n^{\log_2 2} \times \log n^{\log_2 4}$$

$$= n^2(\log n)^2 \rightarrow f_1(n) = f_3(n)$$

So: $f_1(n) = f_3(n) < f_2(n)$

41.

$$n^{\frac{1}{10}} \; \epsilon \; \Omega(\log n)$$

42.

$$f(n) = 2n^2 + 4n, \; g(n) = n^3$$

$$f(n) \le cg(n) \rightarrow 2n^2 + 4n \le c(n^3)$$

$$\xrightarrow{c=1} n = 1, \; 2 \times 1 + 4 \times 1 \le 1 \times 1 \rightarrow 6 \le 1 \; \text{False}$$
$$n = 2, \; 2 \times 4 + 4 \times 2 \le 1 \times 8 \rightarrow 16 \le 8 \quad \text{False}$$
$$n = 3, \; 2 \times 9 + 4 \times 3 \le 1 \times 27 \rightarrow 30 \le 27 \; \text{False}$$
$$n = 4, \; 2 \times 16 + 4 \times 4 \le 1 \times 64 \rightarrow 48 \le 64 \; \text{True}$$

So, we have c = 1 and n = 4.

45.

$$f(n) = 4 \times 2^n + n^3 \; , \; g(n) = n2^n$$

$$f(n) \le cg(n) \rightarrow 4 \times 2^n + n^3 \le cn2^n \xrightarrow{(c=2\,,\;n=4)} 2^7 \le 2^7$$

47.

$$\begin{cases} f(n) = 2\lfloor\sqrt{n}\rfloor - 5 \\ g(n) = n - \lceil\sqrt{n}\rceil \end{cases} \rightarrow \frac{f}{g} \le c \; (for \; all \; n > 1) = \frac{2\lfloor\sqrt{n}\rfloor - 5}{n - \lceil\sqrt{n}\rceil} \le c$$

$$\lim_{n\to\infty} \left(\frac{2\lfloor\sqrt{n}\rfloor - 5}{n - \lceil\sqrt{n}\rceil}\right) = \frac{2\sqrt{n} - 5}{n - \sqrt{n}} = 0$$

Since g is always greater than f, we can conclude that for every c > 1, the equation $f \le cg$ is always true.

48.

$$n^2 + 10n \le c(n^2)$$

$$\xrightarrow{c=2} n = 1, \qquad 1 + 10 \le 2 \times 1 \to 10 \le 2 \quad \text{False}$$

$$n = 2, \qquad 4 + 20 \le 2 \times 4 \to 24 \le 8 \quad \text{False}$$

$\vdots$

$$n = 10, \qquad 100 + 100 \le 2 \times 100 \to 200 \le 200 \quad \text{True}$$

So, we have c = 2 and n = 10.

50.

$$\frac{n(n-1)}{2} \le cn^2$$

$$c = \frac{1}{2} \to \frac{n(n-1)}{2} \le \frac{n^2}{2} \to \frac{n(n-1)}{2} = \frac{n^2}{2} \to n_0 = 0$$

52.

$$n! + 7n^5 \le c \times n^n \xrightarrow{c=2} n! + 7n^5 \le 2 \times n^n$$

$$\xrightarrow{n=6} 6! + 7(6^5) \le 2 \times 6^6 = 2 \times 6 \times 6^5 = 12 \times 6^5$$

$$\to 6! + 7(6^5) \le 12 \times 6^5$$

So, we have c = 2 and $n_0 \ge 6$.

59.

$$3n \lfloor \log n \rfloor = O(n^2)$$

$$\frac{3n \lfloor \log n \rfloor}{n^2} \to (\lfloor s \rfloor \le s \, for \, all \, s > 1) \le \frac{3n \log n}{n^2} \to (\log s \le s \, for \, all \, s > 1) \le \frac{3n^2}{n^2} = 3$$

Thus $3n \lfloor \log n \rfloor$ *is* $O(n^2)$ *because* $3n \lfloor \log n \rfloor \le 3n^2$ *whenever* $n > 1$

61.

$$5n^2 \ge c(n^2)$$

$$\xrightarrow{c=1} n = 1, \quad 5 \times 1 \ge 1 \times 1 \to 5 \ge 1 \quad \text{True}$$

So, we have c $= 2$ and $n_0 \geq 1$.

65.

$$f_{(n)} \geq cg(n) \rightarrow 2n^3 + 7n^2 \geq cn^2 \rightarrow c = 1, \ n \geq 2$$

66.

$$2n^3 + 7n^2 \geq cn^3$$

$$c = 2 \rightarrow 2n^3 + 7n^2 \geq 2n^3$$

$$7n^2 \geq 0 \ \rightarrow \ n \geq 0 \rightarrow n_0 = 0$$

69.

$$0.5n^2 + 3n + 1 \in \ \Theta \ (n^2)$$

$$c_1(n^2) \ \leq 0.5n^2 + 3n + 1 \leq c_2(n^2)$$

$$\overset{c_1=1, \ c_2=2}{\longrightarrow} n = 1, \ \ 1 \times 1 \leq 0.5 + 3 \leq 2 \times 1 \rightarrow 1 \leq 3.5 \leq 2 \ \ \text{False}$$

$$n = 2, \ \ 1 \times 4 \leq 0.5 \times 4 + 3 \times 2 \leq 2 \times 4 \rightarrow 4 \leq 8 \leq 8 \ \ \text{True}$$

$$n = 3, \ \ 1 \times 9 \leq 0.5 \times 9 + 3 \times 3 \leq 2 \times 9 \rightarrow 9 \leq 13.5 \leq 18 \ \ \text{True}$$

So, we have $c_1 = 1$, $c_2 = 2$, and $n = 2$.

70.

$$c_1 n^2 \leq 60n^2 + 5n + 1 \leq c_2 n^2$$

$$c_1 = 1 \quad c_2 = 66$$

$$n = 2 \quad \implies \quad 2^2 \leq 60(2)^2 + 5(2) + 1 \leq 66(2)^2$$

$$4 \leq 261 \leq 264$$

72.

$$\log n! \ = \ \theta(n\log n)$$

$$\implies \log n! = \log n + \log n - 1 + \cdots + \log 1 \leq \underbrace{\log n + \log n + \cdots + \log n}_{n}$$

$$\leq n\log n$$

$$\implies \log n! \in \ O(n\log n)$$

$$\log n! = log\, n + log\, n - 1 + \cdots + log\, 1 \geq \frac{n}{2}\log\frac{n}{2}$$

$$\implies \log n! \geq \frac{n}{2}\log n - \frac{n}{2}\log 2 \geq \frac{1}{4}\,n\log n$$

$$\implies \log n! \in \Omega(n\log n)$$

74.

$$c_1(n^2) \leq 7n^2 - 6n + 2 \leq c_2(n^2)$$

$$7n^2 - 6n + 2 \leq c_2(n^2)$$

$$7n^2 - 6n + 2 \leq 7n^2$$

$$-6n + 2 \leq 0 \rightarrow n \geq \frac{1}{3} \rightarrow n_0 = \frac{1}{3} \text{ and } c_2 = 7$$

$$7n^2 - 6n + 2 \geq c_1(n^2)$$

$$7n^2 - 6n + 2 \geq n^2$$

$$3n^2 - 3n + 1 \geq 0 \rightarrow n_0 = 1 \text{ and } c_1 = 1$$

$$\rightarrow n = 2\,, c_1 = 1\,, c_2 = 7$$

112. $\lim_{n \to \infty} \frac{\frac{1}{2}(n^2 - n)}{n^2} = \frac{1}{2}\lim_{n \to \infty} \frac{(n^2 - n)}{n^2} = \frac{1}{2}\lim_{n \to \infty}(1 - \frac{1}{n}) = \frac{1}{2}$ so

$\begin{cases} f(n) \in O(g(n)) \\ f(n) \in \Omega(g(n)) \text{ are correct.} \\ f(n) \in \theta(g(n)) \end{cases}$

This means that all three $f(n) \in O(g(n))$, $f(n) \in \Theta(g(n))$ and $f(n) \in \Omega(g(n))$ are correct.

113.

$$\lim_{n \to \infty} \frac{f(n)}{g(n)} \neq 0 \rightarrow f(n) \in \Omega(g(n))$$

$$\lim_{n \to \infty} \frac{n^3}{n^2} = n = \infty \rightarrow n^3 \in \Omega(n^2)$$

114. $\lim_{n \to \infty} \frac{\frac{n^2}{2} + 3n}{n^2} \xrightarrow{hop} \frac{n+3}{2n} = \frac{1}{2}\lim_{n \to \infty} 1 + \frac{3}{n} = \frac{1}{2} \rightarrow 0.5n^2 + 3n \in \Theta(n^2)$ is correct.

119.

$$\lim_{n\to\infty} \frac{n^4 + 700n}{n^4} \xrightarrow{hop} \frac{4n^3 + 1400n^2}{4n^3} = \lim_{n\to\infty} 1 + \frac{350}{n} = 1$$

So $n^4 + 700n^2 \in \Theta(n^4)$ is correct.

120.

$$\lim_{n\to\infty} \frac{\frac{12n^5}{\log n} + 7n^2}{n^5} = \frac{12n^5 + 7n^2 \log n}{n^5 \log n} = \frac{12}{\log n} + \frac{7}{n^3} = 0$$

122.

$$\lim_{n\to\infty} \frac{f_{(n)}}{g_{(n)}} = \frac{f'_{(n)}}{g'_{(n)}}$$

$$\lim_{n\to\infty} \frac{2n^3 + 7n^2}{n^3} = \frac{6n^2 + 14n}{3n^2} = \lim_{n\to\infty} 2 + \frac{14}{3n} = 2$$

So $\begin{cases} f(n) \in O(g(n)) \\ f(n) \in \Omega(g(n)) \\ f(n) \in \theta(g(n)) \end{cases}$ are correct.

123.

$$\lim_{n\to\infty} \frac{(n+1)^2}{n^2} = \frac{n^2 + 2n + 1}{n^2} = \frac{2n+2}{2n} = \lim_{n\to\infty} 1 - \frac{1}{n} = 1$$

So $\begin{cases} f(n) \in O(g(n)) \\ f(n) \in \Omega(g(n)) \\ f(n) \in \theta(g(n)) \end{cases}$ are correct.

124.

$$\lim_{n\to\infty} \frac{3n \lfloor \log n \rfloor}{n^2} \to (if \ n \notin \mathbb{Z}) = \frac{3\log n}{n} = 0$$

125.

$$\lim_{n\to\infty} \frac{n}{2^n} = 0 \ \to \ f(n) < g(n) \ \to \ f(n) \in O(g(n))$$

126.

$$\lim_{n\to\infty} \frac{100000n + 1}{2^n} = 0 \to f(n) < g(n) \to 100000n + 1 = O(2^n)$$

129. $T(n) = 6n + 4 \in \Omega(n) \geq 6n$, then for $n_0 = 1$ and $C = 6$, we have $T(n) = 6n + 4 \in \Omega(n)$

130. Suppose there is a C and n_0 so that for $n \geq n_0$ we have:
$T(n) = 3n + 2 \geq Cn^2 \Rightarrow Cn^2 - 3n - 2 \leq 0$.

As you can see, there is not a certain value in this inequality for C. Therefore, $T(n) = 3n + 2 \notin \Omega(n^2)$.

131.

$$\lim_{n\to\infty} \frac{5^n + n^2}{2^n} = \left(\frac{5}{2}\right)^n = \infty$$

132.

$$\lim_{n\to\infty} \frac{\log_a n}{\log_b n} = \lim_{n\to\infty} \frac{\frac{\log_c n}{\log_c a}}{\frac{\log_c n}{\log_c b}} = \frac{\log_c b}{\log_c a} = \log_a b > 0 \Rightarrow \log_a n = \Theta(\log_b n).$$

134.

$$n^3 + 20n + 1 = cn^3$$

$$\to c = 2 \, , \, n_0 = 4$$

$$\to n^3 + 20n + 1 = 4n^3$$

135. We know that n^3 grows faster than n^2, so the statement is true.

136.

$$n^3 + 20n \geq cn^2$$

$$\to c = 1 \, , \, n_0 = 1$$

$$\to \ n^3 + 20n \geq n^2$$

137.

$$f(n) = 5n^2 + 2n + 10 \, , \ g(n) = n$$

$$\lim_{n\to\infty} \frac{f(n)}{g(n)} = \frac{f(n)'}{g(n)'} \ \to$$

$$\lim_{n\to\infty} \frac{5n^2 + 2n + 10}{n} = \lim_{n\to\infty} 10n + 2 = \infty \to f(n) > g(n) \to f(n) \in \Omega\,(g(n))$$

138.

$$\lim_{n\to\infty} \frac{f(n)}{g(n)} = \frac{n^3 + 100n^2 + 40}{10000n + 20} = \lim_{n\to\infty} \frac{3n^2 + 200n}{10000} = n^2 = \infty \to f(n) > g(n) \to f(n) \in \Omega\,(g(n))$$

139.

$$\lim_{n\to\infty} \frac{f(n)}{g(n)} = lim_{n\to\infty} \frac{6\frac{n^3}{logn}}{n^3} = lim_{n\to\infty} \frac{6}{logn} = 0 \to f(n) \in O(g(n))$$

140.

$$\lim_{n\to\infty} \frac{f(n)}{g(n)} = lim_{n\to\infty} \frac{2\sqrt{n}+10}{\log(n+10)} = \frac{2\sqrt{n}+10}{logn} = \infty \to \Omega$$

141.

$$\lim_{n\to\infty} \frac{f(n)}{g(n)} = lim_{n\to\infty} \frac{n^{2^n}+102^n}{n^{2^n}} = \frac{n^{2^n}}{n^{2^n}} = 1 \to \Theta$$

142.

$$f(n) = n\sqrt{n}+n, \ g(n) = n^2 - 10n$$

$$\lim_{n\to\infty} \frac{n\sqrt{n}+n}{n^2-10n} = \lim_{n\to\infty} \frac{n^{\frac{3}{2}}}{n^2} = \lim_{n\to\infty} n^{\frac{3}{2}-2} = \lim_{n\to\infty} n^{\frac{-1}{2}} = \lim_{n\to\infty} \frac{1}{\sqrt{n}} = \lim_{n\to\infty} \frac{1}{\infty} = 0$$

$$f(n) = o(g(n))$$

143.

$$\lim_{n\to\infty} \frac{n\sqrt{n}+n}{n^2-10n} \xrightarrow{hop} \lim_{n\to\infty} \frac{\frac{3}{2}n^{\frac{1}{2}}}{2n-10} = 0 \to f(n) < g(n) \to f(n) = o(g(n))$$

144. $f(n) = \Omega(g(n))$

145. $f(n) = \Omega(g(n))$

146. $f(n) = \Omega(g(n))$

147.

$$\lim_{n\to\infty} \frac{4\times 2^n + n^2}{2^n} = \lim_{n\to\infty} 4\times \left(\frac{n^2}{2^n}\right) = 4 \to f(n) = g(n) \to f(n) \in \Theta(g(n))$$

149.

$$\lim_{n\to\infty} \frac{(logn)'}{(loglog(n^2))'} = \lim_{n\to\infty} \frac{\frac{1}{n}\log_2 e}{\frac{\log_2 e}{2logn}} = \lim_{n\to\infty} \frac{(2logn)'}{n'} = 2\lim_{n\to\infty} \frac{\frac{1}{n}\log_2 e}{1}$$

$$= 2\log_2 e \lim_{n\to\infty} \frac{1}{n} = 2\log_2 e \lim_{n\to\infty} \frac{1}{\infty} = 0$$

$$f(n) = o(g(n))$$

150.

$$\lim_{n\to\infty} \frac{\frac{n^3}{logn}}{n(logn)^3} =$$

$$\lim_{n\to\infty} \frac{(n^2)^{'}}{((logn)^4)^{'}} = \lim_{n\to\infty} \frac{2n}{4(logn)^3 \times \frac{1}{n^4}(log_2e)^{\,4}} = \frac{1}{2(log_2e)^{\,4}} \lim_{n\to\infty} \frac{(n^5)^{'}}{((logn)^3)^{'}}$$

$$= \frac{1}{2(log_2e)^{\,4}} \lim_{n\to\infty} \frac{5n^4}{3(logn)^2 \times \frac{1}{n^3}(log_2e)^{\,3}} = \frac{5}{2(log_2e)^{\,7}} \lim_{n\to\infty} \frac{(n^7)^{'}}{((logn)^2)^{'}}$$

$$= \frac{5}{2(log_2e)^{\,7}} \lim_{n\to\infty} \frac{7n^6}{2logn \times \frac{1}{n^2}(log_2e)^{\,2}}$$

$$= \frac{35}{4(log_2e)^{\,9}} \lim_{n\to\infty} \frac{(n^8)^{'}}{(logn)^{'}} = \frac{35}{4(log_2e)^{\,9}} \lim_{n\to\infty} \frac{8n^7}{\frac{1}{n}log_2e} = \frac{70}{(log_2e)^{\,10}} \lim_{n\to\infty} n^8 = \infty$$

$$f(n) = \Omega(g(n))$$

154.

$$\frac{(n+1)(n+3)}{n+2} = \frac{n^2+4n+3}{n+2} = n+1+\frac{n+1}{n+2}$$

$$T(n) = n+1+\frac{n+1}{n+2} \le n+1+1 = n+2 \le n+2n = 3n \qquad n \ge 1$$
$$\Rightarrow \quad T(n) = O(n)$$

$$T(n) = n+1+\frac{n+1}{n+2} \ge n \quad \Rightarrow \quad T(n) = \Omega(n) \quad \Rightarrow \quad T(n) = \theta(n)$$

156.

$$2+4+6+\cdots+2n = \sum_{i=1}^{n} 2i$$

$$\sum_{i=1}^{n} 2i \in \Theta(n^2) \rightarrow c_1(n^2) \le \sum_{i=1}^{n} 2i \le c_2(n^2)$$

$$\overset{c_1=1,c_2=2}{\longrightarrow} n=1, \qquad 1 \times 1 \le 2 \times 1 \le 2 \times 1 \rightarrow 1 \le 2 \le 2 \quad \text{True}$$
$$n=2, \qquad 1 \times 4 \le 2 \times 1 + 2 \times 2 \le 2 \times 4 \rightarrow 4 \le 6 \le 8 \quad \text{True}$$
$$n=3, \qquad 1 \times 9 \le 2 \times 1 + 2 \times 2 + 2 \times 3 \le 2 \times 9 \rightarrow 9 \le 12 \le 18 \quad \text{True}$$

So, we have $c_1 = 1$, $c_2 = 2$, n = 1.

157.

$$\lim_{n \to \infty} \frac{\frac{(n^2+\log n)(n+3)}{n+n^2}}{n} = \frac{n^3 + 3n^2 + n\log n + 3\log n}{n^3 + n^2} = 1 \to f(n) = \theta(n)$$

158.

$$2 + 4 + \cdots + 2^n = \sum_{i=1}^{n} i^2 = \frac{n(n+1)(2n+1)}{6}$$

$$\lim_{n \to \infty} \frac{\frac{n(n+1)(2n+1)}{6}}{n^3} = \lim_{n \to \infty} \frac{n^3}{n^3} = 1 \implies 2 + 4 + \cdots + 2^n \in \theta(n^3)$$

159.

$$\sqrt{10n^2 + 7n + 3} \to \sqrt{n^2} \to n \to \sqrt{10n^2 + 7n + 3} \in \theta(n)$$

$$c_1 n \leq \sqrt{10n^2 + 7n + 3} \leq c_2 n \xrightarrow{(c_1=1, c_2=4, n=2)} 2 \leq \sqrt{57} \leq 8$$

165.

$$c_2 f(n) = h(n) = c_1 f(n)$$

$$c_4 f(n) = h(n) = c_3 g(n)$$

$$c_2 f(n) + c_4 g(n) = 2h(n) = c_1 f(n) + c_3 g(n)$$

$$\Rightarrow \quad \frac{c_2}{2} f(n) + \frac{c_4}{2} g(n) \leq h(n) \leq \frac{c_1}{2} f(n) + \frac{c_3}{2} g(n)$$

$$c_2' = Max \left\{ \frac{c_2}{2}, \frac{c_4}{2} \right\} \quad , \quad c_1' = Max \{c_1, c_3\}$$

$$\Rightarrow \quad c_2'(f(n) + g(n)) \leq h(n) \leq c_1'(f(n) + g(n))$$
$$\Rightarrow \quad f(n) + g(n) = \theta(h(n))$$

173.

$$c_1 g(n) \leq f(n) \leq c_2 g(n), \ c_3 t(n) \leq h(n) \leq c_4 t(n), \ c_1 g(n) - c_3 t(n) \leq f(n) - h(n) \leq c_2 g(n) - c_4 t(n)$$

$$f(n) = 10n^2 + 4n \ , \ g(n) = n^2 \ , \ h(n) = 5n^2 + 4n \ , \ t(n) = 2n^2$$

$$c_1 g(n) \leq f(n) \leq c_2 g(n)$$

$$\xrightarrow{c_1=1, \ c_2=15} n = 1, \qquad 1 \times 1 \leq 10 + 4 \leq 15 \to 1 \leq 14 \leq 15 \quad True$$

$$c_3 t(n) \leq h(n) \leq c_4 t(n)$$

$$\xrightarrow{c_3=1, c_4=4.5} n = 1, \qquad 1 \times 2 \leq 5 + 4 \leq 4.5 \times 2 \to 2 \leq 9 \leq 9 \quad \text{True}$$

$$c_1 g(n) - c_3 t(n) \leq f(n) - h(n) \leq c_2 g(n) - c_4 t(n)$$

$$\to c_1(n^2) - c_3(2n^2) \leq 10n^2 + 4n - 5n^2 - 4n \leq c_2(n^2) - c_4(2n^2)$$

$$n = 1 \to 1(1) - 1(2) \leq 5 \times 1 \leq 15(1) - 4.5(2) \to -1 \leq 5 \leq 6 \quad \text{True}$$

so, we have

$$f(n) = 10n^2 + 4n, \ g(n) = n^2, \ h(n) = 5n^2 + 4n, \ t(n) = 2n^2$$

181. $f(n) \leq \max\{f(n), g(n)\}$ and $g(n) \leq \max\{f(n), g(n)\}$
$\Rightarrow f(n) + g(n) \leq 2\max\{f(n), g(n)\}$
$\Rightarrow f(n) + g(n) = O(\max\{f(n), g(n)\})$
$\Rightarrow \max\{f(n), g(n)\} \leq f(n) + g(n) \Rightarrow f(n) + g(n) = \Theta(\max\{f(n), g(n)\})$

184.
$$a_k n^k + a_{k-1} n^{k-1} + \cdots + a_1 n + a_0 = \Theta(n^k)$$

Suppose:

$$C = a_k + a_{k-1} + \cdots + a_1 + a_0$$

Then:

$$a_k n^k + a_{k-1} n^{k-1} + \cdots + a_1 n + a_0$$

$$= a_k n^k + a_{k-1} n^k + \cdots + a_1 n^k + a_0 n^k$$

$$= (a_k + a_{k-1} + \cdots + a_1 + a_0) n^k = C n^k$$

Hence:
$$a_k n^k + a_{k-1} n^{k-1} + \cdots + a_1 n + a_0 = O(n^k)$$

Also:

$$a_k n^k + a_{k-1} n^{k-1} + \cdots + a_1 n + a_0 = a_k n^k$$

$$a_k n^k + a_{k-1} n^{k-1} + \cdots + a_1 n + a_0 = \Omega(n^k)$$

Therefore:

$$a_k n^k + a_{k-1} n^{k-1} + \cdots + a_1 n + a_0 = \Theta(n^k)$$

185.

$$1 - f_{1(n)} = O(g_{1(n)}) \Rightarrow f_{1(n)} \leq c_1(g_{1(n)})$$
$$2 - f_{2(n)} = O(g_{2(n)}) \Rightarrow f_{2(n)} \leq c_2(g_{2(n)})$$

$$\xrightarrow{1+2} f_{1(n)} + f_{2(n)} \leq c_1(g_{1(n)}) + c_2(g_{2(n)})$$
$$\Rightarrow f_{1(n)} + f_{2(n)} \leq c_3((g_{1(n)}) + (g_{2(n)}))$$
$$\Rightarrow f_{1(n)} + f_{2(n)} = O((g_{1(n)}) + (g_{2(n)}))$$

186.

$$1 - f_{1(n)} = \Omega(g_{1(n)}) \Rightarrow f_{1(n)} \geq c_1(g_{1(n)})$$
$$2 - f_{2(n)} = \Omega(g_{2(n)}) \Rightarrow f_{2(n)} \geq c_2(g_{2(n)})$$

$$\xrightarrow{1+2} f_{1(n)} + f_{2(n)} \geq c_1(g_{1(n)}) + c_2(g_{2(n)})$$
$$\Rightarrow f_{1(n)} + f_{2(n)} \geq c_3((g_{1(n)}) + (g_{2(n)}))$$
$$\Rightarrow f_{1(n)} + f_{2(n)} = \Omega((g_{1(n)}) + (g_{2(n)}))$$

189.

$$1 - f_{1(n)} = O(g_{1(n)}) \Rightarrow f_{1(n)} \leq c_1(g_{1(n)})$$
$$2 - f_{2(n)} = O(g_{2(n)}) \Rightarrow f_{2(n)} \leq c_2(g_{2(n)})$$

$$\xrightarrow{1 \times 2} f_{1(n)} \times f_{2(n)} \leq c_1(g_{1(n)}) \times c_2(g_{2(n)})$$
$$\Rightarrow f_{1(n)} \times f_{2(n)} \leq c_3((g_{1(n)}) \times (g_{2(n)}))$$
$$\Rightarrow f_{1(n)} \times f_{2(n)} = O((g_{1(n)}) \times (g_{2(n)}))$$

190.

$$1 - f_{1(n)} = \Omega(g_{1(n)}) \Rightarrow f_{1(n)} \geq c_1(g_{1(n)})$$
$$2 - f_{2(n)} = \Omega(g_{2(n)}) \Rightarrow f_{2(n)} \geq c_2(g_{2(n)})$$

$$\xrightarrow{1 \times 2} f_{1(n)} \times f_{2(n)} \geq c_1(g_{1(n)}) \times c_2(g_{2(n)})$$
$$\Rightarrow f_{1(n)} \times f_{2(n)} \geq c_3((g_{1(n)}) \times (g_{2(n)}))$$
$$\Rightarrow f_{1(n)} \times f_{2(n)} = \Omega((g_{1(n)}) \times (g_{2(n)}))$$

194.

$$1 - f_{1(n)} = \Theta(h_{(n)}) \Rightarrow c_1(h_{(n)}) \leq f_{1(n)} \leq c_2(h_{(n)})$$
$$2 - g_{1(n)} = \Theta(h_{(n)}) \Rightarrow c_3(h_{(n)}) \leq g_{1(n)} \leq c_4(h_{(n)})$$
$$\xrightarrow{1+2} c_1(h_{(n)}) + c_3(h_{(n)}) \leq f_{1(n)} + g_{1(n)} \leq c_2(h_{(n)}) + c_4(h_{(n)})$$
$$\Rightarrow c_5((h_{(n)}) + (h_{(n)})) \leq f_{1(n)} + g_{1(n)} \leq c_6((h_{(n)}) + (h_{(n)}))$$
$$\Rightarrow f_{1(n)} + g_{1(n)} = \Theta(h_{(n)})$$

195. $x = 2^{\sqrt{2logn}} \Rightarrow log_n x = \sqrt{2log\ n} \times log_n 2 = \frac{\sqrt{2logn} \times 1}{log_2 n} = \frac{\sqrt{2}}{\sqrt{logn}} = \sqrt{\frac{2}{logn}}$

$\Rightarrow x = n^{\sqrt{\frac{2}{logn}}}$

203.

$$f = o(g(n)) \ and \ f \neq \theta(g(n))$$

$$f(n) = n \ \ and \ \ g(n) = 2^n$$

207.

$$f(n) = 2n^2, g(n) = n^2 + 1$$

208.

$$1 + 2 + \cdots + n = \frac{n(n+1)}{2} = \Theta(n^2)$$

$$\rightarrow \ C_1 n^2 \leq \frac{n(n+1)}{2} \leq C_2 n^2$$

$$\rightarrow \ C_1 = \frac{1}{2} \ , \ C_2 = 2 \ , \ n_0 = 1$$

209. For n =1, we have:

$$1^k + 2^k + \cdots + n^k = n.n^k = n^{k+1}$$

Then:

$$1^k + 2^k + \cdots + n^k = O(n^{k+1})$$

Also:

$$1^k + 2^k + \cdots + n^k = [n/2]^k + \cdots + (n-1)^k + n^k = [(n+1)/2][n/2]^k$$

$$= n^{k+1}/2^{k+1} 1^k + 2^k + \cdots + n^k = \Omega(n^{k+1})$$

$$1^k + 2^k + \cdots + n^k = \Theta(n^{k+1})$$

210.

$$1 + \frac{1}{2} + \frac{1}{3} + \cdots + \frac{1}{n} \ is \ Harmonic \ numbers, \ let \ it \ be \ H_n$$

$$\frac{1}{2} \leq \int_1^2 \frac{dx}{x} \leq 1 \ , \ \ \ \frac{1}{3} \leq \int_2^3 \frac{dx}{x} \leq \frac{1}{2} \ , \ \ \ \frac{1}{4} \leq \int_3^4 \frac{dx}{x} \leq \frac{1}{3} \ , \ \ \ \cdots$$

$$\frac{1}{2} + \frac{1}{3} + \frac{1}{4} + \cdots + \frac{1}{n+1} \leq \int_1^{n+1} \frac{dx}{x} \leq 1 + \frac{1}{2} + \frac{1}{3} + \cdots + \frac{1}{n}$$

$$1 + \int_1^n \frac{dx}{x} \geq H_n \geq \int_1^{n+1} \frac{dx}{x}$$

$$1 + \log n > H_n > \log(n+1)$$

Because $1 + \log n$ and $\log(n+1) \in \Theta(\log n)$ by Squeeze Theorem $H_n \in \Theta(\log n)$.

or

consider the following relationships:

$$(1) \quad \frac{1}{2} + \frac{1}{3} + \cdots + \frac{1}{n} < \log_e^n$$

$$(2) \quad \log_e^n < 1 + \frac{1}{2} + \cdots + \frac{1}{n-1}$$

Using relation 1:

$T(n) = 1 + \frac{1}{2} + \cdots + \frac{1}{n} \Rightarrow T(n) - 1 < \log n \Rightarrow T(n) < \log n + 1$

$T(n) = O(\log n)$

Using relation 2:

$$\log_e^n < T(n) - \frac{1}{n} \Rightarrow \log_e^n + \frac{1}{n} < T(n) \Rightarrow \log_e^n < T(n)$$
$$\Rightarrow T(n) = \Omega(\log_e^n)$$

Therefore:

$$T(n) = \Theta(\log_e^n) = \Theta(\log n)$$

213. To show that $(n+a)^b = \Theta(n^b)$, we want to find constants $c_1, c_2, n_0 > 0$ such that $0 \leq c_1 n^b \leq (n+a)^b \leq c_2 n^b$ for all $n \geq n_0$.

Note that:

$$n + a \leq n + |a| \leq 2n, \quad \text{when } |a| \leq n,$$

and

$$n + a \geq n - |a| \geq \frac{1}{2}n, \quad \text{when } |a| \leq \frac{1}{2}n.$$

Thus, when $n \geq 2|a|$,

$$0 \leq \frac{1}{2}n \leq n + a \leq 2n.$$

Since $b > 0$, the inequality still holds when all parts are raised to the power b:

$$0 \le \left(\frac{1}{2}n\right)^b \le (n+a)^b \le (2n)^b,$$

$$0 \le \left(\frac{1}{2}\right)^b n^b \le (n+a)^b \le 2^b n^b.$$

Thus, $c_1 = (1/2)^b$, $c_2 = 2^b$, and $n_0 = 2|a|$ satisfy the definition.

215. Are not comparable with each other.

220.

$$\sum_{i=3}^{n+1} 1 \to \sum_{i=1}^{n+1} 1 - \sum_{i=1}^{2} 1 \to (n+1) - 2 \to n - 1 \to n$$

221.

$$\sum_{i=3}^{n+1} i \to \sum_{i=1}^{n+1} i - \sum_{i=1}^{2} i \to \frac{(n+1)(n+2)}{2} - 3 \to n^2 + 3n \to n^2$$

222.

$$\sum_{i=3}^{n+1} i(i+1) \to \sum_{i=1}^{n+1} i(i+1) - \sum_{i=1}^{2} i(i+1) \sum_{i=1}^{n+1} i^2 + \sum_{i=1}^{n+1} i + \sum_{i=1}^{2} i^2 + \sum_{i=1}^{2} i$$

$$\to \frac{(n+2)(n+1)(n+3)}{3} - (5+3) \to \frac{n^3 + 6n^2 + 11n - 18}{3} \to n^3 + n^2 \to n^3$$

226.

$$\int_{i=0}^{n-1} (i^2+1)^2 = n + \frac{n(2n-1)(n-1)(n^2-n+3)}{10} - 1$$

$$\to O(n^5)$$

229.

$$\sum_{i=0}^{n} n - i \to n + \sum_{i=1}^{n} n - i \to n + \frac{n^2 - n}{2} \to n^2$$

245.

$$n^{\ln(n)} \to (\ln(n))^n$$

246.

$$(\ln(n))! \to n^{\ln\ln\ln(n)}$$

247.

$$e^{c\sqrt{n}} \text{ and } e^{\sqrt{n}} \implies \lim_{n\to\infty} \frac{\left(e^{\sqrt{n}}\right)^c}{e^{\sqrt{n}}} = \lim_{n\to\infty} \left(e^{\sqrt{n}}\right)^{c-1}$$

$$\begin{cases} c > 1 \to \infty \implies e^{c\sqrt{n}} > e^{\sqrt{n}} \\ c = 1 \to 1 \implies e^{c\sqrt{n}} \in \theta(e^{\sqrt{n}}) \\ c < 1 \to 0 \implies e^{c\sqrt{n}} < e^{\sqrt{n}} \end{cases}$$

257.

θ	ω	Ω	o	O	g(n)	F(n)
no	no	no	yes	yes	n^ε	$\log^k n$
no	no	no	yes	yes	c^n	n^k
no	no	no	no	no	$n^{\sin(n)}$	$\sqrt{n}$
no	yes	yes	no	no	$n^{\frac{n}{2}}$	2^n
yes	no	yes	no	yes	$c^{\log n}$	$n^{\log c}$
yes	no	yes	no	yes	$\log(n^n)$	$\log(n!)$

265. False.

$n \log n \in O(\log_2^{n!}) \Rightarrow 2^{n \log_2^n} \in O(2^{\log_2^{n!}}) \Rightarrow n^n \notin O(n!)$

Or other example: $2 \log_2^n \in O(\log_2^n) \Rightarrow 2^{\log_2^{n^2}} \in O(2^{\log_2^n}) \Rightarrow n^2 \notin O(n)$

272. False.

$f(n) = \frac{1}{n} \Rightarrow f^2(n) = \frac{1}{n^2} \Rightarrow f(n) \notin O(f^2(n))$

Chapter 3
Recurrence Relations

Abstract Recursive relations are useful methods for analyzing recursive algorithms. This chapter provides exercises for developing skills in solving recurrence relations. You may be familiar with how to analyze the time complexity of algorithms. However, to analyze recursive algorithms, we need more sophisticated techniques such as solving recursive relations. In this chapter, we study how to define and solve recurrence relations. This chapter illustrates how to solve recursive relations through a variety of examples and provides 199 exercises about it. To this end, this chapter provides exercises on iteration method, homogeneous and nonhomogeneous linear recurrence equation with constant coefficients, general formula, changing Variables in recurrence relations, recurrence with full history, recurrence with floors and ceilings, master method, recursion tree method, recurrence relations with more than one variable, and generating functions.

3.1 Lecture Notes

Recursive relations are useful methods for analyzing recursive algorithms. This chapter provides exercises for developing skills in solving recurrence relations.

You may be familiar with how to analyze the time complexity of algorithms. However, to analyze recursive algorithms, we need more sophisticated techniques such as solving recursive relations. In this chapter, we study how to define and solve *recurrence relations*. Consider the factorial function

$$n! = \begin{cases} 1 & \text{if } n = 1 \\ n \cdot (n-1)! & \text{if } n > 1 \end{cases}$$

This function can be implemented as following recursive algorithm to compute $n!$:

```
int Factorial(int n)
{
    if n == 1 return 1
    else
    return n * Factorial(n − 1)
}
```

© The Author(s), under exclusive license to Springer Nature Switzerland AG 2022
H. Izadkhah, *Problems on Algorithms*,
https://doi.org/10.1007/978-3-031-17043-0_3

How many multiplications, indicated by $T(n)$, does FACTORIAL perform?

1. No multiplication operation is performed for $n = 1$.
2. Otherwise, 1 multiplication operation is performed.
3. *Plus* the number of multiplications performed in the recursive call, FACTORIAL$(n - 1)$.

This can be expressed as a formula (similar to the definition of $n!$).

$$T(0) = 0$$
$$T(n) = 1 + T(n - 1)$$

We call this as a *recurrence relation*.

Definition 3.1 A *recurrence relation* for a sequence $\{a_n\}$ is an equation that expresses a_n in terms of one or more of the previous terms in the sequence,

$$a_0, a_1, \ldots, a_{n-1}$$

for all integers $n \geq n_0$ where n_0 is a nonnegative integer. A sequence is called a *solution* of a recurrence relation if its terms satisfy the recurrence relation.

For example, $a_n = 5$, $a_n = 3n$, and $a_n = 2^n$ are solutions for the recurrence relation $a_n = 2a_{n-1} - a_{n-2}$. Generally, {initial conditions (other name: boundary conditions) + recurrence relation} uniquely determine the solution for the recurrence relation.

Example 3.1 Consider the following (recursive) algorithm for computing Fibonacci numbers:

```
int Fib(int m,  int n)
{
    if (n ≤ 1)
        return n;
    else
        return Fib(n − 1) + Fib(n − 2)
}
```

This algorithm can be expressed as following recurrence relation

$$Fib(n) = Fib(n - 1) + Fib(n - 2)$$
$$Fib(1) = 1$$
$$Fib(0) = 1$$

The solution to the Fibonacci recurrence is

$$fib_n = \frac{1}{\sqrt{5}} \left(\frac{1 + \sqrt{5}}{2} \right)^n - \frac{1}{\sqrt{5}} \left(\frac{1 - \sqrt{5}}{2} \right)^n$$

To analyze time complexity, let $T(n)$ be the total number of addition steps by this algorithm for computing Fib_n.

$$T(n) = \begin{cases} 0 & n = 0 \\ 0 & n = 1 \\ T(n-1) + T(n-2) + 1 & n \geq 2 \end{cases}$$

The general form of recursive functions that we commonly encounter in algorithm analysis is as follows.

$$T(n) = \alpha T(n-b) + f(n), \quad T(x) = c$$

or

$$T(n) = \alpha T\left(\frac{n}{b}\right) + f(n), \quad T(x) = c$$

Note that it may be necessary to define *several* $T(x)$, initial conditions. To completely describe the sequence, the first few necessary terms are needed in the sequence, called the initial conditions. For example, in the Fibonacci numbers $F(0) = F(1) = 1$ are initial conditions.

In the general form of recursive functions, *f(n)* denotes non-recursive terms. Generally, recurrence relations have two parts: recursive terms and non-recursive terms. For example:

$$T(n) = \underbrace{2T(n-1)}_{\text{recursive}} + \underbrace{n^3 + 2}_{\text{non-recrusive}}$$

The recursive terms indicates that an algorithm is calling itself. Non-recursive terms correspond to the "non-recursive" cost of the algorithm.

We also can use the subscript notation for the recurrence relations, so the Fibonacci numbers are

$$fib_n = fib_{n-1} + fib_{n-2}$$
$$fib_1 = 1$$
$$fib_0 = 1$$

3.1.1   Catalog of Recurrence

Let's look at these functions (and some others), as they appear when expressed as recurrence relations.

3.1.1.1   Linear

Function: $\Theta(n)$.
recurrence: $T(1) = 1, T(n) = T(n-1) + 1$
context: Singly-nested loops, visit everything once.
examples: Naive multiplication, depth-first search.
variations: $T(n) = 1 + 3T(n/3)$, $T(n) = T(n/2) + n$, $T(n) = T(n/5) + T(7n/10 + 6) + n$.

3.1.1.2   Log

Function: $\Theta(\log n)$.
recurrence: $T(1) = 1, T(n) = T(n/2) + 1$
context: Recurse on half of input and throw half away.
examples: Euclid, Repeated Squaring, search in balanced trees.
variations: $T(n) = T(99n/100) + 1, T(n) = T(2n/3) + n/3$.

3.1.1.3   Quadratic

Function: $\Theta(n^2)$.
recurrence: $T(2) = 1, T(n) = T(n-1) + n$
context: Nested loops.
examples: Matrix multiplication, sorting.
variations: $T(n) = T(99n/100) + T(n/100) + 1$, other polynomials: $T(n) = n^2 + T(n-1)$.

3.1.1.4   Exponential

Function: $\Theta(2^n)$.
recurrence: $T(1) = 1, T(n) = 2T(n-1)$
context: Solve both of two slightly smaller versions of problem.
examples: Number of nodes in a binary tree.
variations: $T(n) = 2T(n-1) + 1$, other exponentials: $T(n) = kT(n-1)$.

3.1.1.5   N Log N

Function: $\Theta(n \log n)$.
recurrence: $T(1) = 1, T(n) = 2T(n/2) + n$
context: Scan, divide, recurse on both halves.
examples: Sorting, Fast Fourier Transform.
variations: $T(n) = \log T(n-1) + n$.

3.1.2 Solving Recurrence

3.1.2.1 The Substitution Method

Idea: Make an intelligent guess and prove by induction.

Example 3.2 Solve: $T(n) = 2T(\lfloor n/2 \rfloor) + n$
Guess: $T(n) \le cn \log n$.
Prove:
Base case: Assume constant size inputs take constant time. Assume: $T(\lfloor n/2 \rfloor) \le c\lfloor n/2 \rfloor \log(\lfloor n/2 \rfloor)$
Then:

$$
\begin{aligned}
T(n) &\le 2T(\lfloor n/2 \rfloor) + n \\
&\le 2c\lfloor n/2 \rfloor \log(\lfloor n/2 \rfloor) + n \\
&\le 2c(n/2) \log(n/2) + n \\
&= cn \log(n/2) + n \\
&= cn \log(n) - cn + n \\
&= cn \log(n) - n(c - 1) \\
&\le cn \log(n)
\end{aligned}
$$

So we see we need $c \ge 1$.

We can start with making good guesses. To this end, start with the answer to what you have already seen. Use experience to figure out which aspects of a formula are not important. For example, in $T(n) = 2T(n/2 + 21) + n$, the 21 is probably not that important since $n/2 + 17$ is a lot like $n/2$ for large n.

3.1.2.2 The Iteration Method

The substitution method is great when you have enough experience to guess the answer.

Example 3.3 Solve:

$$
\begin{aligned}
T(1) &= 1 \\
T(n) &= 3T(n/4) + n
\end{aligned}
$$

We have:

$$
\begin{aligned}
T(n) &= 3T(n/4) + n \\
&= 3(3T(n/16) + n/4) + n = 9T(n/16) + 7n/4 \\
&= 9(3T(n/64) + n/16) + 7n/4 = 27T(n/64) + 37n/16 \\
&= 27(3T(n/256) + n/64) + 37n/16 = 81T(n/256) + 175n/64 \\
&= 3^4 T(n/(4^4)) + n \sum_{i=0}^{3}(3/4)^i \\
&= \ldots \\
&= 3^k T(n/(4^k)) + n \sum_{i=0}^{k-1}(3/4)^i \\
&= 3^k T(n/(4^k)) + 4n(1 - (3/4)^k)
\end{aligned}
$$

3.1.2.3   Terminating the Iteration

$T(n) = 3^k T(n/(4^k)) + 4n(1 - (3/4)^k)$.
   For what k does $n/(4^k) = 1$?

$$
\begin{aligned}
n/(4^k) &= 1 \\
n &= (4^k) \\
\log(n) &= k\log(4) \\
k &= \log(n)/2
\end{aligned}
$$

Substituting into the formula above...
$3^k = 3^{\log(n)/2} = n^{\log(3)/2}$
$(3/4)^k = (3/4)^{\log(n)/2} = n^{\log(3)/2-1}$
So,

$$
\begin{aligned}
T(n) &= n^{\log(3)/2} T(1) + 4n(1 - n^{\log(3)/2-1}) \\
&= n^{\log(3)/2} + 4n - 4n^{\log(3)/2} \\
&= 4n - 3n^{\log(3)/2} \\
&= O(n)
\end{aligned}
$$

Example 3.4   Give a solution to

$$
T(n) = T(n-1) + 2n
$$

where $T(1) = 5$.
Solution:

1. Start by unfolding the recursion by a simple substitution of the function values.
   We have

$$
T(n-1) = T((n-1)-1) + 2(n-1) = T(n-2) + 2(n-1)
$$

   By substituting this into the original recursion, we have

$$
T(n) = T(n-2) + 2(n-1) + 2n
$$

   By continuing this, we get the following.

$$
\begin{aligned}
T(n) &= T(n-2) + 2(n-1) + 2n \\
&= T(n-3) + 2(n-2) + 2(n-1) + 2n \\
&= T(n-4) + 2(n-3) + 2(n-2) + 2(n-1) + 2n \\
&\quad\vdots \\
&= T(n-i) + \sum_{j=0}^{i-1} 2(n-j)
\end{aligned}
$$

2. By solving the sum, we get

$$T(n) = T(n - i) + 2n(i - 1) + 2\frac{(i-1)(i-1+1)}{2} + 2n$$

3. We do not want to have the recursive term at the end. To this end, we need to know in what iteration we reach our baseline. That is, for what value of i can we use the initial condition, $T(1) = 5$?
4. It is easy to see that when $i = n - 1$, we get the base case.
5. Substituting $i = n - 1$ into the equation above, we get

$$\begin{aligned} T(n) &= T(n - i) + 2n(i - 1) - i^2 + i + 2n \\ &= T(1) + 2n(n - 1 - 1) - (n - 1)^2 + (n - 1) + 2n \\ &= 5 + 2n(n - 2) - (n^2 - 2n + 1) + (n - 1) + 2n \\ &= n^2 + n + 3 \end{aligned}$$

3.1.2.4 The Table Method

The iteration method is pretty algebra intensive.

$$\begin{aligned} T(0) &= 3 \\ T(1) &= 3\sqrt{2} \\ T(n) &= 2T(n - 2) \end{aligned}$$

n	$T(n)$	
0	3	
1	$3\sqrt{2}$	
2	6	$3(2)$
3	$6\sqrt{2}$	
4	12	$3(2^2)$
5	$12\sqrt{2}$	
6	24	$3(2^3)$
7	$24\sqrt{2}$	

If n is even, $T(n) = 3(2^{n/2})$.
Now check n odd ..., it works!

3.1.3 Linear Homogeneous Recurrences

Definition 3.2 A linear homogeneous recurrence relation of degree k with constant coefficients is a recurrence relation of the form

$$a_n = \alpha_1 a_{n-1} + \alpha_2 a_{n-2} + \cdots + \alpha_k a_{n-k}$$

with $\alpha_1, \ldots, \alpha_k \in \mathbb{R}$, $c_k \neq 0$.

In the above definition:

- Linear indicates that the recurrence relation consists of the sum of the previous terms in the sequence and the coefficients are all constants (not functions depending on n).
- Homogeneous indicates that there is no terms occur that is not multiples of the a_j's.
- Degree k: a_n is expressed in terms of k terms of the sequence.

Example 3.5 The following recurrence relations are examples of linear homogeneous recurrence relations.

$$a_n = 4a_{n-1} + 5a_{n-2} + 7a_{n-3}$$

$$a_n = 2a_{n-2} + 4a_{n-4} + 8a_{n-8}$$

To describe these relationships, we need the initial conditions, which will be equal to the number of degrees of these relationships, $k = 3, 8$, respectively. For example, in the first recurrence relation we have three and in the second relation we have eight initial conditions.

Form $a_n = r^n$ where r is some (real) constant is a solution for linear homogeneous recurrences

$$r^n = \alpha_1 r^{n-1} + \alpha_2 r^{n-2} + \cdots + \alpha_k r^{n-k}$$

We can now divide both sides by r^{n-k}, collect terms, and we get a k-degree polynomial.

$$r^k - \alpha_1 r^{k-1} - \alpha_2 r^{k-2} - \cdots - \alpha_{k-1} r - \alpha_k = 0$$

$$r^k - \alpha_1 r^{k-1} - \alpha_2 r^{k-2} - \cdots - \alpha_{k-1} r - c_k = 0$$

This is called the *characteristic equation* of the recurrence relation.

The roots obtained from solving this relation are called the *characteristic roots* of the recurrence relation, which these roots can be used to find a solution to the recurrence relation.

3.1.3.1   Second Order Linear Homogeneous Recurrences

The general form of a second order linear homogeneous recurrence is as

$$a_n = \alpha_1 a_{n-1} + \alpha_2 a_{n-2}$$

Theorem 3.1 *Let $\alpha_1, \alpha_2 \in \mathbb{R}$ and suppose that $r^2 - \alpha_1 r - \alpha_2 = 0$ is the characteristic polynomial of a second order linear homogeneous recurrence which has two distinct[1] roots, r_1, r_2. Then a_n is a solution if and only if*

$$a_n = C_1 r_1^n + C_2 r_2^n$$

for $n = 0, 1, 2, \ldots$ where C_1, C_2 are constants dependent upon the initial conditions.

Example 3.6 Solve the recurrence

$$a_n = 5a_{n-1} - 6a_{n-2}$$

with initial conditions $a_0 = 1$, $a_1 = 4$.
Solution:

1. The characteristic equation is

$$r^2 - 5r + 6 = 0$$

2. Using the quadratic formula, the roots are:

$$r^2 - 5r + 6 = (r - 2)(r - 3)$$

so $r_1 = 2, r_2 = 3$

3. Using the second-order theorem, we have a solution,

$$a_n = C_1(2^n) + C_2(3^n)$$

4. By plugging the two initial conditions, we get a system of linear equations.

$$a_0 = C_1(2)^0 + C_2(3)^0$$
$$a_1 = C_1(2)^1 + C_2(3)^1$$

$$1 = C_1 + C_2 \quad (1)$$
$$4 = 2C_1 + 3C_2$$

5. Solving for $C_1 = (1 - C_2)$ in (1), we can plug it into the second.

$$4 = 2C_1 + 3C_2$$
$$4 = 2(1 - C_2) + 3C_2$$
$$4 = 2 - 2C_2 + 3C_2$$
$$2 = C_2$$

6. Substituting back into (1), we get

---

[1] We discuss how to handle this situation later.

$$C_1 = -1$$

7. Putting it all back together, we have

$$a_n = C_1(2^n) + C_2(3^n)$$
$$= -1 \cdot 2^n + 2 \cdot 3^n$$

Example 3.7 Find a solution to

$$a_n = -2a_{n-1} + 15a_{n-2}$$

with initial conditions $a_0 = 0$, $a_1 = 1$.
Solution:

After solving, we have

$$a_n = \frac{1}{8}(3)^n - \frac{1}{8}(-5)^n$$

How do we check the answer ourselves?

Example 3.8 What value does the following function return for $n > 1$ as an answer?

```
int g(int n)
{
    if n < 2 return (n)
    else
    return (5 × g(n − 1) − 6 × g(n − 2))
}
```

Solution:

1. The recurrence relation for this algorithm is

$$g(n) = 5g(n-1) - 6g(n-2), \quad g(0) = 0, \ g(1) = 1$$

2. We turn this recurrence relation into a characteristic equation

$$r^2 - 5r + 6 = 0$$

3. We obtain the roots as

$$r^2 - 5r + 6 = (r - 2)(r - 3)$$

so $r_1 = 2, r_2 = 3$
4. Using Theorem 3.1, we have a solution,

$$a_n = C_1(2^n) + C_2(3^n)$$

5. By plugging the two initial conditions, we get the coefficients and putting it all back together, we have

$$g_n = C_1(2^n) + C_2(3^n)$$
$$= -1 \cdot 2^n + 2 \cdot 3^n$$

3.1.3.2 Single Root Case

Theorem 3.1 can only be used if the roots are distinct, i.e., $r_1 \neq r_2$. If the roots are not distinct, i.e., $r_1 = r_2$, we apply the following theorem.

Theorem 3.2 *Let $\alpha_1, \alpha_2 \in \mathbb{R}$ with $\alpha_2 \neq 0$. Suppose that $r^2 - \alpha_1 r - \alpha_2 = 0$ has only one distinct root, r_0. Then $\{a_n\}$ is a solution to $a_n = \alpha_1 a_{n-1} + \alpha_2 a_{n-2}$ if and only if*

$$a_n = C_1 r_0^n + C_2 n r_0^n$$

for $n = 0, 1, 2, \ldots$ where C_1, C_2 are constants depending upon the initial conditions.

Example 3.9 Solve the recurrence relation

$$a_n = 8a_{n-1} - 16a_{n-2}$$

with initial conditions $a_0 = 1, a_1 = 7$.
Solution:

1. The characteristic equation is

$$r^2 - 8r + 16$$

2. Using the quadratic formula, the roots are:

$$r^2 - 8r + 16 = (r - 4)(r - 4)$$

so $r = 4$
3. From Theorem 3.2, the solution is of the form

$$a_n = C_1 4^n + C_2 n 4^n$$

4. Using the initial conditions, we get a system of equations;

$$a_0 = 1 = C_1$$
$$a_1 = 7 = 4C_1 + 4C_2$$

5. By solving the above, we get $C_1 = 1, C_2 = \frac{3}{4}$

6. The final solution, a_n, is

$$a_n = 4^n + \frac{3}{4}n4^n$$

3.1.3.3   General Linear Homogeneous Recurrences

There is a simple generalization of Theorys 3.1 and 3.2 to higher order linear homogeneous recurrences as

Theorem 3.3 *Let $\alpha_1, \ldots, \alpha_k \in \mathbb{R}$. Suppose that the characteristic equation*

$$r^k - \alpha_1 r^{k-1} - \cdots - \alpha_{k-1}r - \alpha_k = 0$$

has k distinct roots, $r_1, \ldots, r_k$. Then a sequence $\{a_n\}$ is a solution of the recurrence relation

$$a_n = \alpha_1 a_{n-1} + \alpha_2 a_{n-2} + \cdots + \alpha_k a_{n-k}$$

if and only if

$$a_n = C_1 r_1^n + C_2 r_2^n + \cdots + C_k r_k^n$$

for $n = 0, 1, 2, \ldots$, where $C_1, C_2, \ldots, C_k$ are constants.

When some roots are not distinct, we use the following theorem.

Theorem 3.4 *Let $\alpha_1, \ldots, \alpha_k \in \mathbb{R}$. Suppose that the characteristic equation*

$$r^k - \alpha_1 r^{k-1} - \cdots - \alpha_{k-1}r - \alpha_k = 0$$

has t distinct roots, $r_1, \ldots, r_t$ with multiplicities $m_1, \ldots, m_t$. Then a sequence $\{a_n\}$ is a solution of the recurrence relation

$$a_n = \alpha_1 a_{n-1} + \alpha_2 a_{n-2} + \cdots + \alpha_k a_{n-k}$$

if and only if

$$
\begin{aligned}
a_n = \ & (C_{1,0} + C_{1,1}n + \cdots + C_{1,m_1-1}n^{m_1-1})r_1^n + \\
& (C_{2,0} + C_{2,1}n + \cdots + C_{2,m_2-1}n^{m_2-1})r_2^n + \\
& \vdots \\
& (C_{t,0} + C_{t,1}n + \cdots + C_{t,m_t-1}n^{m_t-1})r_t^n +
\end{aligned}
$$

for $n = 0, 1, 2, \ldots$, where $C_{i,j}$ are constants for $1 \le i \le t$ and $0 \le j \le m_i - 1$.

3.1.4 Nonhomogeneous

In the recurrence relation of some recursive algorithms, in addition to recursive calls, it is necessary to consider some costs (a non-recursive cost) that are dependent on the input. Such a recurrence relation is called a *linear nonhomogeneous recurrence relation*, which are of the form

$$a_n = \alpha_1 a_{n-1} + \alpha_2 a_{n-2} + \cdots + \alpha_k a_{n-k} + f(n)$$

where $f(n)$ represents a non-recursive cost. By ignoring $f(n)$, we have

$$a_n = \alpha_1 a_{n-1} + \alpha_2 a_{n-2} + \cdots + \alpha_k a_{n-k}$$

which is the *associated homogenous recurrence relation*.

Each solution for a linear nonhomogeneous recurrence relation consists of the sum of a solution for an associated linear homogeneous recursive relation and a particular solution.

Theorem 3.5 *If* $\{P_n\}$ *is a particular solution of the nonhomogeneous linear recurrence relation with constant coefficients*

$$a_n = \alpha_1 a_{n-1} + \alpha_2 a_{n-2} + \cdots + \alpha_k a_{n-k} + f(n)$$

then every solution is of the form $\{P_n + U_n\}$, *where* $\{U_n\}$ *is a solution of the associated homogenous recurrence relation*

$$a_n = \alpha_1 a_{n-1} + \alpha_2 a_{n-2} + \cdots + \alpha_k a_{n-k}$$

No general method exists for solving the nonhomogeneous linear recurrence relation. However, it is possible to solve such relations for special cases.

In particular, if $f(n)$ is a polynomial or exponential function (or more precisely, when $f(n)$ is the product of a polynomial and exponential function), then there is a general solution.

Theorem 3.6 *Suppose that* $\{a_n\}$ *satisfies the linear nonhomogeneous recurrence relation*

$$a_n = \alpha_1 a_{n-1} + \alpha_2 a_{n-2} + \cdots + \alpha_k a_{n-k} + f(n)$$

where $\alpha_1, \ldots, \alpha_k \in \mathbb{R}$ *and*

$$f(n) = (b_t n^t + b_{t-1} n^{t-1} + \cdots + b_1 n + b_0) \cdot s^n$$

where $b_0, \ldots, b_n, s \in \mathbb{R}$.

When s *is not a root of the characteristic equation of the associated linear homogeneous recurrence relation, there is a particular solution of the form*

$$(p_t n^t + p_{t-1} n^{t-1} + \cdots + p_1 n + p_0) \cdot s^n$$

When s is a root of this characteristic equation and its multiplicity is m, there is a particular solution of the form

$$n^m (p_t n^t + p_{t-1} n^{t-1} + \cdots + p_1 n + p_0) \cdot s^n$$

Note 1. In general, if in a nonhomogeneous recurrence relation $f(n)$ is equal to kq^n, where q is not a root of the characteristic equation of the associated linear homogeneous, then we guess the particular solution as $P_n = Cq^n$ and place it in the original recurrence relation.

Note 2. If in a nonhomogeneous recurrence relation $f(n)$ is equal to kn^q and the number 1 is not a root of the characteristic equation of the associated linear homogeneous recurrence relation, then the particular solution will be as follows:

$$P_n = c_0 + c_1 n + \cdots + c_q n^q$$

Note 3. If in a nonhomogeneous recurrence relation $f(n)$ is equal to kn^q and the number 1 is a simple root of the characteristic equation of the associated linear homogeneous recurrence relation, then the particular solution will be as follows:

$$P_n = c_0 n + c_1 n^2 + \cdots + c_q n^{q+1}$$

Example 3.10 Solve

$$a_n = 5a_{n-1} - 6a_{n-2} + 7^n$$

Solution:

$$a_n - 5a_{n-1} + 6a_{n-2} = 0 \quad r_1, r_2 = 2, 3$$

$$U_n = C_1 (2)^n + C_2 (3)^n$$

We guess P_n as follows:

$$P_n = C7^n$$

Then we place P_n in the original recurrence relation.

$$C7^n = 5C7^{n-1} - 6C7^{n-2} + 7^n$$

$$49C = 35C - 6C + 49 \rightarrow C = \frac{49}{20}$$

$$P_n = \frac{49}{20} 7^n$$

$$U_n + P_n = C_1(2)^n + C_2(3)^n + \frac{49}{20}7^n$$

Example 3.11 Solve $a_n = a_{n-1} + 4^n + 2^n$ Solution:

$$a_n - a_{n-1} = 0 \rightarrow a - 1 = 0 \rightarrow a = 1$$

$$U_n = \alpha(1)^n$$

$$P_n = c_1 4^n + c_2 2^n$$

$$c_1 4^n + c_2 2^n = c_1 4^{n-1} + c_2 2^{n-1} + 4^n + 2^n$$

$$c_1 4^n + c_2 2^n = (\frac{c_1}{4} + 1)4^n + (\frac{c_2}{2} + 1)2^n$$

$$c_1 = \frac{4}{3}, \quad c_2 = 2$$

$$U_n + P_n = \alpha(1)^n + \frac{4}{3} \times 4^n + 2 \times 2^n$$

Example 3.12 Suppose the time complexity function of an algorithm is as follows. Specify its order.

$$F(n) = \begin{cases} 1, & n = 1 \\ n + F(n - 1), & n > 0 \end{cases}$$

Solution:

$$F(n) - F(n - 1) = 0 \rightarrow r = 1$$

$$U_n = \alpha(1)^n$$

Considering note 3, we have:

$$P_n = c_0 n + c_1 n^2$$

$$\alpha(1)^n + c_0 n + c_1 n^2 \rightarrow O(n^2)$$

3.1.5   Recurrence Tree

Another way to analyze recurrence relations is to use the recurrence tree. Each node
in the tree is an example of a call. As we move down, the size of the input decreases.
The share of each level in the function is equal to the number of nodes in that level
multiplied by the non-recursive cost in the input size of that level. For example,
consider the following recursive function

$$T(n) = \alpha T\left(\frac{n}{\beta}\right) + f(n), \quad T(\delta) = c$$

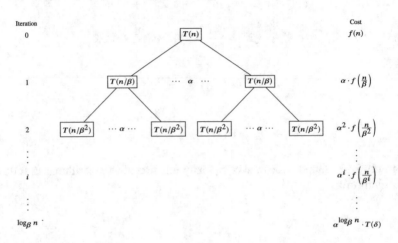

The total value for $T(n)$ is the summation over all levels of the tree:

$$T(n) = \sum_{i=0}^{\log_\beta n} \alpha^i \cdot f\left(\frac{n}{\beta^i}\right)$$

Example 3.13 Solve the following recurrence function by recurrence tree.

$$T(n) = 2T\left(\frac{n}{2}\right) + n, \quad T(1) = 4$$

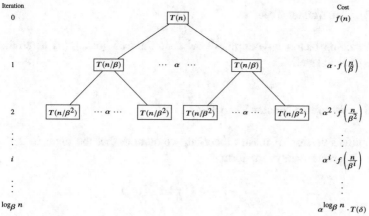

The final solution for $T(n)$ is the summation of the value of all levels. We consider the last level as a special case because its non-recursive cost is different.

$$T(n) = 4n + \sum_{i=0}^{(\log_2 n)-1} 2^i \frac{n}{2^i} = n(\log n) + 4n$$

3.1.6 Master Method

3.1.6.1 Basic Idea

The master method is a general method for solving recurrences where all the subproblems are of the same size. We assume that the input to the master method is a recurrence of the form

$$T(n) = aT\left(\frac{n}{b}\right) + f(n)$$

In this recurrence, there are three constants:

- a is the number of subproblems that we create from one problem, and must be an integer $a \geq 1$.
- b is the factor by which the input size shrinks (it must hold that $b > 1$)
- $f(n)$ is a function indicating the cost of the work done outside the recursive calls.

We gloss over some technicalities: $T(n)$ is usually only defined for integer n, so we really need $T(\lceil \frac{n}{b} \rceil)$, boundary conditions like $T(1) = 0$. This is ok because these details get swallowed up by the big O or Theta that we are eventually going to wrap around things anyway.

Note here, that the Master Theorem does not solve a recurrence relation. We only care about the asymptotic behavior.

3.1.6.2   The Master Tree

a is branching factor, b determines how deep the tree goes, $f(n)$ determines the weight of each level.

3.1.6.3   Simplified Master Theorem

In simplified version of master theorem, we assume that the input to the master method is a recurrence of the form

$$T(n) = aT(\frac{n}{b}) + \Theta(n^k)$$

It means $f(n) = O(n^k)$. In this recurrence, there are three constants:

- a is the number of subproblems that we create from one problem, and must be an integer $a \geq 1$.
- b is the factor by which the input size shrinks (it must hold that $b > 1$)
- k is the exponent of n in the time it takes to generate the subproblems and combine their solutions.

We now state the master theorem, which is used to solve the recurrences.

Theorem 3.7 (Master Theorem). *Let $T(n) = aT(\frac{n}{b}) + \Theta(n^k)$ be a recurrence where $a \geq 1$, $b > 1$. Then*

$$T(n) = \begin{cases} O(n^k \log n) & \text{if } a = b^k \\ O(n^k) & \text{if } a < b^k \\ O(n^{\log_b a}) & \text{if } a > b^k \end{cases}$$

Let us see several examples in this regard.

Example 3.14 $T(n) = 4T(\frac{n}{2}) + n$
The parameters are a $= 4$, b $= 2$, k $= 1$, so $a > b^k$, hence $T(n) = O(n^{\log_2 4}) = O(n^2)$

Example 3.15 $T(n) = T(\frac{n}{3}) + 1$
The parameters are a $= 1$, b $= 3$, k $= 0$, so $a = b^k$, hence $T(n) = O(\log n)$

Example 3.16 $T(n) = 3T(\frac{n}{2}) + n$
The parameters are a $= 3$, b $= 2$, k $= 1$, so $a > b^k$, hence $T(n) = O(n^{\log_2 3}) = O(n^{1.59})$

Example 3.17 $T(n) = 2T(\frac{n}{2}) + n$
The parameters are a $= 2$, b $= 2$, k $= 1$, so $a = b^k$, hence $T(n) = O(n \log n)$

Example 3.18 $T(n) = 2T(\frac{n}{2}) + n^2$
The parameters are a $= 2$, b $= 2$, k $= 2$, so $a < b^k$, hence $T(n) = O(n^2)$

"Fourth" Condition It is important to note that it is not possible to use the master theorem if the non-recursive cost, $f(n)$, is not polynomial. In such cases, we can use the following solution, which is known as the "Fourth" condition.

$$If \ f(n) = \Theta(n^{\log_b(a)}), \ for \ some \ k \geq 0, \ then \ T(n) = \Theta(n^{\log_b(a)} \log^{k+1} n).$$

Example 3.19 $T(n) = 2T(\frac{n}{2}) + O(n \log n)$
The parameters are a = 2, b = 2, but $f(n)$ is not a polynomial. However

$$f(n) = \Theta(n \log n)$$

for k = 1, therefore, by the "Fourth" condition of the master theorem we can say that

$$f(n) = \Theta(n \log^2 n)$$

3.1.6.4 A More General Version of the Master Theorem

We are given the recurrence $T(n) = aT(\frac{n}{b}) + f(n)$.
   [1] If $f(n) = O(n^{\log_b(a)-\epsilon})$ for some constant $\epsilon > 0$, then $T(n) = \Theta(n^{\log_b(a)})$.
   [2] If $f(n) = \Theta(n^{\log_b(a)})$, then $T(n) = \Theta(n^{\log_b(a)} \log n)$.
   [3] If $f(n) = \Omega(n^{\log_b(a)+\epsilon})$ for some constant $\epsilon > 0$, and if $af(n/b) \leq cf(n)$ for some constant $c < 1$ and all sufficiently large n, then $T(n) = \Theta(f(n))$.
   [4] If $f(n) = \Theta(n^{\log_b(a)} \log^k(n))$ for $k \geq 0$, then $T(n) = \Theta(n^{\log_b(a)} \log^{k+1} n)$.

3.1.6.5 The Gaps

There is a polynomial gap on either side. It's not enough for $f(n)$ to be bigger (or smaller) than $n^{\log_b(a)}$, it has to be polynomially so (i.e., the $\epsilon > 0$ matters).
   In one case, we can partially fill this gap. That is, if $f(n)$ is just a log factor bigger than $n^{\log_b(a)}$, we still get a contribution from each level: $T(n) = \Theta(n^{\log_b(a)} \log^{k+1} n)$.

3.1.6.6 Examples

Example 3.20 $T(n) = 9T(\frac{n}{3}) + n$
   a = 9, b = 3, $f(n) = n$, $n^{\log_b(a)} = n^2$. n^2 is polynomially bigger. So, $T(n) = \Theta(n^2)$.

Example 3.21 $T(n) = 3T(\frac{n}{7}) + n^{\frac{3}{4}}$
   a = 3, b = 7, $f(n) = n^{\frac{3}{4}} = n^{0.75}$, $n^{\log_b(a)} \approx n^{0.56}$. $n^{0.75}$ is polynomially bigger. So, $T(n) = \Theta(n^{\frac{3}{4}})$.

Note: As long as $f(n) = \Theta(n^k)$, for $k \geq 0$, Master Theorem definitely applies!

Example 3.22 $T(n) = T(\frac{4n}{5}) + \log^2 n$

$a = 1, b = 5/4, f(n) = \log^2 n, n^{\log_b(a)} = 1$. $f(n)$ is not polynomially bigger. But Rule 4 applies (with $k = 2$). So, $T(n) = \Theta(\log^3(n))$.

Example 3.23 $T(n) = 2T(\frac{n}{4}) + \sqrt{n} \log \log(n)$ $a = 2, b = 4, f(n) = \sqrt{n} \log \log n$, $n^{\log_b(a)} = n^{1/2}$, so $f(n)$ is not polynomially larger than $n^{\log_b(a)}$. So, Master Theorem does not apply.

3.2   Exercises

3.2.1   The Iteration Method

The following give examples of some simple recurrence relations that occur in the study of algorithms. Solve the following recurrences with iteration substitution.

1. T(1) = d, and for all $n > 1$, T(n) = T(n–1) + c.

2. T(0) = 0, and for all $n > 1$, T(n) = 2T(n–1) + 1.

3. T(1) = d, and for all $n > 1$, T(n) = T(n–1) + cn.

4. T(1) = 1, and for all $n > 1$, T(n) = 2T(n–2) + 1.

5. T(1) = d, and for all $n > 1$, T(n) = T(n/2) + 1.

6. T(1) = d, and for all $n > 1$, T(n) = T(n/2) + cn.

7. T(1) = d, and for all $n > 1$, T(n) = 7T(n/2) + cn.

8. T(1) = d, and for all $n > 1$, T(n) = 2T(n/2) + 1.

9. T(1) = d, and for all $n > 1$, T(n) = T(n/2) + cn.

10. T(1) = d, and for all $n > 1$, T(n) = 3T(n/2) + c.

11. T(1) = d, and for all $n > 1$, T(n) = aT(n/b) + cn.

12. T(1) = T(2) = d, and for all $n > 2$, T($\sqrt{n}$) + c.

13. T(2) = d, and for all $n > 1$, solve the recurrence T(n) = 2T($\sqrt{n}$) + lg$_2$ n, using iteration substitution.

14. Hanoi Tower. The algorithm of Hanoi Tower is as follows:

```
void Hanoi(int n, A, B, C)
{
    if(n == 1)  move a disk from A to C
    else {
    Hanoi(n − 1, A, C, B)
    move a disk from A to C
    Hanoi(n − 1, B, A, C)
    }
}
```

Let $T(n)$ denotes the number of calls of the above function per n, we can write the recurrence function corresponding to $T(n)$ as follows

$$T(n) = \begin{cases} 0 & n = 0 \\ T(n − 1) + T(n − 1) + 1 = 2T(n − 1) + 1 & n > 1 \end{cases}$$

Solve the above recurrence relation to obtain the order (time complexity) of the Tower of Hanoi.

15. Consider the following recursive function.

```
int F(int n)
{
    if(n ≤ 2) return 1
    else
    return F(n − 2) ∗ F(n − 2)
}
```

Let $T(n)$ denotes the number of calls of this function per n, we can write the recurrence function corresponding to $T(n)$ as follows

$$T(n) = \begin{cases} 0 & n \leq 2 \\ 2T(n − 2) + 1 & n > 2 \end{cases}$$

Solve the above recurrence relation to obtain the order (time complexity) of this function.

16. Consider the following recursive function.

```
int  F(int  n)
{
    if(n ≤ 2)  return  1
    else
        return  F(n − 2) + F(n − 2)
}
```

Let $T(n)$ denotes the number of calls of the above function per n. Write the recurrence function corresponding to $T(n)$ and then solve it to obtain the order (time complexity) of this function.

17. Consider the following recursive function for computing the power X^n.

```
real  Power(real  X, int  n)
{
    if(n == 1)  return  1
    else
        return  X ∗ Power(X, n − 1)
}
```

Let T(n) denotes the number of multiplications of the above function, so we can write the recurrence function corresponding to T(n) as follows

$$T(n) = \begin{cases} 0 & n = 1 \\ T(n − 1) + 1 & n \geq 2 \end{cases}$$

Solve the above recurrence relation to obtain the order (time complexity) of this function.

3.2.2   Homogeneous Linear Recurrence Equation with Constant Coefficients

Solve the following homogeneous linear recurrence equation using characteristic function of the recurrence relation.

18. T(0) = 1, T(1) = 2, and for all $n > 2$, T(n) = 2T(n–1) – 3T(n–2).

19. T(0) = 0, T(1) = 2, and for all $n > 2$, T(n) = 7T(n–1) – 4T(n–2).

20. Fibonacci numbers. F(1) = F(0) = 1, and for all $n > 1$, F(n) = F(n–1) + F(n–2).

21. $T(1) = 2$, $T(2) = 4$, and for all $n > 2$, $T(n) = 2T(n–1) + 15T(n–2)$.

22. $T(1) = 3$, $T(2) = 5$, and for all $n > 2$, $T(n) = 4T(n-1) + 5T(n-2)$.

23. $T(0) = 2$, $T(1) = 4$, and for all $n > 1$, $T(n) = 3T(n–1) – 2T(n–2)$.

24. $T(0) = 3$, $T(1) = 9$, and for all $n > 1$, $T(n) = 6T(n–1) – 9T(n–2)$.

25. $T(2) = 98$, and for all $n > 2$, $T(n) = 7T(n–1)$.

26. $T(0) = 1$, $T(1) = 3$, and for all $n > 1$, $T(n+2) = 4T(n+1) – 4T(n)$.

27. $T(0) = 2$, $T(1) = 7$, and for all $n > 1$, $T(n) = T(n–1) + 2T(n–2)$.

28. $T(0) = 0$, $T(1) = 1$, and for all $n > 1$, $T(n) = 3T(n–1) + 4T(n–2)$.

29. $T(0) = 1$, $T(1) = 1$, and for all $n > 1$, $T(n) = 2T(n–1) + 3T(n–2)$.

30. $T(1) = 1$, and for all $n > 1$, $T(n) = T(n–1) + T(n–1)$.

31. $T(0) = 2$, $T(1) = 5$, and for all $n > 1$, $T(n) = 4T(n–1) – 3T(n–2)$.

32. $T(0) = T(1) = 1$, and for all $n > 1$, $T(n) = 3T(n–1) + 4T(n–2)$.

33. $T(0) = 1$, $T(1) = 3$, and for all $n > 1$, $T(n) = 3T(n–1) + 4T(n–2)$.

34. $T(0) = 1$, $T(1) = 3$, and for all $n > 1$, $T(n) = T(n–1) + T(n–2)$.

35. $T(0) = T(1) = 1$, and for all $n > 1$, $T(n) = 4T(n–2)$.

36. $T(0) = 1$, $T(1) = 1$, $T(2) = -1$, and for all $n > 2$, $T(n) = 9T(n–1) – 26T(n-2) + 24T(n–3)$.

37. What value does the following function return for $n \geq 2$ as an answer?

```
int  F(int  n)
{
  if (n ≤ 1)  return  1
  else
    return  F(n − 1) + F(n − 2)
}
```

38. What value does the following function return for $n > 2$ as an answer?

```
int F(int n)
{
    if  n < 2  return  (1)
    else
        return  (2 × F(n − 1) + 3 × F(n − 2))
}
```

39. What value does the following function return for $n \geq 2$ as an answer?

```
int  F(int  n)
{
    if  n = 0  return  (1)
    else if  n = 1  return  (3)
    return  (4 × F(n − 1) − 4 × F(n − 2))
}
```

40. Obtain the time complexity of an algorithm with the following complexity function.

$$
T(n) = \begin{cases} 0 & n = 0 \\ 1 & n = 1 \\ 3T(n - 1) + 4T(n - 2) & otherwise \end{cases}
$$

41. Obtain the time complexity of an algorithm with the following complexity function.

$$
T(n) = \begin{cases} 1 & n = 1 \\ T(n - 1) + T(n - 1) & n \geq 2 \end{cases}
$$

3.2.3 Nonhomogeneous Recurrences Equation with Constant Coefficients

Find an explicit formula for each of the sequences defined by the nonhomogeneous recurrence relations with initial conditions.

42. $T(1) = 1$, and for all $n \geq 2$, $T(n) = 3T(n–1) + 2$.

43. $T(n) = 2T(n-1) + 3^n$.

44. $T(0) = 1$, and for all $n > 0$, $T(n) + nT(n–1) = n!$.

45. $T(1) = 3$, and for all $n > 1$, $T(n) = 5T(n–1) + 3$.

46. $T(1) = 5$, and for all $n > 1$, $T(n) = 3T(n–1) + 5n$.

47. $T(1) = 2$, $T(2) = 4$, and for all $n > 2$, $T(n) = 2T(n–1) + 15\,T(n–2) + 2^n$.

48. $T(1) = 3$, $T(2) = 5$, and for all $n > 2$, $T(n) = 4T(n–1) + T(n–2) + 3$.

49. $T(0) = 2$, $T(1) = 4$, and for all $n > 1$, $T(n) = 3T(n–1) – 2T(n–2) + 2^n$.

50. $T(0) = 3$, $T(1) = 9$, and for all $n > 1$, $T(n) = 6T(n–1) – 9T(n–2) + 2^{n+2}$.

51. $T(1) = 1$, and for all $n \geq 2$, $T(n) = 2T(n–1) + 1$.

52. $T(0) = 0$, and for all $n > 0$, $T(n) = 2T(n–1) + 1$.

53. $T(0) = 0$, $T(1) = 12$, and for all $n > 1$, $T(n) – 3T(n–1) = (2n + 1)4^n$.

54. $T(0) = 0$, and for all $n > 0$, $T(n) – T(n–1) = n–1$.

55. $T(1) = 1$, and for all $n > 1$, $T(n) = nT(n–1)$.

56. $T(0) = 0$, and for all $n > 0$, $T(n) = 2T(n–1) + 3n$.

57. Consider the following recursive function.

```
void F(int n)
  {
  if (n == 1){print("ok"); return; }
  for(i = 1; i ≤ n; i + +)
    print("ok");
  else return F(n − 1)
  }
```

Justify why the recurrence function corresponding to this function is as follows

$$T(n) = \begin{cases} 0 & n = 1 \\ n + 2T(n - 1) & n > 1 \end{cases}$$

Solve the above recurrence relation to obtain the order (time complexity) of this function.

3.2.4 General Formula

Find a general formula for the following recurrences.

58. $T(n) - 3T(n–1) + 2T(n–2) = n4^n + 3$.

59. $T(n) = T(n–1) + 4^n + 2^n$.

60. $T(n) = T(n–1) + 2T(n–2)$.

61. $T(n) = 5T(n–1) – 6T(n–2)$.

62. $T(n) = 2T(n/2) + n$.

63. $T(n) = T(n/4) + T(n/2) + n^2$.

64. $T(n) = T(n/2) + 1$.

65. $T(n) = 4T(n–1) – 3T(n–2)$.

66. $T(n) = 3T(n–1) + 4T(n–2)$.

67. $T(n) = 2T(n–1) + 3n$.

68. $T(n) = 4T(n–2)$.

69. $T(n) = T(n–1) + T(n–2)$.

70. $T(n) = 9T(n–1) - 26T(n–2) + 24T(n–3)$.

71. $T(n) = 2T(n–1) + 5$.

72. $T(n) = 3T(n–1)$.

73. $T(n) = T(n/2) + c\log n$.

74. $T(n) = T(n/2) + cn^2$.

75. $T(n) = 2T(n/2) + \log n$.

76. $T(n) = 8T(n/2) + n^2$.

77. $T(n) = 2T(n/2) + n^3$.

78. $T(n) = 2T(9n/10) + n$.

79. $T(n) = 16T(n/2) + (n \log n)^4$.

80. $T(n) = 7T(n/3) + n$.

81. $T(n) = 9T(n/3) + n^3 \log n$.

82. $T(n) = 2T(n/4) + \sqrt{n}$.

83. $T(n) = 3T(n/2) + n \log n$.

84. $T(n) = 5T(n/5) + \frac{n}{\log n}$.

85. $T(n) = T(n-1) + \frac{1}{n}$.

86. $T(n) = T(n-1) + \log n$.

87. $T(n) = T(n-2) + 2 \log n$.

88. Mergesort (simplified): $T(n) = 2T(n/2) + n$.

89. Randomized selection: $T(n) = T(3n/4) + n$.

90. Karatsuba's multiplication algorithm: $T(n) = 3T(n/2) + n$.

91. Ugly divide and conquer: $T(n) = \sqrt{n}\, T(\sqrt{n}) + n$.

92. Randomized quicksort: $T(n) = T(3n/4) + T(n/4) + n$.

93. Deterministic selection: $T(n) = T(n/5) + T(7n/10) + n$.

94. Randomized search trees: $T(n) = 1/4\,T(n/4) + 3/4\,T(3n/4) + 1$.

95. Ham-sandwich trees: $T(n) = T(n/2) + T(n/4) + 1$.

3.2.5 Changing Variables in Recurrence Relations

96. $T(1) = 1$, and for all $n > 1$, solve the recurrence $T(n) = 2\,T(\sqrt{n}) + 1$, by making a change of variables.

97. $T(1) = 1$, and for all $n > 1$, solve the recurrence $T(n) = 2\,T(\sqrt{n}) + n$, using variable substitution.

98. $T(2) = d$, and for all $n > 1$, solve the recurrence $T(n) = 2\,T(\sqrt{n}) + \lg_2 n$, using variable substitution.

99. $T(1) = 3$ and $T(2) = 8$, and for all $n > 0$, solve the recurrence $T(n) = 6T(n/2) - 8\,T(n/4)$, by making a change of variables.

100. $T(1) = 1$ and $T(2) = 3$, solve the recurrence $T(n) = 4T(n/2) - 4T(n/4)$, by making a change of variables.

101. $T(1) = 0$, solve the recurrence $T(n) = T(n/2) + 1$, by making a change of variables.

102. $T(1) = 0$, solve the recurrence $T(n) = T(n/2) + (n-1)$, by making a change of variables.

103. $T(1) = 0$, solve the recurrence $T(n) = 7T(n/2) + 18(n/2)^2$, by making a change of variables.

104. Solve the recurrence $T(n) = 8T(n/9) + n \log n$, by making a change of variables.

105. Solve the recurrence $T(n) = 3T(\sqrt[3]{n}) + \log n$, by making a change of variables.

106. Solve the recurrence $T(n) = 3\sqrt[3]{n^2}\,T(\sqrt[3]{n}) + n \log n$, by making a change of variables.

3.2.6 More Difficult Recurrences

107. Suppose $0 < \alpha, \beta < 1$, where $\alpha + \beta = 1$. Let $T(1) = 1$, and for all $n \geq 1$, $T(n) = T(\alpha n) + T(\beta n) + cn$, for some $c \in \mathbb{N}$. Prove that $T(n) = O(n \log n)$. You may make any necessary assumptions about n.

108. Suppose $0 < \alpha, \beta < 1$, where $\alpha + \beta < 1$. Let $T(1) = 1$, and for all $n \geq 1$, $T(n) = T(\alpha n) + T(\beta n) + cn$, for some $c \in \mathbb{N}$. Prove that $T(n) = O(n)$. You may make any necessary assumptions about n.

109. Suppose $\frac{a}{b} + \frac{c}{d} + \cdots = 1$. Let $T(1) = 1$, and for all $n \geq 1$, $T(n) = T(\frac{a}{b}\,n) + T(\frac{c}{d}\,n) + \cdots + \Theta(n)$, for some $c \in \mathbb{N}$. Prove that $T(n) = O(n \log n)$. You may make any necessary assumptions about n.

110. Suppose $\frac{a}{b} + \frac{c}{d} + \cdots < 1$. Let $T(1) = 1$, and for all $n \geq 1$, $T(n) = T(\frac{a}{b}\, n) + T(\frac{c}{d}\, n) + \cdots + \Theta(n)$, for some $c \in \text{IN}$. Prove that $T(n) = \Theta(n)$. You may make any necessary assumptions about n.

111. Determine the time complexity of $T(n) = T(\frac{3n}{4}) + T(\frac{n}{4}) + n$.

112. Determine the time complexity of $T(n) = T(\frac{n}{2}) + T(\frac{n}{4}) + T(\frac{n}{8})$.

113. Given that we know the Fibonacci numbers F_n for $n \geq 0$ are defined recursively as follows: $F_0 = 0$, $F_1 = 1$, and for $n \geq 2$, $F_n = F_{n-1} + F_{n-2}$. Write the explicit formula F_n.

Let $A(n)$ be the number of different ways of parenthesizing the product of n values. For example, $A(1) = A(2) = 1$, $A(3) = 2$ (they are (xx)x and x(xx)), and $A(4) = 5$ (they are x((xx)x), x(x(xx)), (xx)(xx), ((xx)x)x, and (x(xx))x). Answer questions 114–116.

114. Prove that if $n \leq 2$ then $A(n) = 1$; and otherwise

$$A(n) = \sum_{i=1}^{n-1} (A(i) + A(n - i))$$

115. Show that for all $n \geq 1$, $A(n) \geq 2^{n-2}$.

116. Show that for all

$$A(n) = \frac{1}{n}\binom{2n - 2}{n - 1}$$

117. Solve the following recurrence

$$T(n) = \begin{cases} 1 & n = 1 \\ \frac{3}{2} & n = 2 \\ \frac{3}{2}T(\frac{n}{2}) - \frac{1}{2}T(\frac{n}{4}) - \frac{1}{n} & o.w \end{cases}$$

118. Solve the following recurrence

$$T(n) = \begin{cases} a & n = 0 \\ b & n = 1 \\ \frac{1+T(n-1)}{T(n-2)} & o.w \end{cases}$$

119. The following recurrence relations are defined for integers $n > 2$ and we have $T(0) = T(1) = 1$. Which of these relations does not have a polynomial answer?

a. $T(n) = 2T(n-2) + 1$
b. $T(n) = T(7n/8) + 8n+1$

c. $T(n) = 3T(n/2) + n^2$

d. $T(n) = T(n-1) + n^2$

120. Which of the following is the answer to the recursive relation $T(1) = c$, $T(n) \leq T(\frac{n}{5}) + T(\frac{7n}{10}) + cn$?

a. $T(n) \leq \sum_{i=0}^{\log_{\frac{10}{7}} n} \left(\frac{9}{10}\right)^i n$

b. $T(n) \leq \sum_{i=0}^{\log_{\frac{10}{9}} n} \left(\frac{7}{10}\right)^i n$

c. $T(n) \leq \sum_{i=0}^{\log_5 n} \left(\frac{7}{10}\right)^i n$

d. $T(n) \leq \sum_{i=0}^{\log_{10} n} \left(\frac{9}{10}\right)^i n$

121. For $T(n) = 1$, the answer of the recurrence relation $T(n) = T(n/3) + T(n/6) + n^{\sqrt{\lg n}}$ is ...

a. $\Theta(n^{\sqrt{\lg n}} \lg n)$

b. $\Theta(n^{\sqrt{\lg n}})$

c. $\Theta(n^{\sqrt{n}})$

d. $\Theta(n^{\sqrt{n}} \lg n)$

122. The answer of the recursive function below shows the time complexity of which of the following algorithms?

$$t(1) = 0, \quad t(n) = t(n-1) + \frac{2}{n}$$

a. Binary search

b. Insertion sorting

c. Quicksort

d. Sequential search

3.2.7   Recurrence with Full History

Solve the following recurrences exactly.

123. $T(1) = 1$, and for all $n \geq 2$, $T(n) = \sum_{i=1}^{n-1} T(i) + 1$.

124. $T(1) = 1$, and for all $n \geq 2$, $T(n) = \sum_{i=1}^{n-1} T(i) + 7$.

125. $T(1) = 1$, and for all $n \geq 2$, $T(n) = \sum_{i=1}^{n-1} T(i) + n^2$.

126. $T(1) = 1$, and for all $n \geq 2$, $T(n) = 2\sum_{i=1}^{n-1} T(i) + 1$.

127. $T(1) = 1$, and for all $n \geq 2$, $T(n) = \sum_{i=1}^{n-1} T(n-i) + 1$.

128. $T(1) = 1$, and for all $n \geq 2$, $T(n) = \sum_{i=1}^{n-1}(T(i) + T(n-i)) + 1$.

3.2.8 Recurrence with Floors and Ceilings

Solve the following recurrence relation and find an exact solution for T(n).

129. $T(1) = 1$, for all $n \geq 2$, $T(n) = T(\lfloor \frac{n}{2} \rfloor) + T(\lceil \frac{n}{2} \rceil) + n - 1$.

130. $T(1) = 1$, for all $n \geq 2$, $T(n) = T(\lfloor \frac{n}{2} \rfloor) + 1$.

131. $T(1) = 1$, for all $n \geq 2$, $T(n) = T(\lceil \frac{n}{2} \rceil) + 1$.

132. $T(1) = 1$, for all $n \geq 2$, $T(n) = 2T(\lfloor \frac{n}{2} \rfloor) + 6n - 1$.

133. $T(1) = 2$, for all $n \geq 2$, $T(n) = 4T(\lceil \frac{n}{3} \rceil) + 3n - 5$.

134. $T(1) = 1$, for all $n \geq 2$, $T(n) = T(\lfloor \sqrt{n} \rfloor) + 1$.

135. Let $T(n)$ denote the running time of mergesort on an array containing n elements. The merge function runs in $\Theta(n)$ time.

$$T(n) = \begin{cases} \Theta(n) & n = 1 \\ T(\lfloor \frac{n}{2} \rfloor) + T(\lceil \frac{n}{2} \rceil) + \Theta(n) & o.w \end{cases}$$

Solve this recurrence relation.

3.2.9 The Master Method

To use the master theorem, we simply plug the numbers into the formula. Use the master method to give tight asymptotic bounds for the following recurrences:

136. $T(n) = 9T(n/3) + n$.

137. $T(n) = T(2n/3) + 1$.

138. $T(n) = 3T(n/4) + n \log n$.

139. $T(n) = 3T(n/2) + n^2$.

140. $T(n) = 7T(n/2) + n^2$.

141. $T(n) = 2T(n/2) + n$.

142. $T(n) = 4T(n/2) + n$.

143. $T(n) = 4T(n/2) + n^2$.

144. $T(n) = 4T(n/2) + n^2 \log n$.

145. $T(n) = 2T(n/4) + 1$.

146. $T(n) = 2T(n/4) + \sqrt{n}$.

147. $T(n) = T(7n/10) + n^2$.

148. $T(n) = 2T(n/4) + \log n$.

149. $T(n) = 3T(n/3) + n \log n$.

150. $T(n) = 2T(n/4) + n^{0.51}$.

151. $T(n) = 16T(n/2) + n!$.

152. $T(n) = 8T(n/3) + 2^n$.

153. Can master method be applied to solve the recurrence $T(n) = 2T(n/2) + nlogn$. If not why?

154. Can master method be applied to solve the recurrence $T(n) = 9T(n/3) + \frac{n^2}{logn}$. If not why?

155. Can master method be applied to solve the recurrence $T(n) = 8T(n/2) + n^3logn$. If not why?

156. Determine whether the master method can be used to solve the recurrence $T(n) = 4T(n/2) + n^2logn$? Why or why not? Give an asymptotic upper bound for this recurrence.

Determine whether the master method can be used to solve the following recurrences. If 'Yes', give the running time for that relation. Otherwise, indicate that the master theorem does not apply.

157. $T(n) = 3T(n/2) + n^2$

158. $T(n) = 4T(n/2) + n^2$

159. $T(n) = T(n/2) + 2^n$

160. $T(n) = 16T(n/4) + n$

161. $T(n) = 2T(n/2) + n\log n$

162. $T(n) = 2T(n/4) + n^{0.5}$

163. $T(n) = 0.5T(n/2) + 1/n$

164. $16T(n/4) + n!$

165. $T(n) = \sqrt{n}T(n/2) + \log n$

166. $T(n) = 3T(n/2) + n$

167. $T(n) = 3T(n/3) + \sqrt{n}$

168. $T(n) = 4T(n/2) + cn$

169. $T(n) = 3T(n/4) + n\log n$

170. $T(n) = 3T(n/3) + n/2$

171. $T(n) = 6T(n/3) + n^2\log n$

172. $T(n) = 4T(n/2) + n/\log n$

173. $T(n) = 64T(n/8) - n^2\log n$

174. $T(n) = 7T(n/3) + n^2$

175. $T(n) = 4T(n/2) + n\log n$

176. $T(n) = T(n/2) + n(2 - \cos n)$

Justify why the following equations (177–179) cannot be solved using the master theorem.

177. $T(n) = 2^n T(n/2) + n^n$

178. $T(n) = 2T(n/2) + n/\log n$

179. $T(n) = 0.5T(n/2) + n$

180. Consider the following recursive function.

```
int  F(int  n)
{
    if(n ≤ 1)  return  1
    else
    return  F(n/2) * F(n/2)
}
```

Let T(n) denotes the number of calls or the number of *if* of the above function per n, we can write the recurrence function corresponding to T(n) as follows

$$T(n) = 2T\left(\frac{n}{2}\right) + 1$$

Solve the above recurrence relation to obtain the order (time complexity) of this function.

181. Consider the following recursive function for computing the power X^n by repeated multiplication.

```
real  Power(real  X, int  n)
{
    if(n == 1)  return  X
    else
    return  Power(X, ⌊n/2⌋) * Power(X, ⌈n/2⌉)
}
```

Let T(n) denotes the number of multiplications of the above function, so we can write the recurrence function corresponding to T(n) as follows

$$T(n) = \begin{cases} 0 & n = 1 \\ T(\lfloor\frac{n}{2}\rfloor) + T(\lceil\frac{n}{2}\rceil) + 1 & n \geq 2 \end{cases}$$

Solve the above recurrence relation to obtain the order (time complexity) of this function.

3.2.10 Recursion Tree Method

182. Solve the recurrence $T(n) = 2T(n/2) + 1$.

183. Solve the recurrence $T(n) = 3T(n/4) + cn^2$.

184. Solve the recurrence $T(n) = T(n/3) + T(2n/3) + O(n)$.

185. Solve the recurrence $T(n) = T(n/2) + T(n/4) + n^2$.

186. Solve the recurrence $T(n) = 8T(n/2) + n^2$.

187. Solve the recurrence $T(n) = 2T(n/2) + n^2$.

188. Solve the recurrence $T(n) = T(n/5) + T(7n/10) + n$.

189. Solve the recurrence $T(n) = T(\frac{n}{10}) + T(\frac{9n}{10}) + n$.

190. Solve the recurrence $T(n) = \sqrt{n}T(\sqrt{n}) + n$.

191. Solve the recurrence $T(n) = 1/4T(n/4) + 3/4T(3n/4) + 1$.

3.2.11 Recurrence Relations with More Than One Variable

192. Solve the recurrence

$$T(n, m) = \begin{cases} 2T(m/2, n/2) + mn, & m > 1, n > 1 \\ n, & if\ m = 1 \\ m, & if\ n = 1 \end{cases}$$

193. The binomial coefficient. Solve the recurrence

$$T(n, k) = \begin{cases} T(n-1, k-1) + T(n-1, k), & \\ 1, & k = 1 \\ 0, & k > n \end{cases}$$

194. Solve the recurrence $F(n, m) = F(n–1, m) + F(n, m–1)$.

3.2.12 Generating Functions

Another way to solve a recurrence relation (to find a formula for) is using generating functions. In the following recurrence relations, find the generating function for the relation and use it to find a formula for h_n.

195. $h_n = 4h_{n-1} - 3h_{n-2}$, $h_0 = 2$, $h_1 = 5$.

196. $h_n = 3h_{n-1} + 4h_{n-2}$, $h_0 = h_1 = 1$.

197. $h_n = 2h_{n-1} + 3^n$, $h_0 = 0$.

198. $h_n = 4h_{n-2}$, $h_0 = 1$, $h_1 = 5$.

199. $h_n = h_{n-1} - h_{n-2}$, $h_0 = 0$, $h_1 = 3$.

3.3 Solutions

1.

$$T_n = T_{n-1} + c = (T_{n-2} + c) + c = T_{n-2} + 2c$$

$$= (T_{n-3} + c) + 2c = T_{n-3} + 3c$$

$$= (T_{n-4} + c) + 3c = T_{n-4} + 4c =$$

$$.$$
$$.$$
$$.$$

$$= T_{n-(n-1)} + (n-1)c = T_1 + (n-1)c$$

$$\overset{T_1=d}{\Longrightarrow} d + (n-1)c$$

2.

$$T(0) = 0$$

$$T(1) = 1$$

$$T(2) = 3$$

$$T(3) = 7$$

$$T(4) = 15$$

$$T(5) = 31$$

$$T(6) = 63$$

$$T(n) = 2(T(n-1)) + 1$$

$$T(n) = 2(2T(n-2)) + 1 = 2^2(T(n-2)) + (2+1)$$

$$= 2^n(T(n-1)) + (2^{n-1} + 2^{n-2} + \cdots + 2 + 1)$$

4.

$$T(n) = 2T(n-2) + 1 = 2(2T(n-3) + 1) + 1 = 2^2 T(n-3) + (2+1)$$

$$= 2^2(2T(n-4) + 1) + (2+1) = 2^3 T(n-4) + (4+2+1)$$

$$= 2^3(2T(n-5) + 1) + (4+2+1) = 2^4 T(n-5) + (8+4+2+1)$$

$$= 2^n T(n-2) + (2^{n-1} + 2^{n-2} + 2^{n-3} + \cdots + 2 + 1)$$

6.

$$T(n) = T\left(\frac{n}{2}\right) + cn, \, T(1) = d$$

$$\rightarrow T\left(\frac{n}{4}\right) + \frac{cn}{2} + cn$$

$$\rightarrow T\left(\frac{n}{8}\right) + \frac{cn}{4} + \frac{cn}{2} + cn$$

$$T\left(\frac{n}{\log n}\right) + c\frac{n}{\log n} + \cdots + \frac{n}{8} + \frac{n}{4} + \frac{n}{2} + n$$

$$\rightarrow cn\left(1 + \frac{1}{2} + \frac{1}{4} + \frac{1}{8} + \ldots\right) = cn \rightarrow \theta(n)$$

18.

$$T(n) = 2T(n-1) - 3T(n-2)$$

$$x^2 - 2x + 3 = 0 \rightarrow x_1 = 1 + \sqrt{2}, \, x_2 = 1 - \sqrt{2}$$

$$T(n) = c_1(1 + \sqrt{2})^n + c_2(1 - \sqrt{2})^n$$

$$1 = c_1 + c_2$$

$$2 = c_1\left(1 + i\sqrt{2}\right) + c_2(1 - \sqrt{2})$$

$$c_1 = \frac{1}{2} + \frac{1}{i\sqrt{2}}$$

$$c_2 = \frac{1}{2} - \frac{1}{i\sqrt{2}}$$

$$T(n) = \left(\frac{1}{2} + \frac{1}{i\sqrt{2}}\right)(1 + i\sqrt{2})^n$$

$$T(n) = \left(\frac{1}{2} - \frac{1}{i\sqrt{2}}\right)(1 - i\sqrt{2})^n$$

19.

$$T(n) = 7T(n-1) - 4T(n-2) \Rightarrow T(n) = a^n \rightarrow a^2 - 7a + 4 = 0$$

$$\frac{-b \pm \sqrt{b^2 - 4ac}}{2a} = \frac{-7 \pm \sqrt{7^2 - 4 \times 4}}{2} = \frac{7 \pm \sqrt{33}}{2} = r_1, r_2$$

$$a_n = C_1(r_1)^n + C_2(r_2)^n \Rightarrow a_0 = 0 = C_1\left(\frac{7 + \sqrt{33}}{2}\right)^0 + C_2\left(\frac{7 - \sqrt{33}}{2}\right)^0 \rightarrow C_1 + C_2 = 0$$

$$a_1 = 2 = C_1\left(\frac{7 + \sqrt{33}}{2}\right)^1 + C_2\left(\frac{7 - \sqrt{33}}{2}\right)^1$$

$$\begin{cases} C_1 + C_2 = 0 \\ C_1\left(\frac{7+\sqrt{33}}{2}\right)^1 + C_2\left(\frac{7-\sqrt{33}}{2}\right)^1 = 2 \end{cases}$$

First, multiply the first equation by $\frac{-7+\sqrt{33}}{2}$ and by solving the two equations, the following answer is obtained:

$$C_1 = \frac{2\sqrt{33}}{33}, C_2 = \frac{-2\sqrt{33}}{33}$$

$$a_n = C_1(r_1)^n + C_2(r_2)^n \Rightarrow \frac{2\sqrt{33}}{33}\left(\frac{7 + \sqrt{33}}{2}\right)^n + \frac{-2\sqrt{33}}{33}\left(\frac{7 - \sqrt{33}}{2}\right)^n$$

20.

$$F_n = F_{n-1} + F_{n-2}$$

$$F^2 - F - 1 = 0$$

$$r_1, r_2 = \frac{1 \pm \sqrt{1 - (4 \times 1 \times (-1))}}{2} = \frac{1 \pm \sqrt{5}}{2}$$

$$a_n = C_1(r_1)^n + C_2(r_2)^n$$

$$a_n = C_1\left(\frac{1+\sqrt{5}}{2}\right)^n + C_2\left(\frac{1-\sqrt{5}}{2}\right)^n$$

$$F_0 = C_1 + C_2 = 0$$

$$F_1 = C_1\left(\frac{1+\sqrt{5}}{2}\right) + C_2\left(\frac{1-\sqrt{5}}{2}\right) = 1$$

$$F_n = \frac{1}{\sqrt{5}}\left(\frac{1+\sqrt{5}}{2}\right)^n - \frac{1}{\sqrt{5}}\left(\frac{1-\sqrt{5}}{2}\right)^n$$

25.

$$T(n) - 7T(n-1) = 0 \Rightarrow a - 7 = 0 \ \ a = 7$$

$$a_n = c_1 r_1^{\,n} \rightarrow c_1 7^n$$

$$a_2 = 98 \rightarrow c_1 7^2 = 98 \rightarrow c_1 = 2$$

$$a_n = 2(7)^n$$

26.

$$T(0) = 1, \ T(1) = 3, \ T(n+2) = 4T(n+1) - 4T(n)$$

$$T(n+2) = 4T(n+1) - 4T(n) \Rightarrow \frac{T^{n+2}}{T^n} - \frac{4T^{n+1}}{T^n} + \frac{4T^n}{T^n} = 0$$

$$T^2 - 4T + 4 = 0 \implies r_1, r_2 = 2, 2$$

$$a_n = c_1(r_1)^n + c_2 n(r_2)^n \longrightarrow a_n = c_1(2)^n + c_2 n(2)^n$$

$$\begin{cases} T(0) = 1 \Rightarrow c_1 = 1 \\ T(1) = 3 \Rightarrow 2c_1 + 2c_2 = 3 \Rightarrow c_1 = \frac{1}{2} \end{cases}$$

29.

$$T(n) = 2T(n-1) + 3T(n-2)$$

$$a^2 - 2a - 3 = 0 \Rightarrow (a-3)(a+1) = 0 \Rightarrow r_1 = 3, r_2 = -1$$

$$T(n) = \alpha_1 3^n + \alpha_2(-1)^n \Rightarrow T(0) = 1 \Rightarrow \alpha_1 + \alpha_2 = 1,$$

$$T(1) = 1 \Rightarrow 3\alpha_1 - \alpha_2 = 1 \Rightarrow \alpha_1, \alpha_2 = \frac{1}{2}$$

$$T(n) = \frac{1}{2}(3)^n + \frac{1}{2}(-1)^n$$

35.

$$T(n) = 4T(n-2) \Rightarrow T(n) - 4T(n-2) = 0 \Rightarrow a^2 - 4 = 0 \Rightarrow r_1, r_2 = 2, -2$$

$$\Rightarrow T(n) = c_1(2)^n + c_2(-2)^n$$

$$T(0) = 1 \Rightarrow c_1 + c_2 = 1 \quad , T(1) = 1 \Rightarrow 2c_1 - 2c_2 = 1$$

$$\Rightarrow c_1 = \frac{3}{4}, c_2 = \frac{1}{4} \quad \Rightarrow T(n) = \frac{3}{4}(2)^n + \frac{1}{4}(-2)^n$$

38.

$$F(n) = \begin{cases} 1 & n = 1 \\ 1 & n = 1 \\ 2 \times F(n-1) + 3 \times F(n-2) & n \geq 2 \end{cases}$$

$$F(n) = 2 \times F(n-1) + 3 \times F(n-2)$$

$$\rightarrow F(n) - 2 \times F(n-1) - 3 \times F(n-2) = 0$$

$$\rightarrow a^n - 2a - 3 = 0 \rightarrow r_1 = -1, \ r_2 = 3$$

$$\rightarrow F(n) = c_1(-1)^n + c_2(3)^n$$

$$\rightarrow \begin{cases} F(0) = 1 : c_1 + c_2 = 1 \\ F(1) = 1 : -c_1 + 3c_2 = 1 \end{cases} \rightarrow \begin{cases} c_1 = \frac{1}{2} \\ c_2 = \frac{1}{2} \end{cases}$$

$$\rightarrow F(n) = \frac{1}{2}(-1)^n + \frac{1}{2}(3)^n$$

39.

$$T(n) = 4T(n-1) - 4T(n-2), \ T(0) = 1, T(1) = 3$$

$$a^2 - 4a + 4 = 0 \Rightarrow r = 2$$

$$a_n = \alpha_1 2^n + \alpha_2 n 2^n \implies a_0 = 1 \ , \ \alpha_1 = 1 \ , \ a_1 = 3 \ , \ 2\alpha_1 + 2\alpha_2 = 3 \ , \ \alpha_2 = \frac{1}{2}$$

$$a_n = 2^n + \frac{1}{2}n2^n$$

40.

$$T(n) = \begin{cases} 0 & n = 0 \\ 1 & n = 1 \\ 3T(n-1) + 4T(n-2) & otherwise \end{cases}$$

$$T(n) = 3T(n-1) + 4T(n-2) \implies T(n) = a^n \rightarrow a^2 - 3a - 4 = 0$$

$$r_1 = 4, r_2 = -1$$

$$a_n = C_1(r_1)^n + C_2(r_2)^n \implies a_0 = 0 = C_1(4)^0 + C_2(-1)^0 = C_1 + C_2 = 0$$

$$a_1 = 1 = C_1(4)^1 + C_2(-1)^1 = 4C_1 - C_2 = 1$$

$$\begin{cases} C_1 + C_2 = 0 \\ 4C_1 - C_2 = 1 \end{cases}$$

By solving the above equation we have: $C_1 = \frac{1}{5}, C_2 = \frac{-1}{5}$

$$a_n = C_1(r_1)^n + C_2(r_2)^n = a_n = \frac{1}{5}(4)^n + \frac{-1}{5}(-1)^n$$

41.

$$T(n) = \begin{cases} 1 & n = 1 \\ T(n-1) + T(n-1) & n \geq 2 \end{cases}$$

$$a_n = 2a_{n-1} \rightarrow a_n - 2a_{n-1} = 0 \rightarrow a = 2$$

$$\rightarrow a_n = c_1 r_1^n \rightarrow a_n = c_1(2)^n$$

$$a_1 = 1 \rightarrow 1 = c_1 2^1 \rightarrow c_1 = \frac{1}{2}$$

$$a_n = \frac{1}{2}(2)^n \rightarrow O(2^n)$$

42. T(1) = 1, and for all $n \geq 2$, T(n) = 3T(n - 1) + 2.
Universal solution: T(n) = 3T(n - 1)

1. $a^2 - 3a = 0 \rightarrow r_1, r_2 = 0, 3$
2. $U_n = C_1(3)^n + C_2(0)^n = C_1(3)^n$

Private solution: $P_n = C_0 + C_1 n + \cdots + C_q n^q$

1. $P_n = C$
2. $C = 3C + 2 \implies C = -1$
3. $U_n + P_n = C_1 (3)^n - 1$
4. $T(1) = 1 \implies 3C_1 - 1 = 1$
5. $C_1 = \frac{2}{3}$

$T(n) = \frac{2}{3} (3)^n - 1 = 2 (3)^{n-1} - 1$

43.

$$T(n) = 2T(n-1) + 3^n$$

Homogeneous part:
$T(n) = 2T(n-1) \rightarrow T(n) - 2T(n-1) = 0$
$a - 2 = 0 \rightarrow a = 2$
$U_n = 2(2^n)$
$P_n = c3^n$
$c3^n = 2 c3^{n-1} + 3^n$
$3c = 2c + 3 \rightarrow c = 3$
$P_n = 3 \times 3^n = 3^{n+1}$
$U_n + p_n = \alpha 2^n + 3^{n+1}$

46.

$$T(n) = 3T(n-1) \implies a - 3 = 0 \implies r = 3 \implies U_n = c3^n$$

$$P_n = c_0 + c_1 n$$

$$\implies c_0 + c_1 n = 3 (c_0 + c_1 (n-1)) + 5n \implies c_0 = -\frac{15}{4}, c_1 = -\frac{5}{2} \implies P_n = -\frac{15}{4} - \frac{5}{2}n$$

$$T(n) = U_n + P_n \implies T(1) = U_1 + P_1 = 3c - \frac{15}{4} - \frac{5}{2} = 5 \implies c = \frac{15}{4}$$

$$\implies T(n) = \frac{15}{4}(3)^n - \frac{15}{4} - \frac{5}{2}n$$

47.

$$T(n) - 2T(n-1) - 15T(n-2) = 0$$

$$a^n - 2a^{n-1} - 15a^{n-2} = 0$$

$$a^2 - 2a - 15 = 0 \rightarrow \begin{cases} r_1 = 3 \\ r_2 = -5 \end{cases}$$

$$T\,(n) = c_1(3)^n + c_2(-5)^n$$

$$\rightarrow \begin{cases} T\,(1) = 2 \,:\, 3c_1 - 5c_2 = 2 \\ T\,(2) = 4 \,:\, 9c_1 + 25c_2 = 4 \end{cases} \rightarrow \begin{cases} c_1 = \frac{7}{12} \\ c_2 = -\frac{1}{20} \end{cases}$$

$$\rightarrow T\,(n) = \frac{7}{12}(3)^n + \frac{-1}{20}(-5)^n$$

49.

$$T\,(n) = 3T\,(n-1) - 2T\,(n-2) + 2^n \;\; \&\& \; T\,(0) = 2 \,,\; T\,(1) = 4$$

$$U\,(n) = 3T\,(n-1) - 2T\,(n-2) \Rightarrow \; U\,(n) = a^n \rightarrow a^2 - 3a + 2 = 0$$

$$r_1 = 2, r_2 = 1$$

$$U_n = C_1(r_1)^n + C_2(r_2)^n \Rightarrow U_n = C_1(2)^n + C_2(1)^n = \; U_n = C_1(2)^n + C_2$$

$$P_n = Cn(2)^n$$

$$Cn(2)^n = 3C(n-1)(2)^{n-1} - 2C(n-2)(2)^{n-2} + 2^n$$

$$Cn(2)^2 = 3C(n-1)(2)^1 - 2C(n-2)(2)^0 + 2^2$$

$$4Cn = 6C\,(n-1) - 2C\,(n-2) + 4 \Rightarrow 4Cn = 6Cn - 6C - 2Cn + 4C + 4 \Rightarrow C = 2$$

$$U\,(n) + P\,(n) = C_1(2)^n + C_2 + n(2)^{n+1}$$

$$a_0 = 2 = C_1(2)^0 + C_2(1)^0 + 0(2) = C_1 + C_2 = 2$$

$$a_1 = 4 = C_1(2)^1 + C_2(1)^1 + 1 \times 2^{1+1} = 2C_1 + C_2 + 4 = 4$$

$$\begin{cases} C_1 + C_2 = 0 \\ 2C_1 + C_2 = 0 \end{cases}$$

$$C_1, C_2 = 0$$

$$U\,(n) + P\,(n) = n(2)^{n+1} = n(2)^{n+1}$$

51.

$$T\,(n) = 2T\,(n-1) + 1$$

$$= 2\,(2T\,(n-2) + 1) + 1 \;=\; 2^2 T\,(n-2) + 2 + 1$$

$$= 2^2\,(2T\,(n-3) + 1) + 2 + 1 = 2^3 T\,(n-3) + 2^2 + 2 + 1$$

$$\ldots = 2^k T (n - k) + 2^{k-1} + \cdots + 2^2 + 2 + 1$$

$$= 2^n T (0) + 1 + 2 + 2^2 + \cdots + 2^{k-1}$$

$$= 2^{n-1} + \cdots + 2^2 + 2 + 1 = 2^n - 1 \implies O(2^n)$$

58. (A) To obtain the general answer, we separate the heterogeneous part and we get the homogeneous part by changing the variable $a_n = a^n$ to the following relation.

$$a^n = 3a^{n-1} - 2a^{n-2}$$

(B) Divide the sides by a^{n-2} and obtain the roots of the equation and the general answer

$$a^2 = 3a - 2 \qquad (a - 2)(a - 1) = 0 \rightarrow r_1 = 2 , \qquad r_2 = 1$$

$$U_n = C_1(2)^n + C_2(1)^n$$

$$P_n = \alpha_1 (n4^n + 3)$$

(C) To get the general answer, we place P_n in the main equation

$$\alpha_1 (n4^n + 3) = 3\alpha_1 \left(n4^{n-1} + 3\right) - 2\alpha_1 \left(n4^{n-2} + 3\right) + n4^n + 3$$

Divide the sides by $(n4^n + 3)$ and get the root of the equation

$$\alpha_1 \left(n4^n + 3\right) = 3\alpha_1 \left(4^{-1} + 1\right) \left(n4^n + 3\right) - 2\alpha_1 \left(4^{-2} + 1\right) \left(n4^n + 3\right) + (1) \left(n4^n + 3\right)$$

$$\alpha_1 = 3\alpha_1 \left(4^{-1} + 1\right) - 2\alpha_1 \left(4^{-2} + 1\right) + 1$$

$$\alpha_1 = \left(\frac{3}{4} + 3 - \frac{1}{8} - 2\right) \alpha_1 + 1$$

$$\alpha_1 = \left(\frac{5}{8} + 1\right) \alpha_1 + 1$$

$$\left(\frac{5}{8}\right) \alpha_1 = -1$$

$$\alpha_1 = -\frac{8}{5}$$

(D) To get the general answer, we calculate $P_n + U_n$

$$P_n + U_n = -\frac{8}{5}\left(n4^n + 3\right) + C_1(2)^n + C_2(1)^n$$

65.

$$T(n) = 4T(n-1) - 3T(n-2)$$

$$T(n) = 4T(n-1) - 3T(n-2) \implies T(n) = a^n \rightarrow a^2 - 4a + 3 = 0$$

$$r_1 = 3, r_2 = 1$$

$$a_n = C_1(r_1)^n + C_2(r_2)^n \implies a_0 = 0 = C_1(3)^n + C_2(1)^n$$

66.

$$T(n) = 3T(n-1) + 4T(n-2)$$

$$T(n) - 3T(n-1) + 4T(n-2) = 0$$

$$a^2 - 3a - 4 = 0 \rightarrow (a-4)(a+1) = 0$$

$$r_1, r_2 = +4, -1$$

General formula: $a_n = c_1r_1{}^n + c_2r_2{}^n$

$$a_n = c_1(4)^n + c_2(-1)^n$$

Chapter 4
Algorithm Analysis

Abstract There is usually more than one algorithm for solving problems. The question that arises in such cases is which of these algorithms works best. The algorithms are compared based on runtime and memory usage. Therefore, it is an efficient algorithm that wastes running time and less memory. Because there is enough memory today, it is very important to calculate the execution time, so we will explain the methods of calculating the time in this chapter. This chapter illustrates the methods to analyze algorithms through a variety of examples and provides 130 exercises on algorithm analysis.

4.1 Lecture Notes

There is usually more than one algorithm for solving problems. The question that arises in such cases is which of these algorithms works best. The algorithms are compared based on runtime and memory usage. Therefore, it is an efficient algorithm that wastes running time and less memory. Because there is enough memory today, it is very important to calculate the execution time, so we will explain the methods of calculating the time in this chapter. The algorithms we use to solve problems are usually divided into two main categories:

1. Sequential algorithms
2. Recursive algorithms

To obtain the execution time of a sequential algorithm, we consider the execution time of replacement commands, computational operators, conditionals, etc. (the time of these commands depends on the type of hardware and compiler).

To calculate the execution time of a program, we calculate the following times:

1. Assignment operations, computational operators, if (simple) conditions, etc. have a fixed time.
2. For a number of commands to be repeated, the execution time will be the product of the number of repetitions at the time of execution of the commands. This part of the program is usually represented by loops.

© The Author(s), under exclusive license to Springer Nature Switzerland AG 2022   135
H. Izadkhah, *Problems on Algorithms*,
https://doi.org/10.1007/978-3-031-17043-0_4

3. If the program contains the *if* and *else* structure. Each of which has times T_1 and T_2, in which case the execution time of this piece of program will be equal to the maximum values of T_1 and T_2.
4. The total time of the program is equal to the sum of the program pieces.
   Intuitively, the order time of an algorithm is usually the order of the part of the program that has the most time. Because we always consider the upper limit or the worst case for the growth function.

Analysis of Algorithms principles

1. Arrangement principle: If p_1 and p_2 are two program pieces or several independent program pieces, the time required to run p_1 and p_2 consecutively will be $p_1 + p_2$.
2. Principle of composition: If p_1 and p_2 are two program pieces or several nested program pieces, the time required to run p_1 and p_2 consecutively will be multiplied.
3. Principle of atomic commands: An atomic command means basic commands at the machine level, such as commands, +, −, comparison, replacement (assignment), etc.; These commands at the hardware level require a time limit of $\Theta(1)$.

4.2 Exercises

4.2.1 Iterative Algorithms

1. Give the worst-case running time using big-O notation.

$$x = 0;$$
$$for (i = 0; \ i < n; \ i++)$$
$$x++;$$

2. Give the worst-case running time using big-O notation. Here c is a positive integer constant.

$$x = 0;$$
$$for (i = 0; \ i < n; \ i \ += \ c)$$
$$some \ O(1) \ expressions$$

3. Give the worst-case running time using big-O notation. Here c is a positive integer constant.

$$x = 0;$$
$$for (i = n; \ i > 0; \ i \ -= \ c)$$
$$some \ O(1) \ expressions$$

4. Give the worst-case running time using big-O notation.

$$Algorithm\ Sum(a,\ x)$$
$$\{s := 0;$$
$$\quad for\ i := 1\ to\ n\ do$$
$$\quad\quad s := s + a[i];$$
$$\quad for\ j := 1\ to\ n\ do$$
$$\quad\quad s := s + b[j];$$
$$return\ s;$$
$$\}$$

5. Give the worst-case running time using big-O notation.

$$Algorithm\ Sum(a,\ x)$$
$$\{s := 0;$$
$$\quad for\ i := 1\ to\ n\ do$$
$$\quad\quad s := s + a[i];$$
$$\quad for\ j := 1\ to\ m\ do$$
$$\quad\quad s := s + b[j];$$
$$return\ s;$$
$$\}$$

6. Give the worst-case running time using big-O notation.

$$Algorithm\ Sum(a, x)$$
$$\{s := 0;$$
$$\quad for\ i := 1\ to\ n/2\ do$$
$$\quad\quad s := s + a[i];$$
$$return\ s;$$
$$\}$$

7. Give the worst-case running time using big-O notation.

$$Algorithm\ Sum(a, x)$$
$$\{s := 0;$$
$$\quad for\ i := 1\ to\ n/3\ do$$
$$\quad\quad s := s + a[i];$$
$$return\ s;$$
$$\}$$

8. Give the worst-case running time using big-O notation.

$$Algorithm \ Sum(a, \ x)$$
$$\{s := 0;$$
$$\quad for \ i := 1 \ to \ n/2 \ do$$
$$\quad\quad s := s + a[i];$$
$$\quad for \ j := 1 \ to \ m/3 \ do$$
$$\quad\quad s := s + b[j];$$
$$return \ s;$$
$$\}$$

9. Give the worst-case running time using big-O notation.

$$Algorithm \ Sum(a, x)$$
$$\{s = 0;$$
$$\quad for(i = 1; \ i \leq n; \ i = i * 2)$$
$$\quad\quad s = s + a[i];$$
$$return \ s;$$
$$\}$$

10. Give the worst-case running time using big-O notation.

$$Algorithm \ Sum(a, x)$$
$$\{s = 0;$$
$$\quad for(i = 1; \ i \leq n; \ i = i * 3)$$
$$\quad\quad s = s + a[i];$$
$$return \ s;$$
$$\}$$

11. Give the worst-case running time using big-O notation. Here c is a positive integer constant.

$$x = 0;$$
$$for(i = 0; \ i < n; \ i *= \ c)$$
$$\quad some \ O(1) \ expressions$$

12. Give the worst-case running time using big-O notation. Here c is a constant greater than 1.

$$x = 0;$$
$$for(i = 2; \ i \leq n; \ i = pow(i, c))$$
$$\quad some \ O(1) \ expressions$$

13. Give the worst-case running time using big-O notation. Here *fun* is sqrt or cuberoot or any other constant root.

$$x = 0;$$
$$for(i = n; \ i > n; \ i = fun(i))$$
$$\quad some \ O(1) \ expressions$$

14. Give the worst-case running time using big-O notation.

$$i = 1;$$
$$while(i \leq n)\{$$
$$/ * \;\; Some \;\; Statement \;\; requiring \;\; \theta\,(1) \;\; time * /$$
$$i = i * 2;$$
$$\}$$

15. Give the worst-case running time using big-O notation.

$$i = n;$$
$$while(i \geq 1)\{$$
$$/ * \;\; Some \;\; Statement \;\; requiring \;\; \theta\,(1) \;\; time * /$$
$$i = i/2;$$
$$\}$$

16. Give the worst-case running time using big-O notation.

$$Algorithm \;\; Sum(a, x)$$
$$\{s = 0;$$
$$while(n > 0)\{$$
$$s = s + 1;$$
$$n = n/10\};$$
$$return \;\; s;$$
$$\}$$

17. Give the worst-case running time using big-O notation.

$$Algorithm \;\; Sum(a, x)$$
$$\{s = 0;$$
$$for(i = 1; \;\; i \leq n; \;\; i = i * 2)$$
$$s = s + a[i];$$
$$for(j = 1; \;\; j \leq n; \;\; j = j * 2)$$
$$s = s + a[i];$$
$$return \;\; s;$$
$$\}$$

18. Give the worst-case running time using big-O notation.

$$Algorithm \;\; Sum(a, x)$$
$$\{s = 0;$$
$$for(i = n; \;\; i \geq 1; \;\; i = i/2)$$
$$s = s + a[i];$$
$$return \;\; s;$$
$$\}$$

19. Give the worst-case running time using big-O notation.

$$Algorithm \ Sum(a, x)$$
$$\{s = 0;$$
$$\quad for(i = n; \ i \geq 1; \ i = i/3)$$
$$\quad\quad s = s + a[i];$$
$$return \ s;$$
$$\}$$

20. Give the worst-case running time using big-O notation. Here c is a positive integer constant.

$$x = 0;$$
$$for(i = n; \ i > 0; \ i \ /= \ c)$$
$$\quad some \ O(1) \ expressions$$

21. Give the worst-case running time using big-O notation.

$$Algorithm \ Sum(a, x)$$
$$\{s = 0;$$
$$\quad for(i = n; \ i \geq 1; \ i = i/2)$$
$$\quad\quad s = s + a[i];$$
$$\quad for(j = n; \ j \geq 1; \ j = j/3)$$
$$\quad\quad s = s + b[i];$$
$$return \ s;$$
$$\}$$

22. Give, using Θ notation, the average-case running time for the sequential search algorithm where x is in the array S with n keys.

$$location = 1;$$
$$while(location \leq n \ \&\& \ S[location]! = x)$$
$$\quad location + +;$$
$$if(location > n)$$
$$\quad location = 0;$$
$$return \ location;$$

23. Give the worst-case running time using big-O notation.

$$i := 2;$$
$$while(i < n)do$$
$$\quad begin$$
$$\quad\quad i := i^2;$$
$$\quad\quad x := x + 1;$$
$$\quad end;$$

24. Give the worst-case running time using big-O notation.

```
int Seq_Search(int a[], int n, int x)
{
  int i
  for(i = 0; i < n; i + +)
  if(a[i] == x)
    return(i)
  return(−1)
}
```

25. Give the worst-case running time using big-O notation.

```
void binsearch(int n, const keytype S[], keytype x, index& location)
{
  index low, high, mid;
  low = 1;  high = n;
  location = 0;
  while(low ≤ high && location == 0)
  {
    mid = ⌊(low + high)/2⌋;
    if (x == S[mid])
    location = mid;
    else if(x < S[mid])
      high = mid − 1;
    else
      low = mid + 1;
  }
}
```

26. Give the worst-case running time using big-O notation.

```
/ * Suppose that n > m * /
while(n > 0){
  r = n%m;
  n = m;
  m = r; }
```

27. Give the worst-case running time using big-O notation.

```
x := 0;
for i = 1 to n do
for j = 1 to n do
  x + +;
```

28. Give the worst-case running time using big-O notation.

$$x := 0;$$
$$for \ i = 1 \ to \ n \ do$$
$$for \ j = 1 \ to \ m \ do$$
$$x + +;$$

29. Give the worst-case running time using big-O notation.

$$x := 0;$$
$$for \ i = 1 \ to \ n/2 \ do$$
$$for \ j = 1 \ to \ m/2 \ do$$
$$x + +;$$

30. Give the worst-case running time using big-O notation.

$$x = 0$$
$$for(i = 0; \ i < n; \ i + +)$$
$$for(j = i; \ j < n; \ j + +)$$
$$x + +;$$

31. Give the worst-case running time using big-O notation.

$$S = 0;$$
$$for(i = 1; \ i \le n; \ i + +)$$
$$for(j = 1; \ j \le i; \ j + +)$$
$$S + +;$$

32. Give the worst-case running time using big-O notation.

$$for(i = 1; \ i \le n; \ i + 1)$$
$$for(j = 1; \ j \le n; \ j = j + i)$$
$$x = x + 1;$$

33. Give the worst-case running time using big-O notation.

$$for(i = 1; \ i \le n; \ i + +)$$
$$for(j = 1; \ j \le n; \ j + +)$$
$$\{$$
$$x + +;$$
$$n - -;$$
$$\}$$

34. Give the worst-case running time using big-O notation.

$$x := 0;$$
$$for(i = 1; \ i \le n; \ i = i * 2)$$
$$for(j = 1; \ j \le n; \ j = j * 2)$$
$$x + +;$$

35. Give the worst-case running time using big-O notation.

$$x := 0;$$
$$for(i = 1; \ i \le n; \ i + +)$$
$$for(j = 1; \ j \le n; \ j = j * 2)$$
$$x + +;$$

36. Give the worst-case running time using big-O notation.

$$x := 0;$$
$$for(i = 1; \ i \le n; \ i = i * 2)$$
$$for(j = 1; \ j \le n; \ j = j * 2)$$
$$for(k = 1; \ k \le j; \ j + +)$$
$$x + +;$$

37. Give the worst-case running time using big-O notation.

$$x := 0;$$
$$for(i = n; \ i \ge 1; \ i = i/2)$$
$$for(j = n; \ j \ge 1; \ j = j/3)$$
$$x + +;$$

38. Give the worst-case running time using big-O notation.

$$x := 0;$$
$$for(i = n; \ i \ge 1; \ i - -)$$
$$for(j = n; \ j \ge 1; \ j = j/3)$$
$$x + +;$$

39. Give the worst-case running time using big-O notation.

$$for(i = 1; \ i \le n; \ i + +)$$
$$if \ odd(i) \ then$$
$$for \ j = i \ to \ n \ do$$
$$x = x + 1$$
$$else \ for \ j = 1 \ to \ i \ do$$
$$y = y + 1$$

40. Give the worst-case running time using big-O notation.

$$ISPRIME(N)$$
$$i = 3$$
$$if \ N = 2 \ or \ N = 3$$
$$then \ return \ true$$
$$if \ N \ mod \ 2 = 0$$
$$then \ return \ false$$
$$while \ i^2 \leq N$$
$$do \ if \ N \ mod \ i = 0$$
$$then \ return \ false$$
$$else \ i = i + 2$$
$$return \ true$$

41. Give the worst-case running time using big-O notation.

$$x := 0;$$
$$for \ i = 1 \ to \ n - 1 \ do$$
$$for \ j = i + 1 \ to \ n \ do$$
$$for \ j = 1 \ to \ j \ do$$
$$some \ O(1) \ statements;$$

42. Give the worst-case running time using big-O notation.

$$x := 0;$$
$$for(i = 1; \ i \leq 1000; \ i + +)$$
$$for(j = 1; \ j \leq 100; \ j + +)$$
$$for(k = 1; \ j \leq 3; \ k + +)$$
$$x + +;$$

43. Give the worst-case running time using big-O notation.

$$x := 0;$$
$$for(i = 1; \ i \leq n; \ i + +)$$
$$for(j = 1; \ j \leq 1000; \ j + +)$$
$$for(k = 1; \ j \leq 3; \ k + +)$$
$$x + +;$$

44. Give the worst-case running time using big-O notation.

$$x := 0;$$
$$for(i = 1; \ i \leq n; \ i = i * 3)$$
$$for(j = 1; \ j \leq 1000; \ j + +)$$
$$for(k = 1; \ j \leq 3; \ k + +)$$
$$x + +;$$

45. Give the worst-case running time using big-O notation.

$$x := 0;$$
$$for(i = 1; \ i \le n; \ i = i * 3)$$
$$for(j = 1; \ j \le 1000; \ j = j * 2)$$
$$for(k = 1; \ j \le 3; \ k + +)$$
$$x + +;$$

46. Give the worst-case running time using big-O notation.

$$x := 0;$$
$$for(i = 1; \ i \le n; \ i + +)$$
$$for(j = 1; \ j \le n; \ j + +)$$
$$for(k = 1; \ j \le 3; \ k + +)$$
$$x + +;$$

47. Give the worst-case running time using big-O notation.

$$void \ function(int \ n)\{$$
$$int \ i, \ j, \ count = 0;$$
$$for(i = n/2; \ i \le n; \ i + +)$$
$$for(j = 1; \ j \le n; \ j = j * 2)$$
$$count + +; \}$$

48. Give the worst-case running time using big-O notation.

$$i := n;$$
$$while \ i \ge 1 \ do$$
$$begin$$
$$for \ to \ n \ do$$
$$x := x + 1;$$
$$i := [i/3];$$
$$end.$$

49. Give the worst-case running time using big-O notation.

$$void \ sum(int \ m, int \ n, float \ s[][])$$
$$\{$$
$$int \ i, \ j;$$
$$for(j = 0; j < m; j + +)$$
$$\{$$
$$S[n - 1][j] = 0;$$
$$for(i = 0; i < n - 1; i + +)$$
$$S[n - 1][j] + = S[i][j];$$
$$\}$$
$$\}$$

50. Give the worst-case running time using big-O notation.

$$void \ \ Bubble(elementType \ A[], int \ n)$$
$$\{$$
$$for(int \ \ i = 0; i < n - 1; i + +)$$
$$\{$$
$$for(j = n - 1; j > i + 1; i - -)$$
$$\{$$
$$if(A[i - 1] > A[j])$$
$$exchange(A[j], A[j - 1])$$
$$\}$$
$$\}$$
$$\}$$

51. Analyze the matrix multiplication algorithm given below.

$a_{m \times n} * b_{n \times p}$

$$void \ \ Algorithm \ (element \ A[][], element \ C[][], int \ m, \ int \ n, \ int \ p)$$
$$\{$$
$$for(int \ i = 0; i < m, i + +)$$
$$for(int \ \ j = 0, j < p, j + +)$$
$$\{C[i][j] = 0;$$
$$for(int \ \ k = 0; k < n; k + +)$$
$$C[i][j] = C[i][j] + A[i][j] * B[k][j];$$
$$\}$$
$$\}$$

52. Give the worst-case running time using big-O notation.

$$exporentiate(x, n)$$
$$\{ \quad m := n; \ \ power := 1; \ \ z := x;$$
$$while(m > 0) \ \ do$$
$$\{$$
$$while((m \ \ mod \ \ 2) = 0)do$$
$$\{m := m/2;$$
$$z := z^2;$$
$$\}$$
$$m := m - 1; \ \ power := power * z;$$
$$\}$$
$$return \ \ power;$$
$$\}$$

53. Give the worst-case running time using big-O notation.

$$for \ i = 1 \ to \ n \ do$$
$$for \ j = 1 \ to \ \lfloor i/2 \rfloor \ do$$
$$x := x + 1;$$

54. What is the time complexity of the following function?

$$for(i = 1; \ i \le n; \ i \ += \ c)\{$$
$$for(j = 1; \ j \le n; \ j \ += \ c)\{$$
$$some \ O(1) \ expressions$$
$$\}$$
$$\}$$

55. What is the time complexity of the following function?

$$for(i = n; \ i > n; \ i \ -= \ c)\{$$
$$for(j = i + 1; \ j \le n; \ j \ += \ c)\{$$
$$some \ O(1) \ expressions$$
$$\}$$
$$\}$$

56. What is the time complexity of the following function? Assume that $log(x)$ returns log value in base 2.

$$void \ fun()$$
$$\{$$
$$int \ i, \ j;$$
$$for(i = 1; \ i \le = n; \ i++)$$
$$for(j = 1; \ j \le log(i); \ j++)$$
$$printf("AAA");$$
$$\}$$

57. Give the worst-case running time using big-O notation.

$$s = 0;$$
$$for(i = 0; \ i < n; \ i++)$$
$$for(j = 1; \ j \le i * i; \ j++)$$
$$for(k = 0; \ k < j; \ k++)$$
$$s = s + 1;$$

58. Give the worst-case running time using big-O notation.

```
i := n;
while(i > 1)do
  begin
   i := i div 2;
   j := i;
   while(j > 1)do
      j = j div 2;
  end.
```

59. Give the worst-case running time using big-O notation.

```
i := n;
while(i > 1)do
  begin
   i := i/2;  j = n;
   while(j > 1)do
      j := j/3;
  end.
```

60. Give the worst-case running time using big-O notation.

```
k := 0;
for i = 1 to n do
  begin
   for j := 1 to m do
      k := k + 1;
   j := 1;
   while j < n do
      begin
       k := k + 1;
       j := 2 * j;
      end;
  end;
```

61. Give the worst-case running time using big-O notation.

```
j := n;
while(j ≥ 1)do
  begin
   for i := 1 to n do
      x = x + 1;
      j := [j/2];
  end.
```

62. Give the worst-case running time using big-O notation.

$$for \ i \leftarrow 1 \ to \ n$$
$$for \ j \leftarrow n \ to \ i$$
$$for \ k \leftarrow 1 \ to \ n^2$$
$$sum \leftarrow sum + A;$$

63. Give the worst-case running time using big-O notation.

$$x = 0;$$
$$for(i = 1; \ i \leq n; \ i++)$$
$$\{$$
$$\qquad for(j = 1; \ j \leq n; \ j++)$$
$$\qquad x++;$$
$$\qquad j = 1;$$
$$\qquad while(j < n)$$
$$\qquad \}$$
$$\qquad x++; j = j * 2;$$
$$\qquad \}$$
$$\}$$

64. Give the worst-case running time using big-O notation.

$$for \ i \leftarrow 1 \ to \ n \ do$$
$$\{$$
$$\qquad j \leftarrow i;$$
$$\qquad while \ j \neq 0 \ do$$
$$\qquad \quad j \leftarrow j \ div \ 2$$
$$\}$$

65. Give the worst-case running time using big-O notation.

$$int \ prime(int \ n)$$
$$\{int \ i, temp = 1;$$
$$for(i = 2; (i <= \lfloor \sqrt{n} \rfloor) \&\& (temp) ; \ i++)$$
$$if(num \ \% \ i == 0)$$
$$temp = 0;$$
$$return \ temp;$$

66. Give the worst-case running time using big-O notation.

$$P = 0;$$
$$for(i = 1; \ i < n; \ i++)$$
$$for(j = i + 1; \ j \leq m; \ j++)$$
$$P++;$$

67. Give the worst-case running time using big-O notation.

$$m = 0;$$
$$for(i = 0; \ j < n; \ i++)$$
$$for(j = 0; \ j < m; \ j++)$$
$$for(k = 0; \ k < p; \ k++)$$
$$m++;$$

68. Give the worst-case running time using big-O notation.

$$m = 0;$$
$$for(i = 0; \ j < n; \ i++)$$
$$for(j = i + 1; \ j < n; \ j++)$$
$$for(k = j + 1; \ k < n; \ k++)$$
$$m++;$$

69. Give the worst-case running time using big-O notation.

$$i = n;$$
$$while(i \geq 1)\{$$
$$j = i;$$
$$while(j \leq n)\{$$
$$/* \ Some \ Statement \ requiring \ \theta \ (1) \ time \ */$$
$$j = j * 2; \ \}$$
$$i = i/2;$$
$$\}$$

70. What is the time complexity of *fun()*?

$$int \ fun(int \ n)$$
$$\{$$
$$int \ count = 0;$$
$$for(int \ i = n; \ i > 0; \ i/= 2)$$
$$for(int \ j = 0; \ j < i; \ j++)$$
$$count+ = 1;$$
$$return \ count;$$
$$\}$$

71. What is the time complexity of *fun()*?

```
int fun(int n)
{
int count = 0;
for(int i = 0; i < n; i + +)
  for (int j = i; j > 0; j − −)
    count = count + 1;
return count;
}
```

72. What is the time complexity of the following function?

```
void fun(int n, int arr[])
{
int i = 0, j = 0;
for(; i < n; + + i)
  while(j < n && arr[i] < arr[j])
    j + +;
}
```

73. What is the time complexity of the following function?

```
int unknown(int n){
int i, j, k = 0;
for(i = n/2; i ≤ n; i + +)
  for(j = 2; j ≤ n; j = j * 2)
    k = k + n/2;
return k;
}
```

74. Let $A[1, ..., n]$ be an array storing a bit (1 or 0) at each location, and $f(m)$ is a function whose time complexity is $\theta(m)$. Consider the following program fragment written in a C like language:

```
counter = 0;
for(i = 1; i <= n; i + +)
{
  if(A[i] == 1)
    counter + +;
  else{
    f(counter);
    counter = 0;
  }
}
```

The complexity of this program is ....

75. What is the time complexity of the following function?

```
int  fun1(int  n)
{
    int  i,  j,  k,  p,  q = 0;
    for(i = 1;  i < n;   ++i)
    {
        p = 0;
        for(j = n;  j > 1;  j = j/2)
            ++p;
        for(k = 1;  k < p;  k = k * 2)
            ++q;
    }
    return  q;
}
```

4.2.2   What is Returned?

76. What is the value of count at the end of the procedure in terms of $n > 0$?

$$count = 0; x = 2$$
$$while(x < n)\{$$
$$x = 2x;$$
$$count = count + 1\}$$
$$PRINT count;$$

77. Consider the following recursive function: Get the recursive value for n = 5.

```
int  f(int  n)
{
    if  n == 1  return  1;
    else
    return  f(n − 1) + 2n;
}
```

78. Consider the following recursive function: Get the recursive value for a = 8.

```
int  f(int  a)
{
    if  a ≤ 1  return  (a × a − 5);
    else
    return  (3 × f(a − 2) + 4));
}
```

79. Consider the following recursive function: Get the recursive value for n = 5.

```
int recursive(int n)
{
  if  n == 1  recursive = 1;
  else
  recursive = recursive(n − 1) + recursive(n − 1);
}
```

80. Consider the following recursive function: Get the recursive value *F(6, 3)*.

```
int F(int m,  int n)
{
  if (m == 1 || n == 0 || m == n)
    return 1;
  else
    return  F(m − 1, n) + F(m − 1, n − 1); }
```

81. Consider the following recursive function: Get the recursive value *m(20, 28)*.

$$
m(A, B) = \begin{cases} m(B, A), & if (A < B) \\ A, & if (B = 0) \\ m(B, mod(A, B)), & otherwise \end{cases}
$$

82. Consider the following recursive function: Get the recursive value *F(10, 1)*.

```
int F(int n,  int i)
{
  if (i ≤ n)
    return  (F(n, i + 1) + i);
  else return 0;
}
```

83. Consider the following recursive function: Get the recursive value for z = 6 and y = 5.

```
int f(int y,  int z)
{
  if z = 0 return (0);
  else
  return (f(2y, ⌊z/2⌋) + y(z  mode  2));
}
```

84. What formula does the following function calculate? Get the recursive value *Rec(10)*.

$$int \ Rec(int \ n)$$
$$\{$$
$$if \ n == 1 \ return \ 1;$$
$$else$$
$$return \ n + f(n - 1);$$
$$\}$$

85. Consider the following recursive function:

$$int \ F(int \ n)$$
$$\{$$
$$if \ n \leq 1 \ return \ n * n;$$
$$else$$
$$return \ F(n - 1) + F(n - 2) + n;$$
$$\}$$

(a) Calculate the recursive value F(4).
(b) How many calls are made to calculate the value of F(4)?
(c) How many sum operators are performed to calculate the value of F(4)?
(d) How many multiplication operators are performed to calculate the value of F(4)?
(e) If T(n) is the number of recursive calls to calculate the value of F(n), write a recursive formula for T(n).
(f) If T(n) is the number of sum operators to calculate the value of F(n), write a recursive formula for T(n).
(g) If T(n) is the number of multiplication operators to calculate the value of F(n), write a recursive formula for T(n).

86. Consider the following recursive function:

$$int \ Rec(int \ n)$$
$$\{$$
$$if \ n == 1 \ return \ 1;$$
$$else$$
$$return \ n * Rec(n - 1);$$
$$\}$$

(a) What formula does this function calculate? Express your answer as a function of n.
(b) Get the recursive value Rec(5).
(d) How many multiplication operators are performed to calculate the value of Rec(5)?
(d) If T(n) is the number of recursive calls to calculate the value of Rec(n), write a recurrence relation for T(n).

(d) If T(n) is the number of multiplication operators to calculate the value of Rec(n), write a recurrence relation for T(n).

87. Consider the following code.

```
for i=1 to n do
    for j=i to 2*i do
        output "foobar"
```

Let T(n) denote the number of times 'foobar' is printed as a function of n.

(a) Express T(n) as a summation (actually two nested summations).
(b) Simplify the summation. Show your work.

88. Consider the following code.

```
for i=1 to n/2 do
    for j=i to n-i do
        for k=1 to j do
            output "foobar"
```

Assume *n* is even. Let T(n) denote the number of times 'foobar' is printed as a function of n.

(a) Express T(n) as a summation (actually two nested summations).
(b) Simplify the summation. Show your work.

89. For the following function, get the recursive value *Rec(3, 5)*. What formula does the following function calculate? Express your answer as a function of *a* and *b*.

```
int Rec(int a, int b)
{
    if b == 0 return a;
    else
        return Rec(a, b − 1) + 1;
}
```

90. For the following function, get the recursive value *Rec(3, 5)*. What formula does the following function calculate? Express your answer as a function of *a* and *b*.

```
int Rec(int a, int b)
{
    if b == 1 return a;
    else
        return a + Rec(a, b − 1);
}
```

91. For the following function, get the recursive value Rec(3, 4). What formula does the following function calculate? Express your answer as a function of *a* and *b*.

$$int\ Rec(int\ a, int\ b)$$
$$\{$$
$$if\ b == 1\ return\ a;$$
$$else$$
$$return\ a * Rec(a, b - 1);$$
$$\}$$

92. For the following function, get the recursive value *Rec(8, 3)*. What formula does the following function calculate? Express your answer as a function of *a* and *b*.

$$int\ Rec(int\ a, int\ b)$$
$$\{$$
$$if(a < b)\ return\ a;$$
$$else$$
$$return\ Rec(a - b, b);$$
$$\}$$

93. For the following function, get the recursive value *Rec(8, 3)*. What formula does the following function calculate? Express your answer as a function of *a* and *b*.

$$int\ Rec(int\ a, int\ b)$$
$$\{$$
$$if(a < b)\ return\ 0;$$
$$else$$
$$return\ Rec(a - b, b) + 1;$$
$$\}$$

94. Consider the following recursive function. Get the recursive value *Rec(1), Rec(2), Rec(3), and Rec(4)*. What formula does the following function calculate? Express your answer as a function of *n*.

$$int\ Rec(int\ n)$$
$$\{$$
$$if(n == 0)\ return\ 0;$$
$$else$$
$$return\ 2n + Rec(n - 1) - 1;$$
$$\}$$

95. What formula does the following function calculate? Express your answer as a function of *n* and *m*.

```
int Rec(int m, n)
{
    if (n == 0) return m;
    else
    return Rec(n, m mod n);
}
```

96. What formula does the following function calculate? Express your answer as a function of *n* and *x*.

```
function  f (n : integer ;  x :  real)  :  Real;
begin
    if n <= 1   then  f := x;
    else  f := x * f(n − 1 , x);
end;
```

97. Consider the following recursive function: Get the recursive value for n = 4.

```
function       f(n : intrger)  :  intrger;
    begin
        if  (n < 3) then    f := n;
        else
            f := f(n − 1) + f(n − 2) + f(n − 3);
end;
```

98. Consider the following recursive function: Get the recursive value for a = 2 and b = 5.

```
Function      MP (a, b : Byte)  :  word;
Begin
    if  b = 1    then   Mp := a;
    else   Mp := a * Mp (a, b − 1);
end;
```

99. What does the following function do?

```
procedure    A(var     x : integer ;   j : integer);
Begine
        if  x <= j    then      exit;
    x := x − 1;
    A(x, j);
    writ(x);
end;
```

100. Consider the following recursive function: Get the recursive value for i = 4.

$$
\begin{aligned}
&Function \quad K(i \ : \ integer) \ : \ Integer; \\
&bBegin \\
&\quad if(i <= 0) \ then \quad K := 1; \\
&\quad else \ if \ (i <= 1) \ then \ K := 2; \\
&\quad else \quad K := K(i - 1) + K(i - 2); \\
&end;
\end{aligned}
$$

101. Consider the following recursive function: Get the recursive value for a = 2 and n = 6.

$$
\begin{aligned}
&function \quad DC(a, n); \\
&begin \\
&\quad if \ n = 1 \ then \ return \ (a); \\
&\quad if \ n \ \ is \ even \ then \ return \ (DC(a, \ n/2) * DC(a, n/2)); \\
&\quad return \ (a * DC(a, n - 1)); \\
&end;
\end{aligned}
$$

102. Consider the following recursive function: Get the recursive value A(2, 2).

$$
A(m, n) = \begin{cases} n + 1, & if(m = 0) \\ A(m - 1, 1), & if(m > 0 \ and \ n = 0) \\ A(m - 1, A(m, n - 1)), & if(m > 0 \ and \ n > 0) \end{cases}
$$

103. What value would the following function return for the input x = 95?

$$
\begin{aligned}
&function \ fun(x : integer) : integer; \\
&begin \\
&\quad if \ x > 100 \ then \ fun = x - 10; \\
&\quad else \ fun = fun(fun(x + 11)); \\
&end;
\end{aligned}
$$

104. Consider the following recursive function: Get the recursive value for *x=7 and y=3.

```
void k(int * x, int y) {
    if(*x <= y) return;
    *x = *x - 1;
    k(x, y);
    printf ("%d", *x);
}
```

105. Consider the following C function definition.

```
int Trial (int a, int b, int c)
{
    if((a ≥ b) && (c < b) return b;
    else if (a ≥ b) return Trial(a, c, b);
    else return Trial(b, a, c);
}
```

The function Trial:

(a) finds the maximum of a, b and c
(b) finds the minimum of a, b and c
(c) finds the middle number of a, b and c
(d) None of the above

4.2.3 Recursive Algorithm

106. Give the worst-case running time using big-O notation.

```
int gcd(int a, int b)
{
    if(b == 0) return a;
    else return gcd(b, a mod b);
}
```

107. Give the worst-case running time using big-O notation.

```
int func(int n)
{
    if(n == 1)
        return(1);
    else
        return(n + func(n − 1));
}
```

108. Give the worst-case running time using big-O notation.

```
int func(int n)
{
    if(n == 1)
        return(1);
    else
        return(n * func(n − 1));
}
```

109. Give the worst-case running time using big-O notation.

```
int func(int n)
{
    if(n ≤ 1)
        return(0);
    else
        return(func(n/2) + 1);
}
```

110. Give the worst-case running time using big-O notation.

```
int func(int n)
{
    if(n == 1)
        return(1);
    else
        return(func(n/3) + 1);
}
```

111. Give the worst-case running time using big-O notation.

```
int recursive(int n)
{
    if  n == 1  recursive = 1;
    else
        recursive = recursive(n − 1) + recursive(n − 1);
}
```

112. Give the worst-case running time using big-O notation.

```
int  F(int n)
{
    if(n ≤ 2)  return  1;
    else
        return  F(n − 2) * F(n − 2);
}
```

113. Give the worst-case running time using big-O notation.

```
int  F(int  n)
{
  if(n ≤ c) return  2;
  else
  return  F(n/2) * F(n/2);
}
```

114. Give the worst-case running time using big-O notation.

```
int  T(int  n)
{
  if(n ≤ c) return  2;
  else
  return  T(n − 1) + T(n−1)/(T(n−1)+1);
}
```

115. Give the worst-case running time using big-O notation.

```
int  T(int  n)
{
  if(n ≤ c) return  2;
  else
  return  T(n−1)/(T(n−1)+1);
}
```

116. Give the worst-case running time using big-O notation.

```
int  T(int  n)
{
  if(n ≤ c) return  2;
  else
  return  √(T(n/2) + T(n/2));
}
```

117. Give the worst-case running time using big-O notation.

```
int  F(int  m,  int  n)
{
  if(m == 1 || n == 0 || m == n)
   return  1;
  else
  return  F(m − 1, n) + F(m − 1, n − 1); }
```

118. Give the worst-case running time using big-O notation.

```
Algorithm Hanoy(n, from, help, to)
 {
  if(n == 1) then
   move top disk from (from → to);
  else
   {
    Hanoy(n − 1, from, to, help);
    move top disk from(from → to);
    Hanoy(n − 1, help, from, to);
   }
 }
```

119. Give the worst-case running time using big-O notation.

```
index location(index low, index high)
 {
  index mid
  if(low > high)
   return 0;
  else
   {
    mid = ((low + high)/2);
    if(x == S[mid])
     return mid;
    else if(x < S[mid])
     return location(low, mid − 1)
    else
     return location(mid + 1, high);
   }
 }
```

120. Give the worst-case running time using big-O notation.

```
Algorithm rsearch(i)
 {
 case
   i > n :  return(−1);  // x not found
   a[i] = x :  return(i);  // x is a[i]
   else return(rsearch(i + 1));  // recursive
  end case;
 }
```

121. Obtain the complexity function of the Fibonacci sequence using the recursive method?

```
int  fib(int n)
{
    if (n == 1)
        return 0;
    else
        if (n == 2)
            return (1);
        else
            return(fib(n − 1) + fib(n − 2));
}
```

(a) O(n)
(b) O(n^2)
(c) O(n log n)
(d) $\frac{(1+\sqrt{5})^n}{2}$

122. Give the worst-case running time using big-O notation.

$$f[0] = 0$$
$$if (n > 0)\{$$
$$\quad f[1] = 1$$
$$\quad for(i = 2; i \leq n; i++)$$
$$\quad\quad f[i] = f[i − 1] + f[i − 2];$$
$$\}$$
$$return\ f[n];$$

123. Give the worst-case running time using big-O notation.

```
int  S(int  List[],  int  n)
{
    if (n == 1)
        return(List[1]);
    else
        return(List[n] + S(List, n − 1));
}
```

124. Consider the following functions. What is the time complexity of the function?

```
int  fun1(int n)
{
    if (n ≤ 1)  return n;
    return  2 ∗ fun1(n − 1);
}
```

```
int  fun2(int  n)
{
  if (n ≤ 1) return n;
  return  fun2(n − 1) + fun2(n − 1);
}
```

125. Consider the following C-function:

```
double  foo(int  n)
{
  int i;
  double sum;
  if (n == 0) return  1.0;
  else
  {
    sum = 0.0;
    for (i = 0; i < n; i + +)
      sum+ = foo(i);
    return sum;
  }
}
```

The space complexity of the above function is ....

126. Consider the following C code

```
int  f (int  x)
{
  if (x < 1)return  1;
  else return  (f (x − 1) + g(x));
}
int  g(int  x)
{
  if (x < 2) return  2;
  else return  (f (x − 1) + g(x/2));
}
```

Of the following, which best describes the growth of f(x) as a function of x?

(a) Linear
(b) Exponential
(c) Quadratic
(d) Cubic

4.2.4 Recurrence Relations for Recursive Functions

It's often possible to compute the time complexity of a recursive function by formulating and solving a recurrence relation.

127. Let function T(n) denotes the number of elementary operations performed by the function call *Rec(n)*. Write recurrence relation, T(n), and get the running time of this function in terms of big-O by solving T(n).

> *Rec(n)*
>   *if n* = 1 *then*
>   *do something with O(1)*
>   *else*
>   *Rec(n* − 1);
>   *Rec(n* − 1);
>   *for i* = 1 *to n do*
>   *do something with O(1)*

128. Let function T(n) denotes the number of elementary operations performed by the function call *Rec(n)*. Write recurrence relation, T(n), and get the running time of this function in terms of big-O by solving T(n).

> *Rec(n)*
>   *if (n* = 1) *or (n* = 2) *then*
>   *do something with O(1)*
>   *else*
>   *Rec(n* − 1);
>   *for i* = 1 *to n do*
>   *do something with O(1)*
>   *Rec(n* − 1);

129. Let function T(n) denotes the number of elementary operations performed by the function call *Rec(n)*. Write recurrence relation, T(n), and get the running time of this function in terms of big-O by solving T(n).

> *Rec(n)*
>   *if n* = 1 *then*
>   *do something with O(1)*
>   *else if n* = 2 *then*
>   *do something with O(1)*
>   *else*
>   *for i* = 1 *to n do*
>   *Rec(n* − 1);
>   *do something with O(1)*

130. Let function T(n) denotes the number of elementary operations performed by the function call *Rec(n)*. Write recurrence relation, T(n), and get the running time of this function in terms of big-O by solving T(n).

$$
\begin{aligned}
&Rec(n) \\
&\quad if\ \ n = 1\ \ then \\
&\qquad do\ something\ with\ O(1) \\
&\quad else \\
&\qquad for\ \ i = 1\ to\ n\ do \\
&\qquad\quad Rec(i); \\
&\qquad do\ something\ with\ O(1)
\end{aligned}
$$

4.3   Solutions

1-O(n)

2-O(n)

3-O(n)

4-O(n) + O(n) = O(n)

5-O(n) + O(m) = O(n + m)

6-O(n)

7-O(n)

8-O(n) + O(m) = O(n + m)

9-O(log n)

10-O(log n)

14-O(log n)

15-O(log n)

16-O(log n)

27-O(n^2)

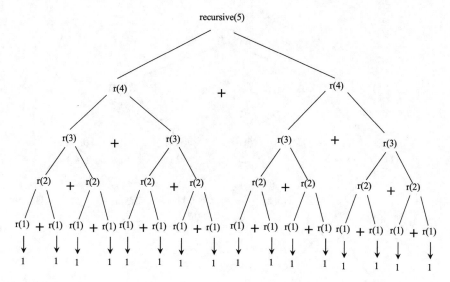

Fig. 4.1 Recursive tree for question 79

29-O(n × m)

31-O(n^2)

34-O(log^2 n)

35-O(n log n)

42-O(1)

43-O(n)

58-O(log^2 n)

60-O(n(m + log n))

62-O(n^3)

79-18. See Fig. 4.1.

80-20. See Fig. 4.2.

82-Computes $1 + \cdots + n = \frac{n(n+1)}{n}$. See Fig. 4.3.

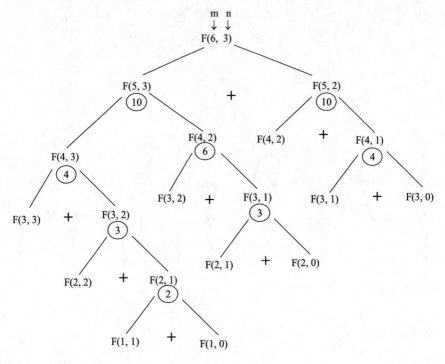

Fig. 4.2 Recursive tree for question 80

85-(a) For calculate F(4), we need to compute F(0), F(1), F(2), and F(3).
F(0)=0, F(1)= 1*1=1, F(2)= F(1) * F(0) + 2 = 3, F(3)= F(2) * F(1) + 3 = 7, F(4)=
F(3) * F(2) + 4 = 14
(b) Fig. 4.4 depicts the number of calls for F(4). According to this figure, the number
of calls is 9.
(c) For n > 1, F(n) performs two sum operators. According to Fig. 4.4, there are four
recursive call where n > 1, therefore, the sum operator is performed eight times.
(d) For n ≤ 1, F(n) performs multiplication operator one times. According to Fig. 4.4,
there are five recursive call where n ≤ 1, therefore, the multiplication operator is performed five times.
(e) $T(n) = T(n-1) + T(n-2) + 1$, $T(0) = T(1) = 1$
(f) $T(n) = T(n-1) + T(n-2) + 2$, $T(0) = T(1) = 0$
(g) $T(n) = T(n-1) + T(n-2)$, $T(0) = T(1) = 1$

89-a + b

90-a * b

92-a mod b

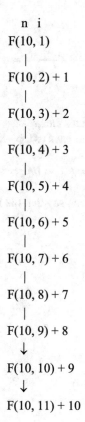

Fig. 4.3 Recursive tree for question 82

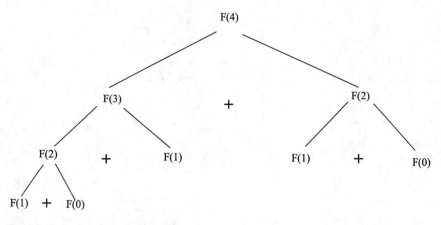

Fig. 4.4 Recursive tree for question 85

93-a div b

94-$1 + 3 + 5 + \cdots + (2n - 1) = \frac{(1+2n-1)\times n)}{2} = n^2$

95-gcd(m, n)

106-O(log n). This function calculates the greatest common divisor of two numbers. It can be shown that the divisor is always twice as large as the remainder.

$$a = bq + r, 0 \leq r \leq b, q \geq 1 \implies r < bq \implies r + r < bq + r \implies 2r < a$$

so, $a \bmod b < (a/2)$. That is, we can say that the function halves a call each time.

107-O(n).

108-O(n).

109-O(log n).

110-O(log n).

111-O(2^n).

112-O($2^{\frac{n}{2}}$).

114-O(3^n).

Chapter 5
Basic Data Structure

Abstract In this chapter, we examine the four basic data structures including arrays, stacks, queues, and linked lists. To this end, this chapter provides 191 exercises on the basic data structures in terms of theoretical and application.

5.1 Lecture Notes

In this chapter, we examine the four data structures: array, stack, queue, and linked list.

5.1.1 Arrays

Data structures are generally divided into two categories: linear and nonlinear. A data structure is called linear whenever its elements form a sequence. In other words, it is a linear list. There are two basic ways to display a linear data structure in memory. One of these methods is to have a linear relationship between elements that are represented by consecutive memory cells. Linear structures are called arrays. An array is one of the most widely used data structures and is often used to implement linear data.

An array is a list of *n* elements or finite sets of data elements of the same type, for example, array elements can only be of the integer type or of any other type. Its elements cannot be of both the integer type and the float type. Finite means that the number of array elements is known to be small or large. Array elements are stored in consecutive memory cells, respectively. So that each element with a index is directly accessible. Direct access means that it can be accessed by specifying the location of each element in the array. Arrays can be one-dimensional, two-dimensional, or multidimensional. A two-dimensional array is also called matrix.

Two types of searching on arrays can be performed, sequential (linear) search and binary search. Several applications of sequential search are:

– performing a single search in an un-sorted list
– Finding largest or smallest element in an un-sorted list

Several applications of binary search are:

- Finding an element in a sorted array
- To find if n is a square of an integer
- Finding the first value greater than or equal to x in a given array of sorted integers
- Finding the frequency of a given target value in an array of integers
- Finding the peak of an array which increases and then decreases
- A sorted array is rotated n times. Search for a target value in the array.

5.1.2 Stack

A stack is a list of elements in which the action of adding or removing an element is performed only from one side of it, which is called the top element. Stacks are used to temporarily store data in many applications.

The stack can be visualized as a set of plates or boxes on top of each other that can only be accessed to the top plate or box, and to access other plates can only be lifted one by one from the top to reach the desired plate. Two specific terms are used for two basic actions in stacks:

- The term Push, which is used to add an element to a stack. But it actually adds to the top of the stack where the top points.
- The term Pop, which is used to remove an element from a stack. Which actually removes the top of the stack where the top points.

5.1.3 Queue

A queue is a linear list of elements in which the deletion of elements can only be done from the beginning of the queue, denoted by front, and the addition operation can only be done from the end of queue, denoted by rear. In other words, the queue length increases from the end and decreases from the beginning. Queues are called first input, first output. Because the first element that enters the queue is the first element that leaves it.

5.1.4 Linked List

Using an array is a method of storing data. Arrays have drawbacks. For example, adding and removing elements in an array is relatively costly. In addition, since each array usually takes up a block of memory space, it is not easy to double or triple the size of an array when additional memory is needed. This is why arrays are called compact lists. Arrays are also called static data structures.

Another way to store data in memory is to put each element in a list, which contains an information field and the address of the next element in the list, which is called the next address, link, or pointer. This way, the consecutive elements in the list do not need to occupy adjacent space in memory. This makes it easy to add and remove list elements. This data structure is called a linked list. In other words, a linked list is a set of nodes scattered in the main RAM that is used to store data.

5.2 Exercises

5.2.1 Arrays

5.2.1.1 Basic Concept

1. Find the number of non-zero elements in an upper triangular or lower triangular square matrix.

2. To store an upper triangular or lower triangular square matrix, denoted by M, it is better to store it as a one-dimensional array, denoted by N, so that not much space is wasted. In fact, zeros are a waste space. After storing the matrix in a one-dimensional array, we need to get a formula for accessing the elements in the array. That is, we must obtain the relation M[i, j] and N[k]. Specify what the relationship should be between M[i, j] and N[k].

3. A one-dimensional array is used to represent a lower triangular matrix with dimensions $n \times n$. What is the minimum size of the array?

4. Tridiagonal matrix is a square matrix in which where elements with a non-zero value appear only on the diagonal and immediately below or above the diagonal. We store the non-zero elements of this matrix line by line in a one-dimensional array. Get its storage formula?

5. *Spars matrix.* A matrix whose small percentage of elements is not zero is called a sparse. In other words, a matrix that uses the position and values of non-zero elements is more useful than using the whole matrix is called sparse. Sparse matrices occur in practical problems such as simulation of city traffic, city maps, numerical answers of differential equations, and more.

Here's how to store sparse matrices:
In this method only row number, column number and non-zero value should be stored. The sparse matrix can be stored in a matrix that has three columns, in which the first column contains a non-zero value row number, the second column contains a non-zero value column number, and the third column contains a non-zero value. In the first row of the new matrix, we write the general characteristics of the sparse matrix

(i.e., number of rows, number of columns and number of non-zero elements). If the number of non-zero elements of the original matrix is n, then the matrix representing sparse matrix will have $n + 1$ rows.

(a) Represent the spars matrix below as described above.

$$\begin{bmatrix} 0 & 0 & 0 & 0 & 1 \\ 9 & 0 & 0 & 0 & 1 \\ 0 & 2 & 0 & 0 & 0 \\ 0 & 0 & 0 & 0 & 0 \end{bmatrix}$$

(b) Let NZ and n denote the number of non-zero elements in a matrix and the dimension of the matrix, respectively. Explain when the method described above will be useful for storing a spars matrix.

6. Using the array, write a program that adds two integers up to a maximum of 150 digits.

5.2.1.2   Linear Search

7. How many comparisons are made in linear search (the following function) for N elements to find an element (in best, average, and worst cases)?

```
int search(int arr[], int N, int x)
{
    int i = 0;
    for(i = 0; i < N; i++)
        if(arr[i] == x)
            return i;
    return -1;
}
```

8. What is the best, average, and worst cases time complexity for linear search?

9. Can linear search be made parallel for execution on multiple CPU cores?

10. How many comparisons are made in the following function for N elements to find the maximum element (in best, average, and worst cases)?

```
int find_max(int arr[], int N)
{
  int i;
  max = S[0];
  for(i = 1; i < N; i++)
    if(S[i] > max)
      max = S[i];
  return max;
}
```

11. Let A is an array of size *n* containing unsorted integer numbers. Give an algorithm to find the largest and smallest numbers in A.

12. The following function is written to find the smallest and largest numbers in an array. Find the maximum and minimum number of comparisons for an array with *N* elements.

```
void find_minmax(int S[], int N)
{
  int i;
  max = S[0];
  min = S[0];
  for(i = 1; i < N; i++)
    if(S[i] < min)
      min = S[i];
    else if(S[i] > max)
      max = S[i];
}
```

13. Let A is an array of size *n* containing integer numbers. Give an algorithm to rearrange the array as smallest number, largest number, 2nd smallest number, 2nd largest number, 3rd smallest number, 3rd largest number and so on.

14. Let A is an array of size 100 containing integer numbers. Give an algorithm to find the missing number in A of 1 to 100?

15. Let A is an array of size *n* containing integer numbers. Give an algorithm to find duplicate numbers in the array.

16. Explain how do you remove duplicate numbers from an array in place.

17. Let A is an array of size n and a number k is given. Give an algorithm to find all pairs of the array whose sum is equal to k?

18. In linear search, it is sometimes observed that an element is searched many times in an array. The point is, if an item is searched again, it should take less time to search. Explain how we can improve the linear search performance in such cases.

19. You are given a range [L, R]. You are required to find the number of integers X in the range such that $GCD(X, F(X)) > 1$ where $F(X)$ is equal to the sum of digits of X in its hexadecimal (or base 16) representation.

20. You have N rectangles (W and H denoting the width and height of a rectangle). A rectangle is golden if the ratio of its sides is in between [1.6, 1.7], both inclusive. Your task is to find the number of golden rectangles.

21. Suppose array A contains a number of integers, you must find the maximum sum that can be obtained by selecting non-empty subset of the array. If there are many of these non-empty subsets, choose the one with the most elements. Print the maximum total and number of elements in the selected subset.

22. You are given a grid of size $N \times N$ that has the following specifications:

- Each cell in the grid contains either a policeman or a thief.
- A policeman can only catch a thief if both of them are in the same row.
- Each policeman can only catch one thief.
- A policeman cannot catch a thief who is more than K units away from the police-man.

Write a program to find the maximum number of thieves that can be caught in the grid.

5.2.1.3 Binary Search

23. Binary search is an efficient algorithm for finding an item from the sorted list of items. This algorithm works by repeatedly splitting the list into half-lists that can contain items, as long as you limit the possible locations to one item.

Given an array A of n elements with values or records $A_0, A_1, A_2, \ldots, A_{n-1}$ sorted such that $A_0 \le A_1 \le A_2 \le \cdots \le A_{n-1}$, and target value x, the following function uses binary search to find the index of x in A.

$$Function \ binary\_search(A[low..high], \ n, \ x)$$
$$low := 0$$
$$high := n - 1$$
$$while \ low \le high \ do$$
$$\quad mid := floor((low + high)/2)$$
$$\quad if \ A[mid] < x \ then$$
$$\quad\quad low := mid + 1$$
$$\quad else \ if \ A[mid] > x \ then$$
$$\quad\quad high := mid - 1$$
$$\quad else :$$
$$\quad\quad return \ mid$$
$$return \ unsuccessful$$

In the list below

$$[2, 4, 7, 9, 11, 13, 14, 15, 16, 20]$$

if we apply the above algorithm, which group of numbers correctly shows the sequence of comparisons used to find the key 9.

a. 11, 4, 7, 9
b. 13, 7, 11, 9
c. 2, 4, 7, 9
d. 20, 13, 7, 9

24. In the previous question, how many comparisons are needed to be done to find out that the number 18 does not exist in the array?

25. Consider the following sorted array.

2	10	21	39	40	46	51	60	71	83	89	95	100

(a) What is the maximum number of key comparisons performed by binary search in this array?
(b) List all the keys in this array that have the largest number of key comparisons in binary search.
(c) Find the average number of key comparisons performed by binary search in a successful search in this array. Suppose each key is searched with the same probability.
(d) Find the average number of key comparisons performed by binary search in an unsuccessful search in this array.

26. Justify the following time complexities for binary search algorithm.

(a) The best case time complexity is O(1).

(b) The worst case time complexity is O(log n).

(c) The average case time complexity is O(log n).

(d) The space complexity is O(1).

27. The table below shows arrays of different sizes. Specify in each row of the table how many comparisons are performed in the worst case to find a particular element by two sequential and binary search algorithms?

n	Sequential search	Binary search
128		
256		
1024		
1,048,576		
2,097,152		

28. Given a bitonic sequence A of N distinct elements, using binary search devise an algorithm to find a given element B in the bitonic sequence in O($log\ N$) time. A bitonic sequence is a sequence of numbers which is first strictly increasing then after a point strictly decreasing.

29. Given a sorted array A of size N. In time complexity $O(log N)$, devise an algorithm to find number of elements which are less than or equal to B.

30. Given an integer A. Without using *sqrt function* from standard library, compute the square root of A. If A is not a perfect square, return floor(sqrt(A)). You are expected to solve this in O($log\ (A)$) (Hint: use binary search).

31. Let A is a matrix of size $N \times M$, containing integer numbers, in which each row is sorted. Give an algorithm to find the median of the matrix A. Note: No extra memory is allowed.

32. There are two sorted arrays A and B of size m and n respectively. Give an algorithm to find the median of the two sorted arrays (the median of the array formed by merging both the arrays).

33. Given an array of integers A of size N and an integer B. Array A is rotated at some pivot unknown to you beforehand, i.e., 2 3 4 6 7 8 9 might become 6 7 8 9 2 3 4).

    Assume array A is sorted in non-decreasing order before rotation and a target value B is given to search. Devise an algorithm to search B, if found in the array, return its index, otherwise return −1.

34. Give an algorithm to find division of two numbers using binary search.

35. *Duplicate elements*. Classic binary search may return any index whose element is equal to the target value, even if there are duplicate elements in the array. For example, in the array [0, 10, 20, 30, 40, 40, 50, 60, 70] and the target 40, it is possible for the algorithm to either return the 5th (index 4) or 6th (index 5) element. However, it is sometimes necessary to return the leftmost element or the rightmost element for a target value that is duplicated in the array. Devise two algorithms, one to return the index of the rightmost element and one to return the leftmost element if such an element exists.

36. Meta binary search (also called one-sided binary search) is a modified form of binary search that incrementally constructs the index of the target value in the array. Explain this type of searching by using an example.

37. *Ternary search*. In this search, for n greater than 1, the search process recursively continued by comparing K with $A[n/3]$, and if K is larger, compare it with $A[2n/3]$ to determine which one-third of the array to continue the search.

(a) Set up a recurrence for the number of key comparisons in the worst case, and then solve the recurrence.
(b) Compare this algorithm's efficiency with that of binary search.

38. When jump search is a better alternative than a binary search?

39. *Unbounded Binary Search*. Given a monotonically increasing function $f(x)$ on the non-negative integers, find the value of x, where $f(x)$ becomes positive for the first time. In other words, find a positive integer x such $f(x-1), f(x-2), ...$ are negative and $f(x+1), f(x+2), ...$ are positive. For example, $f(x) = 3x - 100$ is a monotonically increasing function. It becomes positive for the first time when $x = 34$. Write a program to find the value 'x' where $f()$ becomes positive for the first time.

5.2.1.4 Applications

40. *Union*. To find union of two sorted arrays A and B, follow the following merge procedure:

a. Use two index variables i and j, initial values $i = 0, j = 0$.
b. If $A[i] < B[j]$ then print $A[i]$ and increment i.
c. If $A[i] > B[j]$ then print $B[j]$ and increment j.
d. If $A[i] = B[j]$ print one of them and increment both i and j.
e. This procedure is repeated until one of the arrays is empty.
f. Print remaining elements of the larger array.

Apply this algorithm on the following arrays to find the union of them.

(a) A = {1, 2, 3, 4}, B = {5, 6}

(b) A = {1, 2, 3, 10}, B = {5, 6}
(b) A = {1, 2, 3, 4}, B = {5, 6, 9}
(b) A = {1, 2, 3, 10}, B = {5, 6, 9}

41. *Median*. the middle element is found by ordering all elements in sorted order and picking out the one in the middle (or if there are two middle numbers, taking the mean of those two numbers). Let A and B are two sorted arrays. The most basic approach is to merge both the sorted arrays using an auxiliary array. The median would be the middle element in the case of an odd-length array or the mean of both middle elements in the case of even length array. Give an algorithm to find the median of them. Analyze the algorithm time complexity.

Examples:
Input: A[] = {1, 4, 5}, B[] = {2, 3}
Output: 3
Explanation:
Merging both the arrays and arranging in ascending:
[1, 2, 3, 4, 5]
Hence, the median is 3

Input: A[] = {1, 2, 3, 4}, B[] = {5, 6}
Output: 3.5
Explanation:
Union of both arrays:
{1, 2, 3, 4, 5, 6}
Median = (3 + 4) / 2 = 3.5

42. The following algorithm is given to find the median of two sorted arrays A and B.

   a. The first element of both arrays is compared.
   b. The smaller element among two becomes a new element of the new array S.
   c. This procedure is repeated until one of the arrays is empty and the newly combined array covers all the elements of both the arrays.
   d. Maintain a variable count for the output array and if count equals $(N + M)/2$, then it is the median of the odd length array and if it is even, store the mean of $(N + M)/2$th and $(N + M)/2 - 1$th element.

Apply this algorithm on the following arrays to find the median of them.

(a) A = {1, 2, 3, 4}, B = {5, 6, 7}
(b) A = {1, 2, 3, 10}, B = {5, 6, 7}
(b) A = {1, 2, 3, 4}, B = {5, 6, 9}
(b) A = {1, 2, 3, 10}, B = {5, 6, 9}

43. Devise an algorithm to find the intersection of two sorted arrays.

44. Let A and B are two unsorted arrays of size n and m. Give an algorithm to find their union and intersection.

45. Let A is a sorted array of size n and B is an unsorted array of size m. Give an algorithm to find their union and intersection.

46. Give an algorithm to find common elements in three sorted arrays.

47. Let A, B, and C are three sorted arrays of possibly different sizes. Give an algorithm to merge them so that the output is also sorted.

48. Array A and an element k are given. Give an algorithm to find number of occurrences of k in A.

49. Let A is an array of size n containing integer numbers. Give an algorithm to move all negative numbers to beginning and positive numbers to end with constant extra space.

50. Suppose an array of integers and an integer k are given. Give an algorithm to find a minimum of all the differences between the maximum and minimum elements in all the sub-arrays of size k starting from the left and moving towards the right by one position each time.

51. Suppose an array of integers and an integer k are given. Give an algorithm to find out the minimum cost required to equalize all elements of each subarray of length k. The cost is calculated as the absolute difference between the two.

Example:
Input: A = {1, 2, 3, 4, 6}, k = 3
Output: 7
Explanation:
Subarray 1: Cost to convert subarray {1, 2, 3} to {2, 2, 2} = $|1 - 2| + |2 - 2| + |3 - 2| = 2$
Subarray 2: Cost to convert subarray {2, 3, 4} to {3, 3, 3} = $|2 - 3| + |3 - 3| + |4 - 3| = 2$
Subarray 3: Cost to convert subarray {3, 4, 6} to {4, 4, 4} = $|3 - 4| + |4 - 4| + |6 - 4| = 3$
Minimum Cost = 2 + 2 + 3 = 7

52. Suppose an array of integers and an integer k are given. Give an algorithm to find the greatest common divisor (GCD) of all subarrays of size k.

53. Let A is an array of size n such that the elements are in the range 1 to n. One number from set $\{1, 2, \ldots, n\}$ is missing and one number occurs twice in the array.

Give an algorithm to find these two numbers.

54. Write a program to rotate an array of size n by k elements.

(a) Use temp array
(b) Rotate one by one

55. Using GCD of n and k, try to give an algorithm to rotate an array of size n by k elements.

56. Suppose array A contains distinct numbers sorted in increasing order. This array has been rotated (clockwise) k times. Give an algorithm to find the value of k.

57. Give an algorithm to search a given number in a rotated array.

58. Using reversal operation, write a program to rotate an array of size n by k elements. The idea is as follows:
Let AB are the two parts of the input array where A = arr[0..k−1] and B = arr[k..n−1]. Perform the following steps:

Reverse A to get $A^r B$, where A^r is reverse of A.
Reverse B to get $A^r B^r$, where B^r is reverse of B.
Reverse all to get $(A^r B^r)^r = BA$.

Does this algorithm correctly rotate an array of size n by k elements? Give an example.

59. Suppose an array of integers and an integer k are given. Give an algorithm to find the median of each window of size k starting from the left and moving towards the right by one position each time.

Example:
Input: A = {−2, 4, 18, 6, 1, 5, 5, 2}, k = 3
Output: 4 6 6 5 5 5

60. Let array A of size n and a number k is given. Give an algorithm to split array A from index k, and move the first part of the array to the end.

61. An array contains positive and negative numbers at random so that the number of positive and negative numbers is not necessarily the same. At time O(n) and extra space O(1) provide an algorithm for rearranging the array elements so that the positive and negative numbers are alternated.

62. Let array A of size n containing positive integers and a number k is given. Give an algorithm to find the minimum number of swaps required to put together all the numbers less than or equal to k.

Example:

Input: A = {4, 3, 5, 7, 2, 9}, k = 3

Output: 1

63. Let A is an array of size *n*. Give an algorithm to find the length of the longest sub-sequence such that elements in the subsequence are consecutive integers, the consecutive numbers can be in any order.

Example:

Input: A = {2, 8, 4, 10, 5, 25, 3}

Output: 4

Explanation: subsequence 2, 4, 5, 3

64. Let A is an array of size *n* containing integer numbers. Give an algorithm to find the length of the longest increasing subsequence whose adjacent element difference is one.

65. Let A is an array of size *n* containing integer numbers and an integer *k* is given. Give an algorithm to find length of longest subsequence whose sum is less than or equal to K.

66. Let A is an array of size *n* containing integer numbers. Give an algorithm to print elements of the array which are not a part of the longest increasing subsequence.

67. Explain how to find the second largest element in an array of integers.

68. Let A is an unsorted array of size *n* containing integer numbers and an integer *k* are given. Give an algorithm to find k'th smallest/largest element in the array.

69. Let A is an unsorted array of size *n* containing positive and negative distinct elements. Give an algorithm to find the smallest positive number missing from the array in O(n) time using constant extra space.

70. Let A is an unsorted array of size *n* containing integer numbers. Give an algorithm to find the array mean, mode, and median elements.

71. Let A is an unsorted array of size *n* and an integer *x* is given. Give an algorithm to find floor and ceiling of *x* in A.

Floor of *x* is the largest element that is less than or equal to *x*. If *x* is smaller than the smallest element A, there is no floor *x*.

Ceil of *x* is the smallest element that is greater than or equal to *x*. If *x* is greater than the largest element A, there is no ceil.

72. Let A is an unsorted array of size *n* and an integer *k* is given. Give an algorithm to find the mean of the array after removing k percent elements from the smallest and largest elements of the array.

73. Let A is an unsorted array of size *n*. Give an algorithm to print all unique pairs in the array with equal sum.

74. Given a matrix of integers A of size $n \times m$ and an integer B. Provide an algorithm to search for number B in matrix A. Matrix A has the following properties:

- Numbers in each row are sorted from left to right.
- The first number of each row is greater than or equal to the last number of the previous row.

Return 1 if B is present in A, else return 0.

75. Given an array A consisting of *N* integers. The task is to count the number of triples (A[i], A[j], A[k]), where i, j, and k denote the respective indices, such that one of the integers can be written as the summation of the other two integers.

Example:
Input: A[] = 1, 2, 3, 4, 5
Output: 4
Explanation:
The valid triplets are -
(1, 2, 3) : 1 + 2 = 3
(2, 3, 5) : 2 + 3 = 5
(1, 3, 4) : 1 + 3 = 4
(1, 4, 5) : 1 + 4 = 5

76. *Find the peak.* Given an array of integers, give an algorithm to find a peak element in it. An array element is a peak if it is not smaller than its neighbours. For corner elements, only one neighbor is considered.

5.2.2   Stack

5.2.2.1   Basic Concept

77. Assume we have the characters B, C, D, E, F, and A. Apply the following Push and Pop sequences on a stack and determine the final characters in the stack.

   Push(A); Pop(); Push(B); Push(C); Pop(); Push(D); Push(E); Pop(); Push(F);

78. Suppose a stack can have up to six elements inside. The stack is empty at first. Get the output of the following code?

Set A = 2, B = 5
Push (Stack, A)
Push (Stack, B + 10)
Push (Stack, 3)
Push (Stack, A – B)
Repeat while TOP <> 0
    POP (Stack, item)
    print: item
return

79. Get the final values of A, B, and C (A = 10, B = 2, C = 5).

push (B)
push (A + B)
pop (C)
push (A – B)
push (C)
push (B)
pop (A)
pop (B)
push (A × B)
push (C)
push (A)
pop (B)
pop (C)
pop (A)

80. If we enter a sequence of numbers 1, 2, 3, 4, 5 into a stack, respectively, which of the following outputs from this stack will be possible?

a. 2-3-5-1-4
b. 5-4-3-2-1
c. 1-3-5-4-2
d. 5-1-3-2-4

81. If we use two stacks S1 and S2 instead of one stack S and their Push and Pop operations, what kind of outputs from the stacks are possible with the input data A, B, C, D, E, F?

5.2.2.2 Implementation

82. Implement a stack using an array. The two basic operations of the stack are

– Push (): Add an element to the top of the stack
– Pop (): Remove an element from the top of the stack

To make the best use of a stack, its condition must also be checked. For this purpose, the following functions have been added to the stack:

– Peek(): Get the highest element of the stack without removing it
– isFull (): Check that the stack is full
– isEmpty (): Check that the stack is empty

83. Provide an algorithm to design a stack which supports following operations in O(1) time complexity.

– push—adds an element to the top of stack.
– pop—removes an element from top of stack.
– findMiddle—returns middle element of the stack.
– deleteMiddle—deletes the middle element.

84. Explain the various ways to implement k stacks in an array. Determine which of the these methods effectively implements the k stacks in a single array.

85. *Two stacks in an array (Multiple Stack).* Let STACK[n] is an array of size n used to represent two stacks where STACK[n] divided into two stacks Stack[A] and Stack[B]. Then the value of n is such that the combined size of both the Stack[A] and Stack[B] will never exceed n. Stack[A] will grow from left to right, whereas Stack[B] will grow in opposite direction i.e., right to left.

(a) What is the condition for both stacks to be full in this type of stack?
(b) Implement two stacks in an array.

86. The two stacks are implemented in an array of S[1..m] that grow in opposite directions to each other. What is the time complexity of Push and Pop functions?

87. If we need k stacks in the program, then the array of size n is divided into k parts of size n/k, each part is assigned to one stack. For each stack b[i] indicates the lowest location of the stack and t[i] represents the top of stack i. If t[i] = b[i] means that stack i is empty and if t[i] = b[i + 1] means stack i is full.

(a) In this case, to represent k stacks, we divide S[1..n] into k parts. It is better to divide the array according to our needs, but if we do not already know the needs of each stack, it is better to divide the array into equal parts. In this case, specify the initial values of b[i] and t[i] (their initial address) for all stacks.
(b) Write the code for push and pop functions in multiple stacks.

88. Dividing an array of size n into k parts of size n/k, assigning each part to one stack is inefficient use of array space. A stack push operation may cause the stack to overflow even if there is space in the array. Provide an efficient way.

89. Let A be an array of size n. We are going to define six stacks inside it. One way is to divide the array into 6 parts and define two stacks in each part, one growing from right to left and the other in opposite direction i.e., left to right. Implement this

method and discuss its effectiveness.

90. How to implement three stacks with one array?

91. Implement a stack using a linked list.

92. We want to create four stacks in array S[1..495], get the address of the beginning of each stack?

93. Explain how you can implement stack using priority queue or heap?

5.2.2.3 Infix, Prefix, and Postfix Expressions

94. *Infix to Postfix Conversion.* Let, X is an arithmetic expression written in infix notation. The following algorithm finds the equivalent postfix expression Y.

1. Push "(" into STACK, and add ")" to the end of X.
2. Scan X from left to right and repeat Step 3 to 6 for each character of X until the STACK is empty.
3. If an operand is encountered, add it to Y.
4. If an open parenthesis is encountered, push it onto Stack.
5. If an operator is encountered, then:

   a. Repeatedly pop from the STACK and add to Y each operator on the top of the STACK which has the same precedence as or higher precedence than operator.
   b. Add operator to Stack.

6. If a closed parenthesis is encountered, then:

   a. Repeatedly pop from STACK and add to Y each operator on the top of STACK until an open parenthesis is encountered.
   b. Remove the open Parenthesis.

7. END.

Using this algorithm, converts the following infix to postfix ($\wedge$ is the power operator).

$$A + (B * C - (D/E \wedge F) * G) * H$$

95. *Infix to Prefix Conversion.* Let, X is an arithmetic expression in infix notation. The following algorithm finds the equivalent prefix expression Y.

1. Push ")" into STACK, and add "(" to the end of the X.
2. Scan X from right to left and repeat step 3 to 6 for each character of X until the STACK is empty.

3. If an operand is encountered add it to Y.

4. If a closed parenthesis is encountered push it into STACK.

5. If an operator is encountered, then:

   a. Repeatedly pop from STACK and add to B each operator on the top of STACK which has same or higher precedence than the operator.

   b. Add operator to STACK

6. If an open parenthesis is encountered then:

   a. Repeatedly pop each operator on top of stack from the STACK and add to Y until a closed parenthesis is encountered.

   b. Remove the open Parenthesis.

7. END.

Using this algorithm, converts the following infix to prefix ($\wedge$ is the power operator).

$$A + (B * C - (D/E \wedge F) * G) * H$$

96. *Postfix to Infix Conversion.* The following algorithm converts prefix expression to infix expression.

1. Scan the expression from left to right. If character is operand, push it to stack.

2. If character is operator,

   a. pop operand from the stack, call it is s1.

   b. pop operand from the stack, call it is s2.

   c. perform *s2 operator s1* and push the result to the stack.

3. At the end, pop all from the stack and add it to result.

4. Return the result.

Using this algorithm, converts the following postfix to infix.

$$ABC/ - AK/L - *$$

97. Prefix to Infix Conversion. The following algorithm converts prefix expression to infix expression.

1. Scan the expression from right to left (in reverse order). If character is operand, push it to the stack.

2. If character is operator,

   a. pop operand from the stack, call it is s1.

   b. pop operand from the stack, call it is s2.

   c. perform *s1 operator s2* and push result to the stack.

3. At the end, pop all from the stack and add it to result.

4. Return the result.

Using this algorithm, converts the following prefix to infix.

$$* - A/BC - /AKL$$

98. Postfix to Prefix Conversion. The following algorithm converts prefix expression to postfix expression.

1. Scan the expression from left to right. If character is operand, push it to stack.
2. If character is operator,

   a. pop operand from stack, say it is s1.
   b. pop operand from stack, say it is s2.
   c. perform (operator s2 s1) and push it to stack.

3. Once the expression iteration is completed, initialize result string and pop out from stack and add it to result.
4. Return the result.

Using this algorithm, converts the following postfix to prefix.

$$ABC/ - AK/L - *$$

99. Prefix to Postfix Conversion. The following algorithm converts prefix expression to postfix expression.

1. Scan the expression from right to left (in reverse order). If character is operand, push it to stack.
2. If character is operator,

   a. pop operand from stack, say it is s1.
   b. pop operand from stack, say it is s2.
   c. perform (s1 s2 operator) and push it to stack.

3. Once the expression iteration is completed, initialize result string and pop out from stack and add it to result.
4. Return the result.

Using this algorithm, converts the following prefix to postfix.

$$* - A/BC - /AKL$$

100. In the expression ((2 + 3) * 4 + 5 * (6 + 7) * 8) +9, the operator '+' has priority over '*'. Get the equivalent prefix expression.

101. Using a stack, show the steps for converting the expression a + b * (c/(d + e)) * f to postfix expression.

102. Using a stack, calculate the value of the following postfix expression. $\wedge$ is the power operator.

$$623 + -382/ + *2 \wedge 3+$$

103. Using a stack, convert the following postfix to infix expression.

$$ABC + *DE/-$$

104. If a = 1, b = 2, c = 3, and d = 4, get the value of the following postfix expression.

$$ab * c + dc-$$

105. Convert any of the following expressions into prefix and postfix expressions.

(a) A – B + C – D – E
(b) A + B/C + D * E
(c) (A + B)/(C – D)$\wedge$ E * F; ($\wedge$ denotes the power operator.)
(d) A–(B–(C * (D–E)))
(e) (A + B) * (C + D) – E
(f) A + B * (C + D) – E/F * G + H

106. Calculate the result of the following postfix for a = 7, b = 4, c = 3, and d = 2.

(a) a b + c d – *
(b) a b c + – d *
(c) a b c d///
(d) a b + c – d e * /
(e) a b/c/d/

107. Give an algorithm to convert infix expressions into prefix expressions using two stacks.

108. Convert the following infix expression to postfix expression.

$$((A - (B + C)) * D) \wedge (E + F)$$

109. What is the infix equivalent of the following prefix expression?

$$/ * A - B/ * C - (+DE)FG - HI$$

110. What is the infix equivalent of the following postfix expression?

$$ABC - \wedge D/E+$$

5.2.2.4 Valid Multiple Parentheses

111. Given a string containing just the characters '(', ')', '{', '}', '[', and ']'. Check if the following algorithm is correct for determining if the input string is valid.

1. Define a stack.
2. Foe each *character* from the input do
3. Check whether *character* is open bracket

    a. if is '(', push closed bracket ')' to the stack.
    b. if is '[', push closed bracket ']' to the stack.
    c. if is '{', push closed bracket '}' to the stack

4. Else if *character* is close bracket, one of ')' or '}' or ']'

    a. if stack is empty, return false.
    b. Else if stack is not empty, pop from stack and popped bracket should be same as current bracket else return false.

5. If the stack is empty at the end, return true, otherwise return false.

112. Using stack, devise an algorithm to check if parentheses of an expression are balanced.

113. Consider expression {x + (y–[a + b] * c – [(d + e)])}/(j–(k–[1–n])). We want to use a stack to check if the parentheses, brackets, and braces match correctly. The stack used must have at least how many elements of capacity?

114. Write a program that uses stack to check for pairs of {}, (), and [] in a string containing letters and symbols.

Example:
expression AB(C)[EFG] is an acceptable expression and ({BC}} is an unacceptable expression.

115. Give an algorithm to find out maximum depth of nested parenthesis in a string. Return –1, if parenthesis is unbalanced. For example, in the following expression, maximum depth of nested parenthesis is 4.
"(((X)) (((Y))))"

116. Give an algorithm to check whether a given array can represent preorder traversal of a binary search tree.

117. Give an algorithm to find duplicate parenthesis in a given expression. A set of parenthesis are duplicate if a subexpression is surrounded by multiple parenthesis.

Example:
((x + y) + ((z + p)))

The subexpresthesis "z + p" is surrounded by two pairs of parenthesis.

118. Suppose expression E is given. Give an algorithm to count the valid parenthesis in the expression.

Example 1:
Input: E = ((()
Output: ()

Example 2:
Input: E =)()())
Output: ()()

119. *Parenthesis Checker*- Given an expression string x, provide an algorithm to check whether the pairs and the orders of "{","}","(",")","[","]" are correct in the expression. For example, the algorithm should return 'true' for expression = "([])[][()]()" and 'false' for expression = "()".

120. Given an expression with only '}' and '{'. Minimum number of bracket reversals needed to make an expression balanced.

5.2.2.5 Applications

121. *Stock span problem.* The stock span problem is a financial problem in which we have a series of n daily price quotes for a stock and we need to calculate the span of stock's price for the whole n days.

The span S_i of the stock's price on a given day i is defined as the maximum number of consecutive days just before the given day, for which the price of the stock on the current day is less than or equal to its price on the given day. For example, for an array indicating seven days prices as {90, 70, 50, 60, 50, 65, 75}, the span values for corresponding seven days are {1, 1, 1, 2, 1, 4, 6}.

Example:
Input:
N = 7, price[] = [90, 70, 50, 60, 50, 65, 75]
Output:
1 1 1 2 1 4 6
Explanation:
Traversing the span array, for 90 will be 1, 70 is smaller than 90 so the span is 1, 50 is smaller than 70 so the span is 1, 60 is greater than 50 so the span is 2 and so on. Hence the output will be 1 1 1 2 1 4 6.

122. Only using *push(), pop(), and is-empty()* functions, give an algorithm to check whether stack elements are pairwise consecutive. Pairs can be increasing or decreasing, and if the stack has an odd number of elements, the top element of the stack should be ignored.

Examples:
Input: stack = [2, 3, –4, –5, 13, 12, 7, 8, 22]
Output: Yes
Each of the pairs (2, 3), (–4, –5), (13, 12) and (7, 8) consists of consecutive numbers.

Input: stack = [2, 4, 9, 10, 4, 3]
Output: No
(2, 4) are not consecutive.

123. *Next Greater Element*—Given an array A of size N having distinct elements, write a function to find the next nearest element on the right of the element which is greater than the current element.

Given an array A of size N having distinct elements, the task is to find the next greater element for each element of the array in order of their appearance in the array. If the next greater element of the current element does not exist, the next greater element for the current element is –1.

Example:
Input:
N = 4, arr[] = [3 5 4 6]
Output:
4 6 6 –1
Explanation:
In the array, the next greater element to 3 is 4, 5 is 6, 4 is 6 and for 6 is –1.

124. Explain how data can be moved from one stack to another so that the original order is maintained.

125. Let A is an array of size n. Give an algorithm to sort the elements of the array using stack(s).

126. Let A and B are two stacks. Give an algorithm the to check whether the given stacks are the same.

127. Let S is a stack. Using the following functions, give an algorithm to sort the elements of the stack using recursion. You are not allowed to use other commands, e.g., loops, except as stated below.

– is_empty(S): checks whether stack is empty or not.
– push(S): adds new element to the stack.
– pop(S): removes top element from the stack.
– top(S): returns value of the top element.

128. Write a recursive program to reverse stack elements using the following functions. You are not allowed to use other commands, e.g., loops, except as stated below.

– isEmpty(S)
– push(S)
– pop(S)

129. Give an algorithm to reverse the elements of a stack without using extra space.

130. Design a stack that returns the minimum element in constant time.

131. Design a stack that returns a minimum element without using an auxiliary stack.

132. Give recursive solution to sort a stack.

133. Reverse a string using a stack data structure.

134. Explain how to find the greatest common divisor (GCD) of two numbers using a stack.

135. Special Stack- Design a data-structure that supports all the stack operations such as push(), pop(), isEmpty(), isFull() and an additional operation getMin() which should return minimum element from the data-structure. Your task is to complete all the functions, using stack data-structure.

136. Given a stack including push(), pop(), empty() operations, delete the middle of the stack (ceil(size_of_stack/2.0)) without using any additional data structure.

137. Given a stack, give an algorithm to sort a stack such that the top of the stack has the greatest element.

138. Check if a given array can represent preorder traversal of binary search tree.

139. How to implement stack using min heap?

5.2.3 *Queue*

5.2.3.1 Basic Concept

140. Given a queue, write a recursive function to reverse it. Standard operations allowed:

– enqueue(x): Adds item x to the rear of the queue.
– dequeue(): Removes an item from the front of the queue.
– empty(): Checks if a queue is empty or not.

141. How to efficiently implement k queues in a single array?

142. Explain how a queue can be implemented using two stacks?

143. Give an algorithm to sort the elements of a given queue in another queue in increasing order using a stack.

144. Double Ended Queue (Deque) is a generalized version of the classic queue data structure allowing insert and delete at both ends. This data structure supports the following operations.

– insertFront(): Adds an item at the front of Deque.
– insertLast(): Adds an item at the rear of Deque.
– deleteFront(): Deletes an item from front of Deque.
– deleteLast(): Deletes an item from rear of Deque.
– getFront(): Gets the front item from queue.
– getRear(): Gets the last item from queue.
– isEmpty(): Checks whether Deque is empty or not.
– isFull(): Checks whether Deque is full or not.

Implement double ended queue.

145. Give an algorithm for reversing a queue Q. Only the following standard operations are allowed on queue.

– enqueue(x): Adds item x to the rear of the queue.
– dequeue(): Removes an item from front of the queue.
– empty(): Checks if a queue is empty or not.

146. Suppose a queue of integers and an integer k are given. Give an algorithm to reverse the order of the first k elements of the queue. Only standard operations are allowed on queue.

147. Suppose an array of integers and an integer k are given. Using Deque, give an algorithm to find out the maximum for each and every contiguous subarray of size k starting from the left and moving towards the right by one position each time.

Example:
Input: A = {2, 3, 4, 2, 3, 4, 8}, k = 3
Output: {2, 3, 4}, max = 4
{3, 4, 2}, max = 4
{4, 2, 3}, max = 4
{2, 3, 4}, max = 4
{3, 4, 8}, max = 8

148. How to implement a stack using queue?

149. How to implement a queue using stack?

150. Why and when should we use stack or queue data structures instead of arrays/lists?

151. Suppose a circular queue of capacity $n - 1$ elements is implemented by an array of n elements. Assume that *REAR* and *FRONT* variables are employed for the insertion and deletion operations, respectively. Initially, $REAR = FRONT = 0$. Give the conditions to show queue is full and empty.

5.2.3.2 Priority Queue

152. Explain how to implement priority queue using the following data structures.

(a) Array
(b) Binary Search Tree (BST)
(c) Binary Heap
(d) Fibonacci heap

Analyze the algorithms complexity times.

153. Which of the following data structures is preferred for implementing the priority queue and why?

a. Binary Heap
b. Binary Search Tree

154. Which of the following data structures is preferred for implementing the priority queue and why?

a. Binary Heap
b. Linked List

155. Considering following operations, implement priority queue using linked list.

– push(): Inserts a new data into the queue.

– pop(): Removes the element with the highest priority form the queue.
– peek()/top(): Gets the highest priority element in the queue without removing it from the queue.

156. Implement a priority queue using doubly linked list.

157. How to implement stack using priority queue?

158. Let A is array of size n and an integer k are given. Give an algorithm to find the kth smallest element in the array using priority queue.

5.2.3.3 Applications

159. Let A is a square chessboard of $n \times n$ size. The position of the knight and the position of a target are given in A. Write a function to find out the minimum steps a knight will take to reach the target position.

160. Shortest path in a binary maze. Given a $m \times n$ matrix where each element can either be 0 (indicates the absence of a path) or 1 (indicates the existence of a path). Write a function to find the shortest path between a given source cell to a destination cell.

161. Given the number n, write a function that produces all binary numbers with decimal values from 1 to n.

162. Given a binary tree having positive and negative nodes, give an algorithm is to find the maximum sum level in it.

163. Given a binary tree, write a function to check whether the tree is complete binary tree or not.

164. A binary tree and the two level numbers of it are given. Write a function that prints all nodes between two given levels from left to right.

5.2.4 Linked List

165. What is complexity of push and pop for a stack implemented using a linked list?

166. *Reverse operation.* This operation is one step. How to do this is that the last node points to the first node and is reversed for the entire link list. Write a program that takes a singly linked list and reverses it.

167. The following is a list of basic operations that can be performed on a doubly linked list:

– Insertion: Adds an element to the beginning of the list.
– Deletion: Deletes an element from the beginning of the list.
– Insert Last: Adds an element to the bottom of the list.
– Delete Last: Removes an element from the bottom of the list.
– Insert After: Adds an element after a list item.
– Delete: Deletes an element from the list using a key.
– Display forward: Displays the complete list in the forward method.
– Display backward: Displays the complete list in reverse.

Write a program that performs these operations.

168. The following is a list of important operations in circular linked lists:

– Insert: Inserts an element at the beginning of the list.
– Delete: Deletes an element from the beginning of the list
– Display: Displays the list.

Write a program that performs these operations.

169. Write a program that stores the name and tel number of n people in a singly linked list.

(a) Display all list elements.
(b) Take a name and type the equivalent number.
(c) Modify the print operation to do it with the help of a recursive function.
(d) Add insert and delete operations to this program. Always insert at the end of the list.
(e) Add a sorting function to this program. Modify the insert operation so that the list remains in order after insertion.
(f) Add a function to this program that counts the number of nodes in the list.

170. Write a program that stores the name and tel number of n people in a doubly linked list.

(a) Display all list elements.
(b) Take a name and type the equivalent number.
(c) Modify the print operation to do it with the help of a recursive function.
(d) Add insert and delete operations to this program. Always insert at the end of the list.
(e) Add a sorting function to this program. Modify the insert operation so that the list remains in order after insertion.
(f) Add a function to this program that counts the number of nodes in the list.

171. Write a program that stores the name and tel number of n people in a circular linked list.

(a) Display all list elements.
(b) Take a name and type the equivalent number.
(c) Modify the print operation to do it with the help of a recursive function.
(d) Add insert and delete operations to this program. Always insert at the end of the list.
(e) Add a sorting function to this program. Modify the insert operation so that the list remains in order after insertion.
(f) Add a function to this program that counts the number of nodes in the list.

172. Given two numbers A and B which are represented by two linked lists List1 and List2, respectively. Write a function that returns the sum of these two numbers as linked list representation of the addition of the two input numbers. (Hint: Use Recursion).

Example:
Input:
List1: 6– >7– >4 // represents number 476
List2: 9– >5– >3 // represents number 359
Output:
Resultant list: 5– >3– >8 // represents number 835

173. Given two numbers A and B which are represented by two linked lists List1 and List2, respectively. Write a function that returns the multiplication of these two numbers as linked list representation of the multiplication of the two input numbers. (Hint: Use Recursion).

174. Given two large numbers A and B which are represented by two linked lists List1 and List2, respectively. Write a function that returns the substraction of these two numbers as linked list representation of the difference of the two input numbers. Always subtract the smaller number from the larger number. (Hint: Use Recursion).

175. Write a program that inserts a new element before element x in a linked list with n elements.

176. Write a program that inserts a new element after element x in a linked list with n elements.

177. Write a program that removes element after element x in a linked list with n elements.

178. Write a program that removes element before element x in a linked list with n elements.

179. Write a program that removes element x in a list with n elements.

180. Write a program to sort a linked list of n elements using bubble sort.

181. Write a function that recursively counts the number of nodes in a singly linked list.

182. Write a function that recursively examines the equality of two singly linked lists.

183. Write a function that makes a duplicate copy of a singly linked list.

184. Write a function that swaps the nodes n and m of a doubly linked list.

185. Let A is a singly linked list containing n integer numbers. Write a function to find

(a) maximum element of the linked list.
(b) minimum element of the linked list.
(c) average value of the linked list.

186. Given two polynomial numbers A and B which are represented by two linked lists List1 and List2, respectively. Write a function that adds these lists i.e., adding the coefficients who have same variable powers.

Example
Input:
number 1 $= 3x^2 + 2x^1 + 4x^0$
number 2 $= -4x^1 + x^0$
Output:
$3x^2 - 2x^1 + 5x^0$

187. Given two polynomial numbers A and B which are represented by two linked lists List1 and List2, respectively. Write a function that multiplies these lists.

188. Write a function that receives two sorted linked lists and merges them so that the result is also sorted.

189. Find the output of the following function for lined list A$->$B$->$C.

```
void what(List Node * L)
{
    if(L! = Null){
    what(L-> next);
    cout << L-> info;
    what(L-> next);
    cout << L-> info;
    }
}
```

190. The time complexity of the linear search does not depend on whether a list is implemented by an array or a linked list. Is it also correct for searching a sorted list by binary search?

191. How to apply binary search on a sorted linked list?

5.3 Solutions

13.

1. **Start.**
2. **Initialize arrays A and B.**
3. **Read the variables n, and set i = 0, j = n–1, and k = 0.**
4. **Receive n numbers from the input and store them in array A.**
5. **Arrange the members of array A from small to larger and store them in array B.**
6. **If n is even, consider i from 0 to $\frac{n}{2} - 1$ and j from n–1 to $\frac{n}{2}$ and go to 8, otherwise go to 7.**
7. **If n is odd, consider i from 0 to $\frac{n-1}{2}$ and j from n–1 to $\frac{n-1}{2} + 1$ and go to 8.**
8. **A[k] = B[i] and B[k + 1] = B[j].**
9. **i = i + 1, j = j–1 and k = k + 2 go to 8.**
10. **If n is even, i is equal to $\frac{n}{2} - 1$ and j is equal to $\frac{n}{2}$, go to 13.**
11. **If n is odd, i is equal to $\frac{n-1}{2}$ and j is equal to $\frac{n-1}{2} + 1$, go to 13.**
12. **Finish.**

Example: A[] = [1, 8, 5, 4, 7, 10]
Sorted of array A is B[] = [1, 4, 5, 7, 8, 10].
Answer: A[] = [1, 10, 4, 8, 5, 7]

14.
First algorithm:

Find the missing value between 1 and 100:

(a) First we sort the array in increasing order;
(b) To find the missing number between 1 and 100, we add one to each element starting from the first element;
(c) Then we compare it with the next element, if they are equal we repeat this procedure otherwise we print the missing number.

Example:
input: A = [1, 2, 3, 5, ..., 100]
Output: 4 is missing

A[0] + 1 == A[1] → True
A[1] + 1 == A[2] → True
A[2] + 1 == A[3] → False
answer = A[2] + 1 is missing.

Second algorithm:

1. Calculate the sum of first n natural numbers as sumtotal = n*(n + 1)/2
2. Create a variable sum to store the sum of array elements.
3. Traverse the array from start to end
4. Update the value of sum as sum = sum + array[i]
5. Print the missing number as sumtotal-sum.

15.
Algorithm:

1. Create two nested loops
2. The outer loop will iterate through the array A, from 0 to length of A, which will select an element.
3. The inner loop will be used to compare the selected element A[i] and with all next elements.
4. If A[i] == A[j], this means we've got the duplicated element.
5. Print the element/elements.

As a code:

1. for(int i = 0; i<n; i++):
2.   for(int j = i + 1; j<n; j++):
3.   if(arr[i] == arr[j])
4.   print(arr[j]);

Input: *arr* [ ] = {1, 2, 3, 4, 3, 7, 8, 8, 7}
Output: Duplicate numbers answer: {3, 7, 8}

17.

1. for(int i = 0; i<n; i++)
2.   for(int j = i + 1; j<n; j++)
3.   if(arr[i] + arr[j] == k)
4.   print(arr[i], arr[j]);

18. To solve this question, it is enough to use the shift between indexes.
Algorithm:

1. First we consider an array called A
2. We consider the item we are looking for X
3. Now we start searching for X in array A.
4. After finding X, we move it to the previous index, that is, if it is in index = n, we move it to index = n–1.

5. And move the element in index = n–1 to index = n
6. Save the new array instead of the previous one
7. This will be done in each search until X is passed to index = 0

By doing this, we optimize the program every time to reach the last step, which is the most optimal mode.
Note: Of course, there is another way and that is to transfer X directly to index = 0.

Example:

$$X = 3$$

$$Array\ A[7] = [5, 8, 20, 3, 69, 40, 15]$$

By doing the algorithm for the first time we will see that X is transferred from index = 4 to index = 3

$$Output : Array\ A[7] = [5, 8, 3, 20, 69, 40, 15]$$

If we do it again:

$$Output : Array\ A[7] = [5, 3, 8, 20, 69, 40, 15]$$

28.

1. Start
2. Use the binary search to find bitonic point, if the mid element (firstIndex + (lastIndex–firstIndex)/2) is larger than both adjacent elements, mid is the maximum (bitonic point).
3. Else if the mid element is larger than the next element and smaller than the previous element, maximum (bitonic point) is at the left of the mid.
4. Else if the mid element is smaller than the next element and larger than the previous element, maximum (bitonic point) is at the right of the mid.
5. Now if the searched element is greater than bitonic point, element is not in the array (End).
6. Else if the element to be searched for is less than the bitonic point, we use binary search to search for the element in both halves of the array.
7. Else if the element being searched for is *equall* the bitonic point, we print the index of the bitonic point.
8. End.

Example:
Input: arr [] = {–1, 2, 4, 7, 12, 6, 3, 1}
number = 12
Output: index 4

Explanation: In the bitonic array above, we were looking for the index of element 12, which is equal to 4.

29. Use binary search

   1. Low = first element index
   2. High = n–1 (last element index)
   3. while(low < = high):
   4.   mid = floor((low + high)/2)
   5.   if(B < A[mid]):
   6.   high = mid-1
   7.   else if(B > A[mid]):
   8.   low = mid + 1
   9.   else:
   10.   return mid (mid is index of B)
   11. Print(mid+1) (number of elements)

30.

First Algorithm:

   1. Create a variable i, indicating counter, and take care of some base cases, i.e., when the given number is 0 or 1
   2. Iterate until $i \times i <= A$, where A is the given number. Increment i by 1.
   3. Return $i–1$

Example:
Input: x = 4
Output: 2 (the square root of 4 is 2.)
Input: x = 17
Output: 4
Explanation: The square root of 17 lies in between 4 and 5 so the floor of the square root is 4.

Second Algorithm:

   1. Start iterating from i = 1. If $i \times i = n$, then print i as n is a perfect square whose square root is i.
   3. Else find the smallest i for which $i \times i$ is strictly greater than n.
   3. Now n lies in the interval i – 1 and i so we can use binary search algorithm to find the square root.
   4. Find mid of i – 1 and i and compare $mid \times mid$ with n, with precision up to 5 decimal places.

      a. If $mid \times mid = n$ then return mid.
      b. If $mid \times mid < n$ then recur for the second half.
      c. If $mid \times mid > n$ then recur for the first half.

32.

$$\text{Input : arr1 } [\,] = \{-5,\ 3,\ 6,\ 12,\ 15\}$$

$$arr2\ [\,] = \{-9,\ -13,\ -3,\ 2,\ 11\}$$

output: The median is 3.

1. **Start**
2. Use two index variables i and j, initialize values i = 0, j = 0
3. If $A[i]$ is smaller than $B[j]$ then print $A[i]$ and increment i.
4. If $A[i]$ is greater than $B[j]$ then print $B[j]$ and increment j.
5. If $(m + n)$ is odd so calculate $(m + n)\ /\ 2$ and store the element.
6. Else, store the average of elements $(m + n)/2$ and $(m + n)/2\ -\ 1$
7. Print the median variable in output
8. End.

43.

Input:

$$arr1[i] = \{1,\ 3,\ 6,\ 7,\ 8\}$$

$$arr2[j] = \{2,\ 3,\ 7,\ 9\}$$

Output: $Intersection$: $\{3,\ 7\}$

Algorithm:

1. **Start**
2. Use two index variables i and j, initialize values $i = 0,\ j = 0$
3. If $arr1[i] < arr2[j]$ then increment i.
4. If $arr1[i] > arr2[j]$ then increment j.
5. If $arr1[i] = arr2[j]$ then print one of them and increment both i and j
6. End.

Example:
input:
arr1 = [1, 2, 4]
arr2 = [2, 4]
output:
2, 4
arr1[0] < arr2[0] → True then i++;
arr1[1] < arr2[0] → False
arr1[1] > arr2[0] → False
arr1[1] = arr2[0] → True → print(arr1[1] or arr2[0]) then i++ and j++;

arr1[2] < arr2[1] → False
arr1[2] > arr2[1] → False
arr1[2] = arr2[1] → True → print(arr1[2] or arr2[1]); End;

4

44.

Input:
$$arr1[\,] = \{7,\ 1,\ 5,\ 2,\ 3,\ 6\}$$

$$arr2[\,] = \{9,\ 7,\ 4,\ 1,\ 3\}$$

Output: Union = {1, 2, 3, 4, 5, 6, 7, 9}
Intersection = {1, 3, 7}

Algorithm for union:

1. **Start**
2. Create and initialize union array U as empty.
3. Copy all elements of the first array to U.
4. Do the following for every element x of the second array.
5. If x is not available in the first array, then copy x to U.
6. Print U.
7. End.

Algorithm for intersection:

1. **Start**
2. Create and initialize intersection array i as empty.
3. Copy all elements of the first array to U.
4. Do the following for every element x of the first array.
5. **[1.]** If x is available in second array, then copy x to i.
6. Print i.
7. End.

46.

1. **Start**
2. Give initial 0 value to i, j, k variables as the smallest indexes of array1[], array2[], array3[].
3. If array1[i] = array2[j] = array3[k], a common element has been found, then print that element and increase i, j, k.
4. Else, increase the index of the array in which the smaller element is located.
5. Continue this process until you reach the end of each array.
6. End.

Input:
array1[] = {1, 4, 9, 16, 28, 53, 77}
array2[] = {4, 7, 16, 24, 28, 65, 70, 77}
array3[] = {0, 2, 4, 5, 8, 16, 20, 25, 26, 28, 59, 61, 77}
Output:
{4, 16, 28, 77}

47.

First Algorithm:

1. Create an array of size m+n+k as $arr4[\ ]$
2. Traverse simultaneous all arrays and print the smaller element in currant index in $arr4[\ ]$ then increment the index in main array.
3. If there are same elements print one of them

Example:
Input:

$$arr1[5] = \{1, 2, 6, 7, 8\}$$

$$arr2[5] = \{2, 3, 4, 6, 7\}$$

$$arr3[3] = \{3, 5, 6\}$$

Output:

$$arr4[12] = \{1, 2, 2, 3, 3, 4, 6, 6, 6, 7, 7, 8\}$$

Second Algorithm:
Here we first transform the first two arrays together into an ordered array and call it D, then do the same between D, C

1. If we consider the first array as A whose number of elements is m and the second array as B which the number of elements is n and the third array as C which has k elements and an empty array of size m + n + k as T
2. i = 0 and j = 0
3. if j < n and i < m go to 4 else go to 5
4. If A [i] ≤ B[j] then we add A[i] to D and add i to one, otherwise we add B[j] to D and add j to one and then go to 3
5. If j < n we add B[j] to D and we add one to j otherwise go to 6
6. If i < m, we add A[i] to D and add one to i
7. Now with arrays D, C we go back to 2 with the difference that we use m + n instead of m and k instead of n we put the result in T and finally we return T

Example:
Input:

$$\text{Array A} [4] = [2, 4, 5, 9]$$

$$\text{Array B} [3] = [4, 8, 10]$$

So we have:

$$D [7] = [2, 4, 4, 5, 8, 9, 10]$$

$$\text{Array C}[4] = [9, 10, 13, 18]$$

Output: $T [11] = [2, 4, 4, 5, 8, 9, 9, 10, 10, 13, 18]$

48.

Algorithm:

1. We get array A and element X.
2. Set variable *answer* to zero, *answer* = 0
3. From the left we start examining the elements. If the element is equal to X, we add the *answer* to the number one
4. At the end we print *answer*

Example:
Input:

$$\text{Array } A [\ ] = [2, 6, 20, 9, 6, 6, 21],\ X = 6$$

answer: 1
answer: 2
answer: 3
Output: 3

70.

To find the mean:

(a) initialize sum = 0;
(b) finding the sum by looping in the array and updating sum value until the end of array;
(c) average = $\frac{sum\,Of\,elements}{number\,of\,elements}$;
(d) print the average.

To find the median:

(a) Sort the array in increasing order;

(b) if we have odd number of data values the median index will be $\to n = \lfloor \frac{arraySize}{2} \rfloor$. The median value will be A[n].

(c) if we have even number of data values the median will be the average of two data points in the middle of the data. The median indexes will be $a = \frac{arraySize}{2}$, $b = (\frac{arraySize}{2}) - 1$ and finally the median will be (A[a] + A[b]) / 2.

To find the mode:

(a) we count the occurrence of each integer value in A;
(b) we display the value with the highest occurrence.

Example:
finding mean:
input = A = [4, 2, 1, 1, 2]
output = 2.5
mean: $\frac{4+2+1+1+2}{4} = 2.5$
finding median:
input = A = [4, 2, 1, 1, 2]
output = 2
first we sort it = A = [1, 1, 2, 2, 4];
our list has an odd number of values so the median index is: $\lfloor \frac{5}{2} \rfloor = 2$
A[2] = 2 is median
finding mode:
input = A = [4, 2, 1, 1, 2]
output = 1, 2
first we sort the list = A = [1, 1, 2, 2, 4];
1 and 2 has the most number of occurrences.

83.

Push

1. If Top >= n–1 (bigger than or equal to) then return Full!
2. Else array[++Top] = x

Pop

1. If Top = –1 (is equal to) then return Empty!
2. Else define x=array[Top]
3. Top = Top – 1
4. Print(x)

findMiddle

1. If Top = –1 then return Empty!
2. Else if Top % 2 == 0 (remainder of Top/2) then print(array[Top/2]
3. Else print(array[floor(Top/2)])

86.

Stack1 starts from the leftmost element, the first element in stack1 is pushed at index 0 and the stack2 starts from the rightmost element, the first element in stack2 is pushed at index n–1.

Both stacks grow in opposite direction. To check for overflow, the only thing we need to check is whether there is a space between the top elements of both stacks.

Checking the overflow for stacks:

Push_Stack1:

```
    If top1 < top2 – 1:
        top1 = top1 + 1
        Arr[top1] = item
    Else:
        Print('stack overflow')
Push_stack2:
    If top1 < top2 – 1:
        top2 = top2 – 1
        Arr[top2] = item
    Else:
        Print('stack overflow')
Pop_stack1:

    If top1 > 0:
        x = Arr[top1]
        top1 = top1 – 1
    Else:
        Print('stack overflow')
Pop_stack2:

    If top2 < n:
        x = Arr[top2]
        top2 = top2 + 1
    Else:
        Print('stack overflow')
```
So the time complexity for push and pop is:
Push time complexity is: $O(1)$
Pop time complexity is: $O(1)$

94. Using the algorithm, convert the following infix to postfix ($\wedge$ is the power operator).

$A + (B * C - (D/E \wedge F) * G) * H$

Postfix: ABC*DEF∧/G*-H*+

Answer:

Input	Stack	Output
A		A
+	+	A
(	+(	A
B	+(	AB
*	+(*	AB
C	+(*	ABC
–	+(–	ABC*
(	+(–(	ABC*
D	+(–(	ABC*D
/	+(–(/	ABC*D
E	+(–(/	ABC*DE
∧	+(–(/∧	ABC*DE
F	+(–(/∧	ABC*DEF
)	+(–(/∧)	ABC*DEF∧/
*	+(–*	ABC*DEF∧/–*
G	+(–*	ABC*DEF∧/G
)	+(–*)	ABC*DEF∧/G*–
*	+*	ABC*DEF∧/G*–
H	+*	ABC*DEF∧/G*–H
		ABC*DEF∧/G*–H*
		ABC*DEF∧/G*–H*+

99.
Answer:

Step 1: push L	L		
Step 2: push K	L	K	
Step 3: push A	L	K	A
Step 4: s1 = pop A	L	K	
Step 5: s2 = pop K	L		
Step 6: push s1s2/	L	AK/	
Step 7: s1 = pop AK/	L		
Step 8: s2 = pop L			
Step 9: push s1s2-	AK/L-		
Step 10: push C	AK/L-	C	
Step 11: push B	AK/L-	C	B
Step 12: s1 = pop B	AK/L-	C	
Step 13: s2 = pop C	AK/L-		
Step 14: push s1s2/	AK/L-	BC/	
Step 15: push A	AK/L-	BC/	A
Step 16: s1 = pop A	AK/L-	BC/	
Step 17: s2 = pop BC/	AK/L-		
Step 18: push s1s2-	AK/L-	ABC/-	
Step 19: s1 = pop ABC/-	AK/L-		
Step 20: s2 = pop AK/L-			
Step 21: push s1s2*	ABC/- AK/L-*		
Step 22: result = pop ABC/-AK/L-*			

Result = ABC/- AK/L-*

100. In the expression $((2 + 3) * 4 + 5 * (6 + 7) * 8) + 9$, the operator + has priority over *. Get the equivalent Prefix expression.

$$(((((2 + 3) \times (4 + 5)) \times (6 + 7)) \times 8) + 9)$$

Prefix: $+ * * * + 2 3 + 4 5 + 6 7 8 9$

101.

Postfix: abcde+/*f*+

Input	Stack	Output
a		a
+	+	A
b	+	ab
*	+*	ab
(	+*(	ab
c	+*(	abc
/	+*(/	abc
(	+*(/(	abc
d	+*(/(	abcd
+	+*(/(+	abcd
e	+*(/(+	abcde
)	+*(/(+)	abcde+
)	+*(/)	abcde+/
*	+*	abcde+/*
f	+*	abcde+/*f
		abcde+/*f*
		abcde+/*f*+

102.

		2					
3		8	4				
2	5	3	3	7	2		
6	6	1	1	1	7	49	52

103.

C		E	
B	(B+C)	D	(D/E)
A	A	(A*(B+C))	(A*(B+C))

((A*(B+C))−(D/E))

107.

First Algorithm:

1. First, reverse the infix expression given in the problem.
2. Scan the expression from left to right.
3. Whenever the operands arrive, push it into stack1.

4. If the operator arrives and the stack2 is found to be empty, then simply push the operator into the stack2.
5. If the incoming operator has higher or same priority than the TOP of the stack2, push the incoming operator into the stack.
6. If the incoming operator has lower priority than the TOP of the stack2, first, pop operand from stack1 say it is s1, pop operand from stack1 say it is s2, pop operator from stack2 say it is op, from the TOP of the stacks. push op s1 s2 into stack1, then check if the incoming operator has lower or higher or same priority than the new TOP of stack2 and do the relatives.
7. When we reach the end of the expression, pop operand from stack1 say it is s1, pop operand from stack1 say it is s2, pop operator from stack2 say it is op, from the TOP of the stacks. push op s1 s2 into stack1.
8. If the operator is ')', then push it into the stack.
9. If the operator is '(', pop operand from stack1 say it is s1, pop operand from stack1 say it is s2, pop operator from stack2 say it is op, from the TOP of the stacks. Do this till it finds ')' opening bracket in the stack.
10. If the top of the stack is ')', push the operator on the stack.

Second Algorithm:

1. Start.
2. For convert infix to prefix, first should reversing string and go to 3.
3. Read the character from the infix expression until the end of the file.
4. Test the character.
5. If the character is an operand, push it into the stack2, go to 8 otherwise go to 6.
6. If the character is a left parenthesis, push it into the stack1, go to 8 otherwise go to 7.
7. If the character is a right parenthesis then pop entries from the stack1 and push them into the Stack2 until a left parenthesis is popped. Discovered both left and right parenthesis, go to 8.
8. If the character is an operator, pop from the stack1 and push to stack2 when the operator that have stack priority greater than or equal to the infix priority of the read character, then push the read character to the stack1.
9. If the end of the expression, pop all entries that remain in the stack1 and push them to stack2.
10. Then reverse the strings of the stack2 and print it.
11. End.

108.

$$(((A - (B + C)) * D) \wedge (E + F))$$

Postfix: $ABC + -D * EF + \wedge$

109.
Infix = $((A*(B-((C*((D+E)-F))/G)))/(H-I))$

110.

$$ABC - \wedge D/E+ \;\rightarrow\; A(B-C) \wedge D/E+ \;\rightarrow\; (A \wedge (B-C))D/E+ \;\rightarrow$$

$$((A \wedge (B-C))/D)E+ \;\rightarrow\; ((A \wedge (B-C))/D)+E \;\rightarrow$$

$$\left(\frac{A^{B-C}}{D}\right)+E$$

111.

The algorithm has problem in third step.
Correct form:
3. Check if character is open bracket

(a) '(', if yes then push opened bracket to stack.
(b) '[', if yes then push opened bracket to stack.
(c) '{', if yes then push opened bracket to stack.

112.

1. **Start**
2. **Get string from the input.**
3. **Declare a character stack.**
4. **Then traverse the expression string and go to 5.**
5. **If the current character is a starting bracket ('(', '[' , '{'), push it into the stack.**
6. **If the current character is a closing bracket (')', ']' , '}'), should compare the top element of stack and current character. If the top element of stack is not matching with the current character it means that is not balanced and return false. If the top element of stack is matching with the current character it means that is balanced and pop the top element of the stack (starting bracket) and go to 4.**
7. **After complete traverse, if there is some starting bracket left in stack, it means is not balanced.**
8. **End.**

115.

1. Declare variable count equal 0.
2. Scan the expression from left to the right.
3. Whenever a '(' arrives push it in to the stack.
4. When a ')' arrives and stack is not found empty, pop a '(' from the TOP of the stack and add 1 to the count.
5. If ')' arrives but stack is empty return –1.
6. If something else than '(' or ')' arrives, pass it.
7. If we reach end of the expression but stack isn't empty return –1 else print count.

118.

1. Create 2 empty stacks
2. Traverse the expression:
   (a) If the current character was '(' push it into the stack1.
   (b) If the current character was ')' pop from stack1 and if the popped character is the matching starting bracket append stack2 "("+")"
3. Do step 2 until there isn't any pairs
4. Pop all elements of stack2

Explanation:
Input: E = (()())(

Stack1

(

stack2

()
()
()

Output:()()()

144.

Algorithm of insertFront():

1. Start.
2. Create an empty array deque of size N.
3. Initialize front = −1, rear = −1 and x.
4. If front is zero (front == 0) and rear is N−1 (rear == N−1) or front is rear + 1 (front == rear + 1), it shows that queue is Full.
5. Else if front is −1 and rear is −1, it shows that queue is Empty. On that case front = rear = 0 and deque[front] = x.
6. Else if front is zero (front == 0), front = N−1 and deque[front] = x.
7. Else front = front−1 and deque[front] = x.
8. Finish.

Algorithm of insertLast():

1. Start.
2. Create an empty array deque of size N.
3. Initialize front = –1, rear = –1 and x.
4. If front is zero (front == 0) and rear is N–1(rear == N–1) or front is rear + 1(front == rear + 1), it shows that queue is Full.
5. Else if front is –1 and rear is –1, it shows that queue is Empty. On that case front = rear = 0 and deque[rear] = x.
6. Else if rear is N–1 (rear == N–1), rear = 0 and deque[rear] = x.
7. Else rear = rear + 1 and deque[rear] = x.
8. Finish.

Algorithm of deleteFront():

1. Start.
2. Create an empty array deque of size N.
3. Initialize front = –1, rear = –1 and x.
4. If front is zero (front == 0) and rear is N–1(rear == N–1) or front is rear + 1(front == rear + 1), it shows that queue is Full.
5. Else if front is –1 and rear is –1, it shows that queue is Empty.
6. Else if front is rear (front == rear), front = rear = –1.
7. Else if front is N–1(front == N–1), print deque[front] and front = 0.
8. Else print deque[front] and front = front + 1.
9. Finish.

Algorithm of deleteLast():

1. Start.
2. Create an empty array deque of size N.
3. Initialize front = –1, rear = –1 and x.
4. If front is zero (front == 0) and rear is N–1(rear == N–1) or front is rear + 1(front == rear + 1), it shows that queue is Full.
5. Else if front is –1 and rear is –1, it shows that queue is Empty.
6. Else if front is rear (front == rear), front = rear = –1.
7. Else if rear is zero(rear == 0), print deque[rear] and rear = N–1.
8. Else print deque[rear] and rear = rear–1.
9. Finish.

145.

we can use an **auxiliary queue** for that purpose.

1. create a queue and add items with **enqueue(x)** function to the rear of the queue.
2. for reversing the queue, create a auxiliary queue.
3. With **dequeue()** function remove an item from the front of the queue and add that item to the rear of our auxiliary queue with **enqueue(x)**.
4. Do this until our main queue is empty and **empty()** function id True.

5. Now our auxiliary queue is reverse of the main queue.

172.

1. Start.
2. Calculate sizes of two given linked lists.
3. Initialize two pointers first as first node of the first linked list and second as first node of the second linked list and these two linked lists travel in this linked list in the while loop.
4. Read variables addition, divide-by-10, reminder, carry.
5. To get the sum list, traverse these two linked lists.
6. While first is not null or second in not null, go to 7 otherwise go to 12.
7. If sizes of two linked list are the same, then calculate sum using recursion. Hold all nodes in recursion call stack till the rightmost node, calculate the sum of rightmost nodes and forward carry to the left side. Store the sum of first and second and carry in addition. Then divide addition by 10 and store it in the divide-by-10 and go to 6 otherwise go to 8.
8. If divide-by-10 is zero, create a new node in new linked list then store addition in it.
9. Else find remaining addition by 10 and store it in the reminder and store divide-by-10 in carry. Then create a new node in new linked list then store reminder in it and go to 5.
10. If sizes of two linked list are not same, find the differences between two linked lists. Then for solving this problem, as much as the difference between two linked lists, put zero in any empty place in which linked list that has a smaller size. In that case the sizes of two linked list are same and go to 5.
11. At the end before put null, if carry is not zero create a new node in new linked list and store carry in it and the put null.
12. End.

Chapter 6
Hash

Abstract Hashing is an efficient method to store and retrieve elements. The hash table uses key-value pairs to store data. These tables are implemented in some way in almost all programming languages due to their speed and efficiency. Hash tables are called objects in JavaScript and are called dictionaries in Python, and maps in Java, Scala, and Go. A hash table is a data structure that stores data in an interconnected manner. In a hash table, data is stored in the form of arrays in which each data value has its own unique array index value. In this data structure, if we know the desired data index, access to it will be very fast. Hence, this data structure becomes a structure in which insertion and search operations are performed very quickly, regardless of the size of the data. Hash tables use an array as a storage medium and use the hash technique to create an index through which an element is inserted or searched. This chapter provides 48 exercises on the hash methods.

6.1 Lecture Notes

Hashing is an efficient method to store and retrieve elements. The hash table uses key-value pairs to store data. These tables are implemented in some way in almost all programming languages due to their speed and efficiency. Hash tables are called objects in JavaScript and are called dictionaries in Python, and maps in Java, Scala, and Go.

A hash table is a data structure that stores data in an interconnected manner. In a hash table, data is stored in the form of arrays in which each data value has its own unique array index value. In this data structure, if we know the desired data index, access to it will be very fast.

Hence, this data structure becomes a structure in which insertion and search operations are performed very quickly, regardless of the size of the data. Hash tables use an array as a storage medium and use the hash technique to create an index through which an element is inserted or searched. The basic operations of a hash table are as follows.

© The Author(s), under exclusive license to Springer Nature Switzerland AG 2022   219
H. Izadkhah, *Problems on Algorithms*,
https://doi.org/10.1007/978-3-031-17043-0_6

1. Search: Looks for an element in the hash table.
2. Insert: Inserts an element in the hash table.
3. Delete: Deletes an element from the hash table.

In searching process, whenever an element needs to be inserted in the hash table, the hash code of the element is calculated and the position of the index is determined using the hash code as the array index. If the calculated hash code is not found in the array, the linear search method is used to find it.

Whenever an element is to be searched, compute the hash code of the key passed and locate the element using that hash code as index in the array. Use linear searching to retrieve the element ahead if the element is not found at the computed hash code.

In insert operation, whenever an element needs to be inserted in the hash table, the hash code of the element is calculated and the position of the index is determined using the hash code as the array index. If the calculated hash code is not found in the array, the linear search method is used to find it.

In delete operation, whenever we want to delete an element from a hash table, we compute the hash code of the key and find the index using the hash code as the array index. In cases where the calculated hash code is not present in the array, linear exploration is used to find the desired position. When the corresponding index is found, a dummy item is stored in it to keep the hash table function intact.

6.2   Exercises

6.2.1   Basic

1. Let the hash function is $x \ mod \ 10$. How many of the following statements are true for input $\{4322, 1334, 1471, 9679, 1989, 6171, 6173, 4199\}$?

   a. 9679, 1989, 4199 hash to the same value.
   b. 1471, 6171 hash to the same value.
   c. All elements hash to the same value.
   d. Each element hashes to a different value.

2. The keys 23, 42, 37, 26, and 41 are inserted into an empty hash table of length 5 using open addressing with hash function $h(k) = k \ mod \ 5$ and linear probing. What is the resultant hash table?

3. The keys 24, 32, 47, 51, 60, 27, and 76 are inserted into an empty hash table of length 7 using chaining with hash function $h(k) = k \ mod \ 5$ and linear probing. What is the resultant hash table?

4. If array T[0..7] and elements A-E are mapped to the addresses of the table below by linear probing, what is the average number of comparisons for successful (S) and unsuccessful (U) searches?

Element	A	B	C	D	E
Address	3	5	4	3	1

5. Which one of the following hash functions on integers will distribute keys most uniformly over 10 buckets numbered 0 to 9 for i ranging from 0 to 2020?

a. $h(i) = i^2 \bmod 10$
b. $h(i) = i^3 \bmod 10$
c. $h(i) = (11 \times i^2) \bmod 10$
d. $h(i) = (12 \times i) \bmod 10$

6.2.2 Applications

6. Let A and B are two arrays of size n and m, respectively, that elements in both array are distinct. Give four algorithms using the following methods to specify whether B is a subset of A or not.

(a) Use two loops
(b) Use sorting and binary search
(c) Use sorting and merging
(d) Use hashing

Compute their time and auxiliary complexities.

For example:
Input: A = {20, 1, 7, 11, 9, 4}, B = {4, 20, 1}
Output: Yes, B is a subset of A

7. Let A and B are two arrays of size n and m, respectively, that arrays can have duplicate elements. Give an algorithm to specify whether B is a subset of A or not.

For example:
Input: A = {20, 1, 7, 1, 9, 4}, B = {4, 20, 1}
Output: Yes, B is a subset of A

8. Let A and B are two linked lists. Give two algorithms as follows to find union and intersection of the lists.

(a) Use mergesort
(b) Use hashing

For example:
Input:
List1: $6 -> 11 -> 0 -> 15$
List2: $3-> 0 -> 1 -> 6$
Output:
Intersection List: $0 -> 6$
Union List: $1 -> 4 -> 15 -> 0 -> 11 -> 6$

Compute their time and auxiliary complexities.

9. Let A is an array of size n and s is an integer number. Give an algorithm that finds two elements in the array whose sum is exactly s. Print "No" if no such numbers are found. Note: Consider both the presence of duplicate elements and the absence of duplicate elements.

Input: $A = \{4, -1, 0, -1, 3\}, s = -2$
Output: $-1, -1$

10. Let A is an array of size n and s is an integer number. Give an algorithm that finds the number of pairs in the array whose sum is exactly s. Print "No" if no such numbers are found.

Example:
Input: $A = \{-1, 3, 5, 1\}, s = 4$
Output: 2
Pairs with sum 4 are $(1, 3)$ and $(5, -1)$

11. Let a, b, c, and d are distinct elements in a given array of size n. The task is to find two pairs (x, y) and (p, q) such that $x + y = p + q$. If there are a number of pairs, print them all.

Example:
Input: $\{2, 3, 6, 2, 3, 8, 7\}$
Output: $(3, 6)$ and $(2, 7)$

12. Let A be an array of size n. Give an algorithm to find the length of the longest sub-array whose sum equals to k.

Input: $A = \{11, -6, -2, -12, -3, 3, 6, 21\}; k = -20$
Output: 5
Sub-array: $\{-6, -2, -12, -3, 3\}$

13. Let A be an array of size n. Give an algorithm to find the duplicates in A. If there are no duplicates, print "No".

Example:
Input: {5, 5, 10, 10, 20, 20, 10, 20, 40, 2}
Output: 5 10 20

14. Let A be an array of size n. Give an algorithm to determine the most frequent element in the array. If there are multiple answers, print them all.

Example:
Input: A = {2, 5, 2, 3, 4, 3}
Output: 2 and 3
2 and 3 appear two times in array which is maximum frequency.

15. Let A be an array of size n. Give an algorithm to find the first element that occurs k times in the array. Print "No", if there are no such number.

Examples:
Input: {2, 8, 4, 0, 4, 8, 7}, k = 2
Output: 8

16. Let A be an array of size n. Give an algorithm to find the top three duplicated numbers. Print "No", if there are no such number.

Example:
Input: A = {9, 7, 5, 9, 19, 9, 17, 19, 17, 17, 19, 5, 9}
Output: Three largest elements are 9 19 17

17. Let A and B are two arrays of size n and m, respectively, that elements in both array are distinct (no duplicates). Give an algorithm to check if the arrays are disjoint or not?

Input: A = {5, 9, 11, 8, 1}
B = {6, 1, 7, 3}
Output: No disjoint, 1 is common in the two arrays.

18. Let A and B are two arrays of size n and m, respectively. Give an algorithm to count the minimum numbers that should be removed from the arrays so that both arrays become disjoint or don't contains any elements in common.

Example:
Input: A = {3, 6, 9}
B = {6, 10, 2, 5}
Output: 1

19. Let A and B are two arrays of size n and m, respectively. Give an algorithm find numbers which are present in A, but not present in B.

20. Let A and B are two arrays of size n. Give an algorithm to check whether the arrays are equal (contain the same set of elements) or not. Note: the arrangements of elements may be different though.

21. Let A be an array of size n containing distinct elements and a range [a, b]. Give an algorithm to find the missing elements in array A. An element is said missing element when is in range, but not in array.

Example:
Input: A = {8, 11, 13, 15},
a = 10, b = 15
Output: 12, 14

22. Let A is an array of size n. Give an algorithm to find the maximum number that is the product of the two elements in the array. Elements are within the range of 1 to 10^5. Print -1, If no such element exists.

Example:
Input : A = {12, 8, 5, 40, 38}
Output: 40
Explanation: 40 is the product of 8 and 5.

23. Let A and B are two linked lists of size n and m containing distinct elements and a value k. Give an algorithm to count all pairs from both lists (one element from each list) whose sum is equal to k.

Example:
Input: A = 5- >3- >7- >9
B = 10- >4- >7- >5
k = 14
Output: 2
The pairs are: (7, 7) and (9, 5)

24. Let A is an array of size n with distinct elements. For a given integer number k, give an algorithm to find the pairs in the array such that $a\%b = k$.

25. Given a string, write a program to find out maximum occurring character in the string e.g., if input string is "book" then program should print 'o'.

26. Let A is an array of size n. Give an algorithm to find out the first non-repeating element.

Example:
Input: 3 −1 −1 3 2 6 3
Output: 2

27. Let A is an array integers containing distinct elements. Give an algorithm to find out the largest p such that $x + y + z = p$, where all are distinct elements of A.

28. Let A is an array of size n containing only one repetitive element. Give an algorithm to find out the repetitive element.

29. Let A is an array of size n where every element occurs p times, except one element which occurs n times ($p > n$). Give an algorithm to find out the element that occurs n times.

30. Let A is an array of size n. Give an algorithm to count minimum number of partitions from the array such that no subset contains duplicate elements.

Examples:
Input: A = {5, 6, 7, 8}
Output :1
Explanation: A single subset can contain all values and all values are distinct

Input: A = {5, 6, 7, 7}
Output: 2
Explanation: [{1, 2, 3} , {3}] or [{1, 3} , {2, 3}]

31. Let A is an array of integers with any orders of elements and k is a given number. Give an algorithm to find out the number of subarrays that have sum equal to k.

Example: Input: A = {13, 5, −5, −23, 13},
$k = -10$
Output: 3
Subarrays: A[0..3], A[1..4], A[3..4]

32. Let A is an array of integers with any orders of elements and k is a given number. Give an algorithm to determine if array A can be divided into pairs such that sum of every pair is divisible by k.

Example:
Input: A = {9, 6, 2, 3}, k = 4
Output: Yes
The pairs are (9, 3) and (6, 2).

33. Let A, B, C are three arrays with any orders of elements and k is a given number. Give an algorithm to specify if there are three elements x, y, z, each belongs to one of the arrays, such that $x + y + z = k$.

34. Let A denotes an array of integers and k is a given positive integer. Give an algorithm to find out the indexes whose elements sum is k.

Example:
Input: A = {2, 4, 8, 3, 4, 5}, sum = 15
Output: indices 1 and 3; indices 2 and 4

35. Let A denotes an array of integers and k is a given positive integer. Give an algorithm to find out the longest subarray whose elements sum is divisible by k.

Example:
Input: A = {3, 3, 6, 2, 4, 1}, k = 5
Output: 4
The subarray is {3, 6, 2, 4} with sum 15, which is divisible by 5.

36. Let A denotes an array of integers. Give an algorithm to find the elements with the most repetition and the least repetition in the array.

Example:
Input: A = [1, 3, 4, 6, 4, 1, 1, 2, 2, 2, 5]
Output: 2
Most repetition element is 5.
Least repetition elements are 1 and 2.

37. Let S is a string of lowercase alphabets. Give an algorithm to check whether the string is isogram or not. A string is Isogram if no letter occurs more than once.

38. Suppose a dictionary of words with the same length and a pattern of the same length as the dictionary words are given. Give an algorithm to find out all words in the dictionary that match the given pattern.

Example:
Input:
dictionary = {qaa, pgq, xyz, xzz}
pattern = Coo
Output: qaa, xzz
Explanation: in the pattern indexes 1 and 2 are the same characters, so only qaa and xzz have the same characters at index 1 and 2.

39. Suppose a dictionary of words with different length and a pattern of arbitrary size are given. Give an algorithm to find out all words in the dictionary that match the given pattern.

40. Let A is an array of size n and integer number k is given. Give an algorithm to determine if there's a triplet in A which sums up to k.

Example:
Input: A = [3, 4, 7, 6, 1, 8], k = 8
Output:
{3, 4, 1}

41. Let A is an array of integers and integer number k is given. Give an algorithm to find out k elements which have the highest frequency in the array. If two numbers are repeated identically, the larger number must be selected.

42. Let A is an array of integers and integer number k is given. Give an algorithm to find the smallest number that is repeated exactly k times.

43. Let A and B are two arrays of size n and m, respectively. Give an algorithm to count common elements in the two arrays.

44. Let A is an array of integers and window number k is given. Give an algorithm to count distinct elements in every window of size k in the array.

Example:
Input:
k = 3
A = {2, 2, 1, 4, 3, 1, 5}
Output: 2 4 4 4 5

45. If both of the following conditions are met, the two strings are called *K-anagrams*.

1. Both have same number of characters.
2. A string can be changed to another string by changing at most K characters.

Suppose two strings of lowercase alphabets and a value k are given. Give an algorithm to determine whether two strings are K-anagrams of each other.

46. Let A and B are two unsorted arrays of size n and m, respectively. Give an algorithm to find union of these two arrays.

Example:
Input:
A = {10, 20, 30}
B = {30, 50}

Output:
{10, 20, 30, 50}

47. Let A and B are two unsorted arrays of size n and m, respectively. Give an algorithm to find intersection of these two arrays.

48. Let A and B are two sorted arrays of size n and m, respectively and a value sum is given. Give an algorithm to count all pairs from both arrays whose sum is equal to sum.

6.3   Solutions

1. $h(9679) = 9679\%10 = 9$
$h(1989) = 1989\%10 = 9$
$h(4199) = 4199\%10 = 9$
$h(1471) = 1471\%10 = 1$
$h(6171) = 6171\%10 = 1$
As we can see, 9679, 1989 and 4199 hash to same value 9. Also, 1471 and 6171 hash to same value 1.

2.

key	23	42	37	26	41
h(k)	3	2	2	1	1

First 32 enters hole 3, then 42 enters slot 2. Then 37 collides with 42 and because slot 3 is full, it enters slot 4. Then 26 enters slot 1 and finally 41 collides with 26 and enters slot zero.

3.

	E		A	C	B	D		

$$S = \frac{1+1+1+1+4}{5} = \frac{8}{5}, U = \frac{1+2+1+5+4+3+2+1}{8} = \frac{19}{8}$$

6(4).

1. Create a Hash Table for all the elements of array A.
2. Traverse array B and search for each element of B in the Hash Table. If element is not found then return 0.
3. If all elements are found then return 1.

8.

1. Start traversing both the lists.
   (a) Store the current element of both lists with its occurrence in the map.
2. For Union: Store all the elements of the hash table in the resultant list.
3. For Intersection: Store all the elements only with an occurrence of 2 as 2 denotes that they are present in both the lists.

9.

Use a hash table to check for the current array value A[i], if there is a value of $A[i] + x$ in the hash table the value of s is obtained. x is already stored in the hash table. Let's look at the following example.

$A[] = \{-1, -2, 1, -5, 0\}$

$s = -3$

Now start traversing:

Step 1: For '−1' there is no valid number '−2' in the hash table so store '−1' in hash table.

Step 2: For '−2' there is a valid number '−1' in the hash table. Store '-2' in hash table, and answer is $-2, -1$.

Step 3: For '1' there is no valid number '−3' in the hash table so store '1' in hash table.

Step 4: For '−5' there is no valid number '2' in the hash table so store '5' in hash table.

Step 5: For '0' there is no valid number '−3' in the hash table so store '0' in hash table.

1. Initialize an empty hash table HS.
2. Do following for each element A[i] in A[]
   (a) if there is a value of $A[i] + x$ (x is already stored in the hash table) that the value of s is obtained then print the pair (A[i], x)
   (b) Insert A[i] into HS.

14. We create a hash table and store elements and their frequency counts as key-value pairs. Finally, we traverse the hash table and print the key with the maximum value.

17.

1. Create an empty hash table.
2. Store every element of array A in the hash table.

3. Iterate through array B and check if any element is present in the hash table. If present, then returns false (no disjoint), else ignore the element.
4. If all elements of array B are not present in the hash table, return true.

19. Store all elements of array B in a hash table. One by one check all elements of array A and print all those elements which are not present in the hash table.

20. Store all elements of array A and their counts in a hash table. Then we traverse array B and check if the count of every element in B matches with the count in A.

21. Create a hash table and insert all array items into the hash table. Once all items are in hash table, traverse through the range and print all missing elements.

26. The x + y + z = p can be restated as finding x, y, z, p such that x + y = p - z.

1. Store sums of all pairs (x + y) in a hash table
2. Traverse through all pairs (p, z) again and search for (p - z) in the hash table.
3. If a pair is found with the required sum, then make sure that all elements are distinct array elements and an element is not considered more than once.

46.

1. Initialize an empty hash set hs.
2. Iterate through the first array and put every element of the first array in the set hs.
3. Repeat the process for the second array and update hs.
4. Print the set hs.

47.

1. Initialize an empty set hs.
2. Iterate through the first array and put every element of the first array in the set hs.
3. For every element x of the second array, do the following:
   search x in the set hs. If x is present, then print it.

48. Store all first array elements in hash table. For elements of second array, subtract every element from sum and check the result in hash table. If result is present, increment the count.

Chapter 7
Tree

Abstract A tree is a hierarchical data structure consisting of vertices (nodes) and edges that connect them together. Trees are similar to graphs, but the key difference is that they do not exist a cycle in a tree, unlike graphs. There are several types of trees, including the N-ary tree, Balanced tree, Binary tree, Binary search tree, AVL tree, and Red-Black tree. This chapter provides 227 exercises on the tree data structure which address different aspects of this data structure.

7.1 Lecture Notes

A tree is a hierarchical data structure consisting of vertices (nodes) and edges that connect them together. Trees are similar to graphs, but the key difference is that they do not exist cycle in a tree, unlike graph. There are several types of trees, including

- N-ary tree
- Balanced tree
- Binary tree
- Binary search tree
- AVL tree
- Red Black tree

7.2 Exercises

7.2.1 Tree

1. A tree is non-linear and a hierarchical data structure consisting of a collection of nodes such that each node of the tree stores a value, a list of references to nodes (the "children"). The basic terminologies in tree data structure are as follows:
Parent Node, Child Node, Root Node, Degree of a Node, The degree of a tree, Leaf

© The Author(s), under exclusive license to Springer Nature Switzerland AG 2022
H. Izadkhah, *Problems on Algorithms*,
https://doi.org/10.1007/978-3-031-17043-0_7

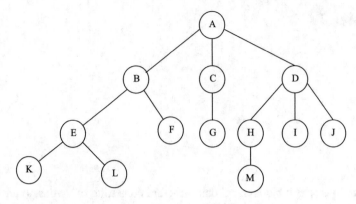

Fig. 7.1 A sample tree

Node or External Node, Ancestor of a Node, Descendant, Sibling, Depth of a node, Height of a node, Level of a node, Internal node, Neighbor of a Node.

Define all of these and also give examples using the tree shown in Fig. 7.1.

2. Unlike array, linked list, stack, and queue which are linear data structures, tree is hierarchical (or non-linear) data structure. True/False?

3. Explain how a tree can be implemented using linked lists.

7.2.2   Binary Tree

A tree whose elements have at most two children is called a binary tree, called the left and right child.

4. For binary tree:

(a) Show the maximum number of nodes at level 'l' of a binary tree is 2^l.
(b) Show the maximum number of nodes in a binary tree of height 'h' is $2^h - 1$.
(c) Show in a binary tree with n nodes, the minimum number of levels (heights) is $Log_2(n + 1)$?

5. Because there are zero or two children in each node, the number of leaf nodes is always one more than the nodes with two children in a binary tree. True/False?

6. The following are common types of binary trees.

Full Binary Tree, Complete Binary Tree, Perfect Binary Tree, Balanced Binary Tree, A degenerate (or pathological) tree.

Define all of these as well as give examples.

7. Regarding binary tree:

(a) How many different unlabelled binary trees can be made with three nodes?
(b) How many different unlabelled binary trees can be there with n nodes?
(c) How many labelled binary trees can be there with n nodes?

8. Show for each binary tree, if n_0 is the number of leaves and n_2 is the number of nodes of degree 2, then we have:

$$n_0 = n_2 + 1$$

9. If in a binary tree, the total number of nodes is 17 and the number of nodes with degree 2 is 6, get the number of nodes with degree 1.

10. Regarding implementation of binary tree:

(a) Explain how a binary tree can be stored in an array?
(b) Explain how a binary tree can be implemented using a linked list?

11. In a complete binary tree with n nodes, what is the maximum distance between two nodes?

12. Given a binary tree, give an algorithm to check whether

(a) it is a left/right skewed binary tree or not. In a left skewed binary tree (it is also called a left side dominated tree), all the nodes are having a left child or no child at all and in a right skewed binary tree (it is also called a right side dominated tree) all the nodes are having a right child or no child at all.
(b) it is a full binary tree or not.
(c) it is complete or not.
(d) it is pure binary.
(e) it is an even-odd tree or not.

13. Inorder, preorder, and postorder are three ways for traversing binary trees. Write them for Fig. 7.2.

14. Write inorder, preorder, and postorder traversals for Fig. 7.3.

15. Write inorder, preorder, and postorder traversals for Fig. 7.4.

16. Assume to construct the binary tree, you are given two traversal sequences. Which of the following is correct?

Fig. 7.2 A sample tree

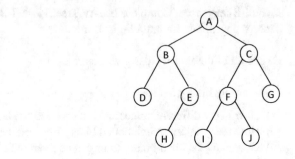

Fig. 7.3 A sample tree

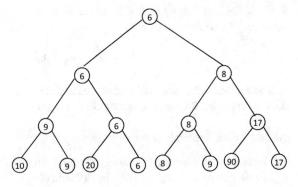

Fig. 7.4 A sample tree

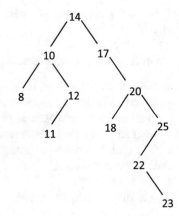

a. {Inorder and Preorder}, {Inorder and Postorder}, and {Inorder and Level-order} can uniquely identify a tree.

b. {Postorder and Preorder}, {Preorder and Level-order}, and {Postorder and Level-order} can not uniquely identify a tree.

17. The inorder and preorder traversal of a binary tree are as follows

Inorder : *DBHEAIFJCG*

$$Preorder: ABDEHCFIJG$$

Draw the desired binary tree.

18. Consider the following traversals of a binary tree

$$inorder = \{dbeafcg\}$$

$$preorder = \{abdecfg\}$$

The postorder of the binary tree made from these traversals is:

a. d e b f g c a
b. e d b g f c a
c. e d b f g c a
d. d e f g b c a

19. True or False? If the inorder and preorder sequences of a binary tree exist, then:

a. A binary tree can be made but it will not be unique.
b. It is not possible to build a binary tree.
c. A binary tree can be made and it will be unique.
d. A binary tree can be made but two different trees will be produced.

20. If the inorder traversal of a full binary tree is as follows, what is its preorder traversal?

$$dbeagcf$$

21. The inorder and preorder traversals of a binary tree are as follows

$$inorder: GFHKDLAWRQPZ$$

$$perorder: ADFGHKLPQRWZ$$

Draw the desired binary tree.

22. The inorder and postorder traversal of a binary tree are as follows

$$inorder: GFHKDLAWRQPZ$$

$$postorder: FGHDALPQRZWK$$

Draw the desired binary tree.

23. The depth of a binary tree equal to the arithmetic expression $(-a) * b * c - d/e * g + h$ is equal to ....

24. Build a binary tree equivalent to $(x - y) * z + (y - w) - x$.

25. The number of nodes in a binary tree that represents a mathematical expression is 14. The operators of this expression are binary or unary. Which of the following is true?

a. This expression must have odd numbers of binary operators.
b. This expression must have a unique pair of operators.
c. This expression cannot have a binary operator.
d. There is at least one unique operator in this expression.

26. In a binary tree, what is the time complexity of the preorder, postorder, and inorder?

27. There is a tree with n nodes in which there are 5 nodes of degree 4, 7 nodes of degree 3 and 4 nodes of degree 2. How many leaves does this tree have?

28. Write a recursive algorithm to determine the following:

(a) The number of nodes or the number of elements in a binary tree,
(b) The number of leaves of a binary tree,
(c) Number of non-leaf nodes,
(d) The sum of the contents of all nodes in a binary tree,
(e) The depth of the binary tree,
(f) The smallest element of the tree.

29. Draw a binary tree for the following math expressions. Then traverse the tree and print their prefix and suffix phrases:

(a) $(A + B) - C$
(b) $A + (B - C)$
(c) $A * (B + C) - (D - (E/F))$
(d) $(A/B - D) * (C + E)$

30. Two binary trees are similar when they are both empty, or if they are non-empty, their left subtree is similar and their right subtree are similar. Write an algorithm that determines whether two binary trees are similar.

31. Check whether the two binary trees are Identical or Not. Two trees are identical if they are identical structurally and nodes have the same values.

32. What does the following function do on a given binary tree?

```
int fun(node node *tree)
{
  if(tree == NULL)
    return 0;
  if(tree− > left == NULL && tree− > right == NULL)
    return 0;
  return 1 + fun(tree− > left) + fun(tree− > right);
}
```

a. Counts leaf nodes
b. Counts internal nodes
c. Determines height (the number of edges in the path from the root to the deepest node) of the tree
d. Returns diameter (the number of edges on the longest path between any two nodes)

33. What does the following function do on a given binary tree?

```
int fun(node node *tree)
{
  if(tree == NULL)
    return 0;
  if(tree− > left == NULL && tree− > right == NULL)
    return 1;
  return fun(tree− > left) + fun(tree− > right);
}
```

a. Counts leaf nodes
b. Counts internal nodes
c. Determines height (the number of edges in the path from the root to the deepest node) of the tree
d. Returns diameter (the number of edges on the longest path between any two nodes)

34. What does the following function do on a given binary tree?

```
int fun(node node *tree)
{
  if(tree == NULL)
    return 0;
  if(tree− > left == NULL && tree− > right == NULL)
    return 1;
  return fun(tree− > left) + fun(tree− > right) + 1;
}
```

35. Explain breadth first traversal of a binary tree (Level Order Binary Tree Traversal).

36. About diameter of a binary tree:

(a) Define diameter of a binary tree.
(b) Draw a complete binary tree with 12 nodes and find its diameter.
(c) Give an algorithm to find the diameter of a binary tree.
(d) Analyze its time complexity.

37. Implement inorder, preorder, and postorder traversals of a binary tree with and without stack.

38. *Threaded binary tree*. The idea is to effectively manage the space of the tree, because in a binary tree of the *2n* total connections, the number of $n + 1$ connections is NULL values which results in wastage of storage space.

A binary tree can be threaded in several ways. This threading depends on the tree traversal method. For example, a threaded binary tree is defined by the inorder traversal as follows:

"For each node, the empty connection on the right points to the next node in the inorder traversal, and the empty connection on the left points to the previous node in the inorder traversal."

There are two types of threaded binary trees.

– Single Threaded
– Double Threaded

Using a sample binary tree shows how to convert a given binary tree to threaded binary tree.

39. Explain inorder traversal in a threaded binary tree.

40. Convert a binary tree to threaded binary tree (use queue).

41. How to insert and delete in single binary threaded tree.

42. How to insert and delete double binary threaded tree.

43. Given a binary tree and a key, write a function that prints all the ancestors of the key in the given binary tree.

44. Which of the following statements is true about binary tree?

a. Every binary tree is either complete or full.
b. Every complete binary tree is also a full binary tree.
c. Every full binary tree is also a complete binary tree.
d. No binary tree is both complete and full.
e. None of the above

45. The sequence of keys 11, 10, 24, 23, 28, 26, 16, 51, 96, 61, 41, 30 indicates postorder traversal of a given binary search tree. The inorder traversal of this tree is

Fig. 7.5 A sample tree

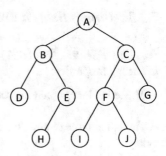

a. 10, 11, 15, 23, 24, 26, 28, 30, 41, 51, 61, 96
b. 10, 11, 16, 23, 41, 51, 61, 96, 24, 26, 28, 30
c. 30, 16, 10, 11, 26, 23, 24, 28, 41, 61, 51, 96
d. 96, 51, 61, 41, 28, 24, 23, 26, 11, 10, 16, 30

46. Threaded binary trees make inorder traversal faster without stack and recursion.
True/False?

47. Consider Fig. 7.5. Make a threaded binary tree by inorder traversal.

48. Explain how to convert any m-ary tree (general tree) to a binary tree.

49. Convert general tree shown in Fig. 7.6 to a binary tree.

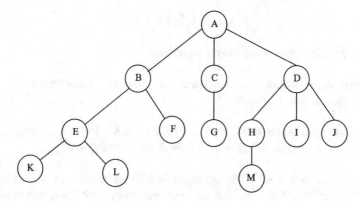

Fig. 7.6 A sample tree

7.2.3   Binary Search Tree

50. *Binary Search Tree (BST)*. BST is a binary tree data structure which has the following properties:

- The left subtree of a node can only contain nodes with keys smaller than the node key.
- The left subtree of a node can only contain nodes with keys greater than the node key.
- The left and right subtree each must also be a binary search tree.

Give three examples of such tree.

51. Suppose the following numbers enter a binary search tree from the left, respectively. Get the resulting tree?

$$5, 10, 8, 3, 14, 1, 13, 4, 16$$

52. Suppose the following numbers enter a binary search tree from the left, respectively. Get the resulting tree height?

$$3, 5, 8, 2, 1, 9, 12, 4, 6, 7$$

53. Suppose the following numbers enter a binary search tree from the left, respectively. Which node is deeper?

$$3, 5, 8, 2, 1, 9, 12, 4, 6, 7$$

54. How to sort an array using binary search tree?

55. 16 elements are stored in a binary search tree, how many comparisons are needed to search for a desired element?

56. If you enter the numbers 44, 10, 8, 60, 82, 23, 65, 45, 15 from left to right in an empty search binary tree. What is the height of the resulting tree?

57. Suppose the numbers 1–1000 are stored in a binary search tree and we want to find the number 363. Which of the following sequences (left to right) cannot indicate the order of access to tree elements in this search? Draw the tree of all options.

 a. 925, 202, 911, 240, 912, 245, 363
 b. 924, 220, 911, 244, 898, 258, 362, 363
 c. 2, 252, 401, 398, 330, 344, 397, 363
 d. 2, 399, 387, 219, 266, 382, 381, 278, 363

58. Employing an example, explain how to

Fig. 7.7 A sample tree

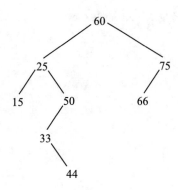

(a) search a key in a BST.
(b) insert a key in a BST.
(c) delete a key in a BST.

59. In a BST, explain

(a) how to delete a leaf node.
(b) how to delete a node that has only one child
(c) how to delete a node that has two children

60. In the binary search tree, the operations like search, minimum and maximum can be performed fast. True/False?

61. n elements with the keys $k_1, k_2, \ldots, k_n$ are entered in an empty binary search tree. Which of the following is true?

a. The resulting tree is unique and is not dependent on the order of the inserted keys.
b. The smallest key can not be the root.
c. The height of the tree is the minimum if the keys are inserted in ascending order.
d. None of the above options are correct.

62. Consider binary search tree shown in Fig. 7.7.

(a) Delete node 44.
(b) Delete node 75.
(c) Delete node 25.

63. Consider binary search tree shown in Fig. 7.8.

(a) Delete the node with key 10.
(b) Delete the node with key 20.

64. The worst-case time complexity of search and insertion operations is $O(h)$ where h is the height of the binary search tree. True/False?

Fig. 7.8 A sample tree

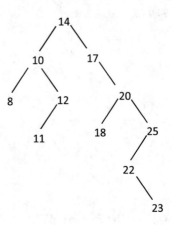

65. What is the worst case time complexity for search, insertion, and deletion operations in a binary search tree?

   a. O(n) for all
   b. O(Log n) for all
   c. O(Log n) for search and insert, and O(n) for delete
   d. O(Log n) for search, and O(n) for insert and delete

66. Suppose we want to delete a node from a BST tree that has a left and right
child. Which of the following is true about inorder successor sequences required in
a deletion operation?

   a. Inorder successor is always a leaf node.
   b. Inorder successor is always either a leaf node or a node with empty left child.
   c. Inorder successor may be an ancestor of the node.
   d. Inorder successor is always either a leaf node or a node with empty right child.

67. Inorder traversal of BST always produces sorted output. True/False?

68. We can construct a BST with only Preorder or Postorder or Level Order traversal. Note that we can always get inorder traversal by sorting the only given traversal.
True/False?

69. Number of unique BSTs with n distinct keys is *Catalan Number*. True/False?

70. Consider a set of n information elements, $A_1, A_2, \ldots, A_n$. Suppose we want to
find all the duplicate elements in this set and delete them. Using BST, devise an
algorithm to perform this operation.

71. If we want to remove duplicate data from the list, what data structure do we use
for that list?

Fig. 7.9 A sample tree

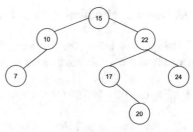

Fig. 7.10 A sample tree

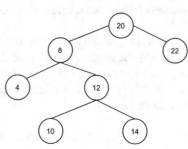

a. Binary search tree
b. Heap tree
c. Stack
d. Queue

72. Consider binary tree shown in Fig. 7.9.

a. It is a binary search tree.
b. It is a complete tree.
c. It is a full tree.
d. None of the above.

73. Write a program that creates a threaded binary search tree and then traverse it.

74. How to find a node with minimum value in a binary search tree.

75. Using inorder traversal, give an algorithm to check if a binary tree is BST or not.

76. True/False? In binary tree, inorder successor of a node is the next node in inorder traversal of the binary tree. Consider binary search tree shown in Fig. 7.10.

(a) inorder successor of 8 is 10.
(b) inorder successor of 10 is 12.
(c) inorder successor of 14 is 20.

77. Find kth smallest element in BST (Order Statistics in BST).

78. Devise an algorithm to merge two BSTs with limited extra space.

79. Suppose a binary tree is given. Provide an algorithm that transforms the binary tree into a binary search tree while preserving its original structure.

80. Explain how to find the second largest element in a BST.

81. The second largest element is the second last element in inorder traversal and the second element in the reverse inorder traversal. True/False?

82. In a BST, all the keys to the left of a key must be smaller and all the keys to the right must be larger. According to this definition, in a binary search tree the values of all keys are distinct. Suppose we change the definition of the binary search tree so that it is possible to add duplicate keys. Explain how to insert and delete in this new binary search tree.

83. Give an example for left-rotation and right-rotation in BST tree.

– Left-Left rotation
– Right-Right rotation
– Right-Left rotation
– Left-Right rotation

84. Two binary search trees with the same number of elements can be converted to each other by rotation. True/False?

85. If L_x means rotation to the left x times and R_x means turning to the right x times. What sequence of rotations turns the left tree into the right tree?

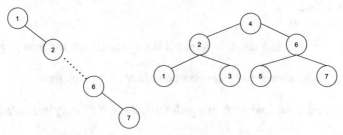

a. L_3, L_2, L_1, L_1, L_5
b. L_5, L_2, L_4, L_4, L_4
c. R_6, R_2, L_3, L_2, L_1
d. R_4, R_3, R_2, R_2, R_6

86. What is the balanced BST and how to construct it from a BST.

87. Devise an algorithm to convert a sorted array to a balanced BST.

88. Devise an algorithm to merge two balanced binary search trees.

89. Find median of BST in $O(n)$ time and $O(1)$ space.

90. How to remove BST keys outside the given range.

91. Write a function to count BST subtrees that lie in given range.

92. Write a function to sum the k smallest elements in BST.

93. Write a function to convert a BST to threaded binary search tree.

94. Explain how to delete a given key from threaded binary search tree.

95. Why Red-Black Trees?

96. Explain search, deletion, and insertion operations in Red-Black Tree.

97. In a Red-Black tree that is initially empty, we perform the following insertion and deletion. Draw the resulting tree after each step.

(a) insertion- A, F, E, D, C, B, G, H, I
(b) deletion- A, F, I, H, B, C, G

98. If the numbers 1, 2, 3, 8, 6 are entered in an empty Red-Black tree, how many rotations are performed?

99. The height of a Red-Black tree is always $O(log\ n)$ where n is the number of nodes in the tree. True/False?

100. Time complexity of search in a Red-Black tree is always $O(logn)$. True/False?

101. Time complexity of delete in a Red-Black tree is always $O(logn)$. True/False?

102. Time complexity of insert in a Red-Black tree is always $O(logn)$. True/False?

103. Show that two Red-Black trees with n nodes can be converted to each other with a maximum of $2n - 2$ rotations.

104. Compare Red-Black tree with AVL?

105. The AVL trees are more balanced compared to Red-Black trees, but they may cause more rotations during insertion and deletion. So if your application involves frequent insertions and deletions, then Red-Black trees should be preferred. True/False?

106. If an application requires fewer insertion and deletion operations and the search is a more frequent operation, the AVL tree should be preferred to the red-black tree. True/False?

107. Because each Red-Black tree is a special case of a binary tree, the Red-Black tree search process is similar to a binary tree. True/False?

108. Give a Red-Black tree that is not structurally AVL tree.

109. Most of the self-balancing BST library functions like map and set in C++ (or TreeSet and TreeMap in Java) use Red-Black tree. True/False?

110. MySQL uses the Red-Black tree for indexes on tables. True/False?

111. A BST of n nodes is balanced if height is in $O(logn)$. True/False?

112. There are many versions of balanced BSTs, including AVL trees, 2-3 trees, and Red-Black trees. True/False?

113. AVL tree is a self-balancing binary search tree where the difference between height of the left and right subtrees cannot be more than one for all nodes. True/False?

114. The height of an AVL tree is always $O(log\ n)$ where n is the number of nodes in the tree. True/False?

115. Why AVL Trees?

116. Explain search, deletion, and insertion operations in AVL Trees?

117. How many nodes are there in an AVL tree with a height of 4?

118. How many AVL trees can be made with keys 1, 2, and 3?

119. An AVL tree with n nodes has height $O(log\ n)$. True/False?

120. Left-Left rotation is performed when a new node is inserted at the left child of the left subtree. True/False?

121. Right-Left rotation is performed when a new node is inserted at the right child of the left subtree. True/False?

122. The valid values of the balance factor are -1, 0, and $+1$. True/False?

123. Let BF denotes the balanced factor for a node. In insertion of a node we have

a. If BF(node) $= +2$ and BF(node $- >$ left-child) $= +1$, perform LL rotation.
b. If BF(node) $= -2$ and BF(node $- >$ right-child) $= 1$, perform RR rotation.
c. If BF(node) $= -2$ and BF(node $- >$ right-child) $= +1$, perform RL rotation.
d. If BF(node) $= +2$ and BF(node $- >$ left-child) $= -1$, perform LR rotation.

124. Let BF denotes the balanced factor for a node. In deleting from the right subtree we have

a. If BF(node) $= +2$ and BF(node $- >$ left-child) $= +1$, perform LL rotation.
b. If BF(node) $= +2$ and BF(node $- >$ left-child) $= -1$, perform LR rotation.
c. If BF(node) $= +2$ and BF(node $- >$ left-child) $= 0$, perform LL rotation.

125. Let BF denotes the balanced factor for a node. In deleting from the left subtree we have

a. If BF(node) $= -2$ and BF(node $- >$ right-child) $= -1$, perform RR rotation.
b. If BF(node) $= -2$ and BF(node $- >$ right-child) $= +1$, perform RL rotation.
c. If BF(node) $= -2$ and BF(node $- >$ right-child) $= 0$, perform RR rotation.

126. Let $N(h)$ denotes the minimum number of nodes in a tree of height h. Show that

$$N(h) = N(h-1) + N(h-2) + 1 \quad N(0) = 1, \ N(1) = 2, \ N(2) = 4$$

127. AVL trees are binary search trees in which the difference between the height of the left and right subtree is either -1, 0, or $+1$. True/False?

128. To make sure that the given tree remains AVL after every insertion, some re balancing are needed. True/False?

129. Because only a small number of pointers change in rotation operations (left and right rotations), these operations require a constant time. True/False?

130. The time complexity of the AVL deletion operation is similar to that of the BST deletion operation, both of which are $O(h)$, which is h the height of the tree. True/False?

131. Let n denotes the node of a tree. Since AVL tree is balanced, the height is $O(\log n)$. Hence, the time complexity of AVL deletion operation is $O(\log n)$. True/False?

132. AVL trees keep themselves balanced by performing rotations. True/False?

133. After each operation in the AVL tree, if the balance becomes -2 or 2, we must make at least one rotation to rebalance the tree. True/False?

134. Write a recurrence relation for the number of nodes in AVL Tree.

135. Write insertion time complexity in the following data structures.

Unsorted Array, Sorted Array, Unsorted Linked List, Sorted Linked List

Balanced Binary Search Tree

136. Explain search and insert operations in 2-3 Trees.

137. The perfectly balanced property enables the 2-3 tree operations (insertion, deletion, and search) to have a time complexity of $O(log(n))$. True/False?

138. Explain search and insert operations in 2-3-4 Trees.

139. 2-3-4 trees allow nodes to have 2, 3, or 4 children, they can always be perfectly balanced. True/False?

140. Give search, insertion, and deletion operations time complexity for 2-3-4 trees and AVL tree.

141. Give search, insertion, and deletion operations time complexity for AVL tree, 2-3 tree, and B-tree.

142. Let n denotes the total number of keys in a 2-3-4 tree. If $n \geq 1$, the height is $\leq logn$. True/False?

143. In 2-3-4 trees, all leaves are at the same level. True/False?

144. Like 2-3 trees, the worst case time complexity of 2-3-4 trees in search, insertion, and deletion operations is $O(log\ n)$, where n denotes the number of nodes. True/False?

145. Although a 2-3-4 tree with n nodes has usually less height than a 2-3 tree with n nodes, that is not the advantage that it has over the 2-3 tree. The advantage is the greater ease in keeping the tree balanced. True/False?

146. A ScapeGoat tree is a self-balancing binary search tree like AVL tree, Red-Black tree, Splay tree. True/False?

147. The worst case time complexity of binary search tree operations such as search, deletion, insertion occurs when the tree is skewed, which is $O(n)$, where n denotes the number of nodes. True/False?

148. Give search, insertion, and deletion operations time complexity in Splay tree.

149. Treap (cartesian tree) data structure combines binary tree and binary heap. Explain it and give search, insertion, and deletion operations time complexity in Treap tree.

150. Another name of Treap data structure—randomized binary search tree. True/ False?

151. The expected time complexity of search, insertion and deletion operations in Treap is $O(log\ n)$, where n denotes the number of nodes. True/False?

152. A treap provides the following operations:

Insert (X) in O(logN).
Adds a new node to the tree.

Search (X) in O(logN).
Looks for a node with the specified key value X.

Erase (X) in O(logN).
Looks for a node with the specified key value X and removes it from the tree.

Build (X_1, ..., X_N) in O(N).
Builds a tree from a list of values. This can be done in linear time (assuming that X_1, ..., X_N are sorted). We will just use N serial calls of Insert operation, which has O(NlogN) complexity.

Union (T_1, T_2) in O(Mlog(N/M)).
Merges two trees, assuming that all the elements are different. It is possible to achieve the same complexity if duplicate elements should be removed during merge.

Intersect (T_1, T_2) in O(Mlog(N/M)).
Finds the intersection of two trees (i.e. their common elements).

   In addition, due to the fact that a treap is a binary search tree, it can implement other operations, such as finding the kth largest element or finding the index of an element. Implement these operations.

153. An order-statistics tree is an augmented version of a binary search tree that supports the additional operations like *Rank(x)*, which returns the rank of x (i.e., the number of elements with keys less than or equal to x) and *FindByRank(k)*, which returns the kth smallest element of the tree. Implement these operations.

154. Implement insertion and deletion operations in interval tree.

155. Give search, insertion, and deletion operations time complexity for interval tree.

156. Extend the intervalSearch() to print all overlapping intervals instead of just one.

157. Compare interval tree versus segment tree.

158. What is difference between B-tree and B^+ tree.

159. How a binomial tree works?

160. A binomial heap is the collection of binomial trees, and every binomial tree follows the min-heap property. True/False?

161. We know that a recurrence for the min number of nodes in an AVL tree with height h can be written as: $N(h) = N(h-1) + N(h-2) + 1$. We would like to know why $N(h) = F(h+2) - 1$ (Fibonacci sequence).

162. True or False?

- The worst case time complexity of search operation in binary tree and binary search tree is O(n).
- The worst case time complexity of search operation in AVL tree is O(log n).

7.2.4 Heap

163. A heap is a special tree data structure in which the tree is a complete binary tree. True/False?

164. Binary heaps are a common way to implement priority queues. True/False?

165. A binary tree is called a min-heap if (1) the data in each node is less than (or equal to) the data of the children in that node, (2) the binary tree is complete. True/False?

166. A binary tree is called a max-heap if (1) the data in each node is greater than (or equal to) the data of the children in that node, (2) the binary tree is complete. True/False?

167. Explain how to insert a new element into a min-heap.

168. Explain how to insert a new element into a max-heap.

169. If the input numbers are as follows, build the corresponding min-heap.

$$24, 43, 12, 58, 92, 45$$

170. If the input numbers are as follows, build the corresponding max-heap.

$$24, 43, 12, 58, 92, 45$$

171. What is the height of a heap with n elements?

a. $n - 1$
b. $\lceil log\, n \rceil$
c. $\lceil log(n + 1) \rceil$
d. $\lfloor log\, n \rfloor$

172. What is the number of leaves of a heap with n elements?

a. $n - 1$
b. $n - 2$
c. $\lceil n/2 \rceil$
d. $\lfloor n/2 \rfloor$

173. Which of the following arrays is not a min-heap?

a. 2, 3, 4, 8, 7, 6, 5, 9, 10
b. 2, 5, 8, 10, 13, 12, 22, 50, 11
c. 3, 4, 5, 6, 9, 8, 50, 10, 12, 7, 11
d. 2, 3, 2, 3, 3, 2, 3, 3, 3, 3, 3, 2

174. Which one of the following array represents a max-heap, which is implemented using an array?

a. 26, 13, 17, 14, 11, 9, 15
b. 26, 13, 17, 14, 11, 9, 15
c. 26, 15, 17, 14, 11, 9, 13
d. 26, 15, 13, 14, 11, 9, 17

175. The array [40, 30, 20, 10, 15, 16, 17, 8, 4] shows a max-heap tree. After inserting the value 35 to this heap, the new heap will be

a. 40, 30, 20, 10, 35, 16, 17, 8, 4, 15
b. 40, 35, 20, 10, 30, 16, 17, 8, 4, 15
c. 40, 30, 20, 10, 15, 16, 17, 8, 4, 35
d. 40, 35, 20, 10, 15, 16, 17, 8, 4, 30

176. Consider the following array of integers.

$$\{99, 29, 70, 37, 32, 35, 12, 15, 17, 21, 16, 19, 110\}$$

The minimum number of interchanges needed to convert this array into a max-heap is ….

177. The minimum number of swaps required to convert the array [89, 19, 40, 17, 12, 10, 2, 5, 7, 11, 6, 9, 70] into a max-heap is ….

178. Which of the following arrays forms a heap?

 a. {23, 17, 14, 6, 13, 10, 1, 12, 7, 5}
 b. {23, 17, 14, 6, 13, 10, 1, 5, 7, 12}
 c. {23, 17, 14, 7, 13, 10, 1, 5, 6, 12}
 d. {23, 17, 14, 7, 13, 10, 1, 12, 5, 7}

179. How many min-heaps can be built with seven different numbers?

180. Write a program that reads numbers from the input and stores them in a min-heap (as an array) and then print the elements of this array.

181. Write a program that reads numbers from the input and stores them in a max-heap (as an array) and then print the elements of this array.

182. K'th largest element in an array can be efficiently found using heaps. True/False?

183. Sorting an almost sorted array can be efficiently performed using heaps. True/False?

184. Merging K sorted arrays can be efficiently performed using heaps. True/False?

185. Given an array representation of min-heap, convert it to max-heap in $O(n)$ time.

186. Given an array representation of BST, convert it to max-heap.

187. Given an array representation of BST, convert it to min-heap.

188. Given a max-heap, devise an algorithm to find the minimum element in the heap.

189. Give an algorithm to find kth largest element in a max-heap.

190. Give time complexity of building a heap.

191. In a binary max-heap containing n numbers, the smallest element can be found in time

 a. O(n)
 b. O(log n)
 c. O(loglog n)

d. O(1)

192. Suppose two max-heaps are given each the size of n elements. What is the minimum possible time complexity to create a max-heap of the elements of these two max-heaps?

a. O(n log n)
b. O(n loglog n)
c. O(n)
d. O(n log n)

193. Which of the following min-heap operation has the worst time complexity?

a. Inserting an item in the heap
b. Merging with another heap
c. Deleting an item from the heap
d. Decreasing value of a key

194. Consider a max-heap which is represented by an arrays as [25, 14, 16, 13, 10, 8, 1]. The content of the array after two deletion operations is

a. 14, 13, 8, 12, 10
b. 14, 12, 13, 10, 8
c. 14, 13, 12, 8, 10
d. 14, 13, 12, 10, 8

195. The following array is the implementation of a heap tree. We want to put the number 94 in it. After adding the number 94, how many moves must be made for the tree to remain heap?

| 99 | 89 | 81 | 84 | 73 | 74 | 72 | 67 | 69 |

196. You are given an array containing 1023 elements, representing a min-heap. The minimum number of comparisons required to find the maximum in this heap is ....

197. Insertion into heap takes $O(log\ n)$ time, but for sorted array it is $O(n)$. True/False?

198. Extract min or (max) takes $O(log\ n)$ in heap, but for sorted array it takes $O(n)$. True/False?

199. Find min or (max) takes O(1) in heap, but in array it takes $O(n)$ linear search. True/False?

200. Devise an algorithm to check if a given array represents a binary heap?

201. Using the following methods, write a program for printing k largest elements in an given array of any order.

(a) Use bubble k times
(b) Use temporary array
(c) Use sorting
(d) Use max-heap
(e) Use order statistics
(f) Use min-heap

202. Let a binary search tree is a complete binary tree. Write a program that converts a binary search tree into a special max-heap with the condition that all the values in the left subtree of a node must be less than all the values in the right subtree of the node.

203. Suppose the level order traversal of a complete binary tree is given. Write a function to determine whether the given binary tree is a min-heap.

204. Assume k linked lists each of size n are given, where each list is sorted in non-decreasing order. Write a function to merge these lists into a single non-decreasing sorted linked list and print the sorted linked list as output. (Hint: Use min-heap)

205. A max-heap with n elements is implemented as an array. Explain what is the most appropriate way to find the minimum element in this data structure?

206. Consider array A of size n, which is used to store the elements of a complete binary tree. Suppose we have an efficient algorithm to check if this binary tree is a heap. What is the complexity of this algorithm in the worst case? Write this algorithm.

207. Complete the following table in terms of time complexity (Big-O time).

	Search	Insertion	Deletion
Binary Tree			
Binary Search Tree			
AVL Tree			
Heap Tree			

208. A k-ary min/max heap is a generalization of binary heap ($k = 2$) in which instead of 2 children, each node has k children, from left to right at each level of the tree. The insertion and deletion operations in k-ary min/max heap are similar to a binary min/max heap tree. Which one of the following represents a 3-ary max heap?

a. 2, 4, 6, 7, 9, 10
b. 10, 7, 4, 1, 8, 6
c. 10, 4, 7, 9, 6, 2
d. 10, 6, 7, 9, 4, 2

209. Heapsort uses binary heap (min or max heap) to sort an array in $O(n\log n)$ time. True/False?

210. Priority queues can be efficiently implemented using binary heap data structure because this data structure supports *insert()*, *delete()*, *extractmax()*, and *decreaseKey()* operations in $O(log\ n)$ time. True/False?

211. We can implement a priority queue by a max-heap. Assume the level order traversal of the heap containing five elements is $11, 9, 6, 4, 3$. The level order traversal of the heap after the insertion of the two new elements '2' and '8' in that order is

a. $11, 9, 8, 6, 4, 3, 2$
b. $11, 9, 8, 3, 4, 2, 6$
c. $11, 9, 8, 2, 3, 4, 6$
d. $11, 9, 8, 4, 3, 2, 6$

212. The priority queues are especially used in graph algorithms such as Dijkstra's algorithm (finding shortest path) and Prim's algorithm (finding minimum spanning tree). True/False?

213. Why is binary heap preferred over BST for priority queue?

214. Binomial heap and fibonacci heap are variations of binary heap. True/False?

215. Explain deletion, extract min and decrease key operations in fibonacci heap.

216. What is difference between binary heap, binomial heap, and fibonacci heap.

217. Complete the following table in terms of time complexity (Big-O time).

	Search	Insertion	Deletion
Binary Heap			
Binomial Heap			
Fibonacci Heap			

7.2.5 Applications

218. Suppose there is a sequence of n values $a_1, a_2, \ldots, a_n$. The goal is to respond quickly to the queries of the form: given i and j, find the smallest value in $a_i, \ldots, a_j$.

(a) Design a data structure that uses $O(n^2)$ space and responds the queries in $O(1)$ time.
(b) Design a data structure that uses $O(1)$ space and responds the queries in $O(log\ n)$ time.

219. *Disjoint set*—A disjoint-set data structure is a data structure representing a dynamic collection of sets $S = \{S_1, \ldots, S_r\}$. In other words, a disjoint set is a col-

lection of sets where no item can be in more than one set. It is also called a *union-find* data structure as it supports union and find operation on subsets. Let's begin by defining them:

Find: Determines in which subset a particular element is located and returns the representative of that particular set.

Union: Merges two different subsets into a single subset.

The disjoint-set also supports one other important operation called MakeSet, which creates a set containing only a given element in it.

(a) One way to store a set is as a directed tree, where each set is represented by a tree data in which each node holds a reference to its parent and the representative of each set is the root of that set's tree. Explain how we can implement *find* and *union* by a tree.

(b) Implement union-find by a hash table to implement a disjoint set.
(c) Use linked list to implement a disjoint set.

220. The tree to implement a disjoint set creates highly unbalanced trees. We can enhance it in two ways.

– Union by Rank
– Path Compression

Explain these approaches. Analyze their running times.

221. *Children Sum Property*—A binary tree follows the children sum property if for every node, the value of the node is equal to the sum of the values at its left and right subtree.

(a) Write a function that returns true if a binary tree satisfies children sum property.
(b) Write a function to convert a given binary tree to a binary tree that satisfies children sum property without changing the structure of the tree and only by incrementing the data values.

222. Given a binary tree, write a function to print the count of nodes whose immediate children are its factors, for example, 3 and 5 are factors of 15.

223. Write a function to determine if a binary tree is height-balanced? A non-empty binary tree T is balanced if, for every node, the difference between heights of left subtree and right subtree is not more than 1.

224. Suppose a binary tree is given. Write a function to

(a) count leaves in it.

(b) find its height.

(c) find the preorder traversal of the tree without using recursion.

(d) convert it into its mirror.

(e) count number of nodes in it.

(f) find the inorder traversal of the tree without using recursion.

(g) find the size of the largest independent set in it. A subset of all tree nodes is an independent set if there is no edge between any two nodes of the subset.

(h) find its minimum depth.

(i) check whether the given binary tree is a prefect binary tree or not. A binary tree is a perfect binary tree in which all internal nodes have two children and all leaves are at same level.

(j) remove all the half nodes (which has only one child).

(k) check whether the given tree follows the max heap property or not.

(l) find the deepest node in it.

(m) find its diameter. The diameter of a tree (sometimes called the width) is the number of nodes on the longest path between two end nodes.

(n) return the tilt of the whole tree. The tilt of a tree node is defined as the absolute difference between the sum of all the values of the left sub-tree and the sum of all the values of the sub-tree node on the right. The tilt of the tree is defined as the sum of all nodes' tilt.

(o) find the minimum elements in each level of the binary tree.

(p) find two subtrees with the same structure and value in a given tree. For each duplicate subtree, only return the root node of each one.

(q) find the minimum distance between two given node values.

(r) find all the nodes that are at distance k from the given target node. No parent pointers are available.

(s) find if two given binary trees are identical or not.

(t) find all nodes that are at distance k from root (root is considered at distance 0 from itself). Nodes should be printed from left to right. If k is more that height of tree, nothing should be printed.

(u) evaluate the a full binary expression tree consisting of basic binary operators $(+, -, *, /)$ and some integers.

225. Write a function to convert a binary tree to binary search tree so that the main structure of the binary tree remains intact.

226. Write a function to check whether the two given binary trees are isomorphic. If one of the trees can be obtained from the other with a series of flips, that is, by swapping the left and right children of several nodes, they are also called isomorphic.

227. Regarding BST:

(a) Given two BSTs, return elements of both BSTs in sorted form.

(b) Given a BST and a range [low, high]. Find all the numbers in the BST that lie in the given range.

(c) Given a BST and a value k, the task is to delete the nodes having values greater than or equal to k.

7.3 Solutions

4. For binary tree:

(a) In a binary tree, level of the root is 0 and we have one node in the root level and the number of nodes in the next level is twice the number of nodes in the previous level. So in that case, we can say that the number of nodes in the level 'l' is twice the number of nodes in the previous level and is obtained as follows:

$$1, \ 2, \ 4, \ \ldots, 2^0, 2^1, 2^2, \ldots.$$

The powers are level of the binary tree. So we have:

$$2^0, 2^1, 2^2, \ldots, 2^l$$

(b) Height of a binary tree with a single node is considered as 1. A binary tree has maximum nodes if all levels have maximum nodes. So maximum number of nodes in a binary tree of height 'h' is $1 + 2 + 4 + \cdots + 2^{h-1}$. Now we should prove $1 + 2 + 4 + \cdots + 2^{h-1} = 2^h - 1$:

$$1 + 2 + 4 + \cdots + 2^{h-1} = 2^h - 1 \rightarrow \ \ 1 + 2 + 4 + \cdots + 2^{h-1} + 1 = 2^h$$

$$2^0 + 2^1 + 2^2 + \cdots + 2^{h-1} + 1 = 2^0 + 2^0 + 2^1 + 2^2 + \cdots + 2^{h-1}$$

$$= 2 \times 2^0 + 2^1 + 2^2 + \cdots + 2^{h-1} = 2^{1+0} + 2^1 + 2^2 + \cdots + 2^{h-1}$$

$$= 2 \times 2^1 + 2^2 + \cdots + 2^{h-1} = 2^2 + 2^2 + \cdots + 2^{h-1} = \cdots$$

$$= 2 \times 2^{h-2} + 2^{h-1} = 2 \times 2^{h-1} = 2^h$$

(c) If we consider the convention where the height and level of a root node is considered as 1 and number of nodes in level 'l' or height 'h' is:

$$n = 2^0 + 2^1 + 2^2 + \cdots + 2^{l-1}$$

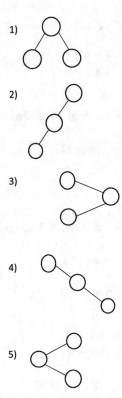

Fig. 7.11 Binary trees

Then we have:

$$\log_2(n+1) = \log_2\left(2^0 + 2^1 + 2^2 + \cdots + 2^{l-1} + 1\right)$$

$$= \log_2\left(2^0 + 2^0 + 2^1 + 2^2 + \cdots + 2^{l-1}\right) = \cdots = \log_2\left(22^{l-1}\right) = \log_2\left(2^l\right) = l$$

Then above formula for minimum possible height and level becomes $\log_2(n+1)$.

5. True,

In fact, a tree with all its nodes having 0 or 2 children is a full tree. In a full tree, the number of leaves is one more than the number of nodes with 2 children.

7. Regarding binary tree: (a) Five trees (see Fig. 7.11)
(b)

$$n = 1 \quad \rightarrow \quad Tree = 1 \quad n = 3 \quad \rightarrow \quad Tree = 5$$

$$n = 2 \quad \rightarrow \quad Tree = 2 \quad n = 4 \quad \rightarrow Tree = 14$$

$$T(n) = \frac{(2n)!}{[(n+1)!n!]} \qquad (\text{n'th Catalan Number})$$

(c)

Number of Labelled Trees = (Number of unlabelled trees) $\times$ n!

$$T(n) = [(2n)! / (n+1)!n!] \times n!$$

9.

Total number of nodes: 17

Number of nodes with degree 2: 6

Relationship between the number of leaf nodes and nodes with degree 2:

$$n_0 = n_2 + 1$$

$$n_0 = 6 + 1 = 7 \rightarrow n_t = n_0 + n_1 + n_2 \quad \rightarrow \quad 17 = 7 + n_1 + 6 \quad \rightarrow \quad n_1 = 4$$

13.

Inorder: D – B – H – E – A – I – F – J – C – G

Preorder: A – B – D – E – H – C – F – I – J – G

Postorder: D – H – E – B – I – J – F – G – C – A

14.

Inorder: 10 – 9 – 9 – 6 – 20 – 6 – 6 – 6 – 8 – 8 – 9 – 8 – 90– 17 – 17

Preorder: 6 – 6 – 9 – 10 – 9 – 6 – 20 – 6 – 8 – 8 – 8 – 9 – 17– 90 – 17

Postorder: 10 – 9 – 9 – 20 – 6 – 6 – 6 – 8 – 9 – 8 – 90 – 17 –17 – 8 – 6

15.

Inorder: 8 10 11 12 14 17 18 20 22 23 25

Preorder: 14 10 8 12 11 17 20 18 25 22 23

Postorder: 8 11 12 10 18 23 22 25 20 17 14

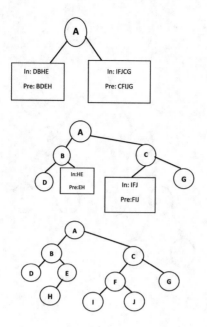

Fig. 7.12 Answer to 17

17: Inorder : DBHEAIFJCG/Preorder : ABDEHCFIJG

In the preorder, the first character is the root, and in the inorder, the letters on the left of the root, are its left branch, and the letters on the right of the root are its right branch, and we use this feature for all steps (see Fig. 7.12).

24. See Fig. 7.13

25. d

$n_0 = n_2 + 1$

$n = n_0 + n_1 + n_2$

$n = 14$

$13 = 2n_2 + n_1$

Because n_2 is multiplied by 2 it's will always be even, and we know that the result is odd so there is odd number of operators in this expression which at least can be 1.

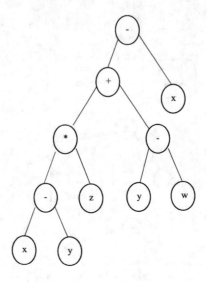

Fig. 7.13 Answer to 24

Fig. 7.14 Answer to 27

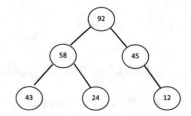

27. See Fig. 7.14

32.
(b) Explanation: The function counts internal nodes.

1. If root is NULL or a leaf node, it returns 0.
2. Otherwise returns, 1 plus count of internal nodes in left subtree, plus count of internal nodes in right subtree.

33. This function counts the number of leafs.

34.
This function counts the number of tree nodes.

37.

1. First initialize current as root.
2. Then while current is not NULL

If the current does not have left child

Fig. 7.15 Answer to 52

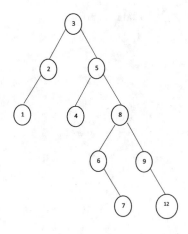

1. Print current
2. current=current->right

Else

1. We should find rightmost node in current left subtree or current==node whose right child

If we found right child is equal to current

1. we update the right child as Null of that node whose right child is current
2. second print current data
3. at the last go to the right

Else

1. make current as the right child of that rightmost node we found
2. finally current=current->left

44. c.
52. The height of the tree is 5 (See Fig. 7.15).

53.
The deepest are 7 and 12.

54. Inorder traversal of the BST.

63. (a) Delete the node with key 10. See Fig. 7.16

(b) Delete the node with key 20 (See Fig. 7.17).

64. True

Fig. 7.16 Answer to 63(a)

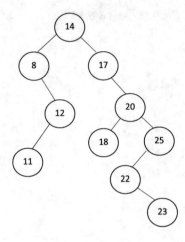

Fig. 7.17 Answer to 63(b)

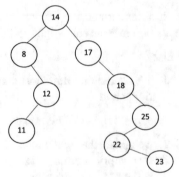

65. a. For skewed BSTs.

66. C, because inorder successor is not always a leaf and is not either a leaf or a node with an empty right or left child.

67. True

69. True

71. a.

72. a.

111. True

119. True

Fig. 7.18 Answer to 169

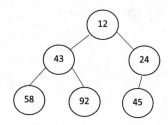

122. True

127. True

132. True

163. True

165. True

169. See Fig. 7.18

173. c. 9 is not smaller than 7.

175.
The correct array is (b). See Fig. 7.19

Step 4 array: {40, 35, 20, 10, 30, 16, 17, 8, 4, 15}
177.

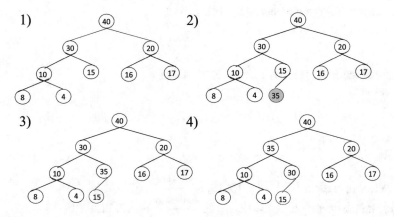

Fig. 7.19 Answer to 175

Fig. 7.20 Answer to 177

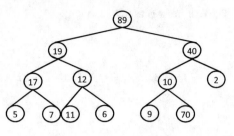

Fig. 7.21 Answer to 178

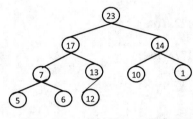

Fig. 7.22 Answer to 187

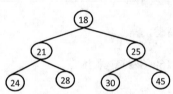

Only element 70 doesn't follow the rule. so, it must be shifted to its correct position (See Fig. 7.20).

Step 1: swap (10, 70)
Step 2: swap (40, 70)

178. Answer: (c) is a max-heap (See Fig. 7.21).

187.

Inorder of the BST: {18, 21, 25, 24, 28, 30, 45}
See Fig. 7.22.

194.
Answer: (c) is correct.
See Fig. 7.23.

first delete operation: 12, 14, 16, 13, 10, 8 $\overset{max-heap}{\longrightarrow}$ 16, 14, 12, 13, 10, 8
second delete operation: 8, 14, 12, 13, 10 $\overset{max-heap}{\longrightarrow}$ 14, 13, 12, 8, 10

Fig. 7.23   Answer to 194

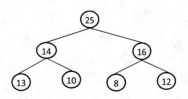

195. Two moves
First replace the 73 and 94 then 80 and 94.

207. Complete the following table in terms of time complexity (Big-O time)

	search	Insertion	deletion
Binary Tree	O(n)	O(n)	O(n)
Binary search tree	O(n)/O(h)	O(n)/O(h)	O(n)/O(h)
AVL Tree	$O(\log_2 n)$	$O(\log_2 n)$	$O(\log_2 n)$
Heap Tree	O(log n)	O(log n)	O(log n)

209. True

Chapter 8
Search

Abstract Searching algorithms are used to retrieve a given element from data structures. According to the search strategy, these algorithms are generally classified into two categories: sequential search and interval search. This chapter provides 76 exercises on the search methods including linear search, binary search, Ternary search, binary search tree, Fibonacci search, exponential search, and interpolation search.

8.1 Lecture Notes

Searching algorithms are used to retrieve a given element from data structures. According to the search strategy, these algorithms are generally classified into two categories:

1. Sequential search: In this type of search, all elements of a list or array are examined sequentially to find a specific element, for example: linear search.
2. Interval search: These algorithms are a class of algorithms designed for searching sorted data structures, dividing a list into several intervals at every step. These types of search algorithms are much more efficient than linear search because they repeatedly target the middle of the list and split the search space in half, for example: binary search.

8.2 Exercises

8.2.1 Preliminary

1. *Log n* Behavior
Suppose the element we are looking for exists in the array. In the range 1–8, how many comparisons are used to guess a number? Table 8.1 shows a sample scenario for the range 1–8.

As can be seen from the table, the maximum number of comparisons for guessing a number is 3. Assume a person chooses a number between 1 and 10. How many comparisons can you guess? Draw a table similar to Table 8.1 to do this.

© The Author(s), under exclusive license to Springer Nature Switzerland AG 2022   269
H. Izadkhah, *Problems on Algorithms*,
https://doi.org/10.1007/978-3-031-17043-0_8

Table 8.1 Sample behavior for guessing number from 1 to 8

Number	First round	Second round	Third round	Fourth round
1	Is it 4?	Is it 2?	Must be 1!	
	Too high	Too high	You win	
2	Is it 4?	Is it 2?		
	Too high	You win		
3	Is it 4?	Is it 2?	Must be 3!	
	Too high	Too low	You win	
4	Is it 4?			
	You win			
5	Is it 4?	Is it 6?	Must be 5!	
	Too low	Too high	You win	
6	Is it 4?	Is it 6?		
	Too low	You win		
7	Is it 4?	Is it 6?	Is it 7?	
	Too low	Too low	You win	
8	Is it 4?	Is it 6?	Is it 7?	Must be 8!
	Too low	Too low	Too low	You win

2. A bartender offers the following $10,000 bet to any patron: "I will choose a secret number from 1 to 1,000,000 and you will have 20 chances to guess my number. After each guess, I will either tell you Too Low, Too High, or You Win. If you guess my number in 20 or fewer questions, I give you $10,000. If none of your 20 guesses is my secret number you must give me $10,000." Would you take this bet? You should, because you can always win.

3. Devise an algorithm to guess number in range [low, high]?

4. The best running time for search algorithms is logarithmic that are very efficient. True/False?

5. How do you find an entry when you search for a thick dictionary?

6. Explain how we can perform searching in a pre-sorted data.

8.2.2   Linear Search

7. A straightforward solution to the search method is linear search. This method examines the data from the beginning of a list to the last item in the list or until it finds x. Write the code of this search. Give the running time of linear search.

8. Given an array containing n integer numbers, give an algorithm to find either the *min* or *max* of the set in $n - 1$ comparisons each. Do you make better than $2n - 3$ comparisons when you need to find both?

9. Find the *min* and *max* elements from a set of n elements using no more than $3n/2 - 1$ comparisons.

10. Linear search method used when

  a. the size of the array is low.
  b. the size of the array is large.
  c. the array is unordered.
  d. never used

11. ...... is the best case runtime of linear search algorithm on an ordered set of elements.

  a. $O(1)$
  b. $O(n)$
  c. $O(\log n)$
  d. $O(n^2)$

12. What is the recurrence relation for the linear search recursive algorithm?

  a. $T(n - 2) + c$
  b. $2T(n - 1) + c$
  c. $T(n - 1) + c$
  d. $T(n + 1) + c$

13. *Sentinel Linear Search*—Sentinel is a useful technique for linear searching. This technique is a type of programming technique rather than an algorithmic one. Discuss about this technique.

8.2.3 Binary Search

14. How many of the following are binary search applications?

  – Finding the lower/upper bound in an ordered sequence
  – Union of intervals
  – Debugging
  – Searching in unordered list

a. 1
b. 2
c. 3
d. 4

15. How many of the following are binary search applications on arrays?

– To find if the number n is a square of an integer
– To find the first value greater than or equal to k in a sorted array
– To find the frequency of a given target value in an array of integers
– To find the peak of an array which increases and then decreases
– A sorted array is rotated n times. Search for a target value in the array.

a. 1
b. 3
c. 4
d. 5

16. Binary search can be categorized into which of the following?

a. brute force technique
b. divide and conquer
c. greedy algorithm
d. dynamic programming

17. Implement binary search as following

(a) recursive
(b) iterative
(c) different idioms for conditionals
(d) another method other than the above methods

18. The idea of binary search is to always delete half of the elements of the searched array. In this case, if the search key is not equal to the middle element of the array, remove one of the two sets of elements to the left or right of the middle element from further investigation. True/False?

19. Here is an iterative implementation adapted from Bentley's book, which includes his bug. Find the bug of this code.

```
//This algorithm returns location of x in given array arr[low..high] if present,
//otherwise -1
int binarySearch(int arr[], int low, int high, int x)
{
  while(low ≤ high){
    int mid = (high − low)/2;
    //Check if x is present at mid
    if(arr[mid] == x)
      return mid;
    //If x greater, ignore left half
    if(arr[mid] < x)
      low = mid + 1;
    //If x is smaller, ignore right half
    else
      high = mid − 1;
  }
  //if we reach here, then element was not present
  return − 1;
}
```

20. When instead of
$$int\ mid = (low + high)/2;$$

we must use
$$int\ mid = low + (high − low)/2;$$

or
$$int\ mid = high − (high − low)/2;$$

21. *Uniform Binary Search*—Uniform binary search is an improvement on the binary search algorithm and is used when multiple searches are performed on an array or many arrays of the same size. In normal binary search, arithmetic operations are performed to find midpoints. In a uniform binary search, midpoints are pre-calculated and entered in the lookup table. The look-up table usually works faster than calculations to find the midpoint. Implement this type of search.

22. *Jump Search*—Like binary search, jump search is employed to search the sorted arrays. The basic idea is to explore fewer elements (than a linear search) by jumping forward with fixed steps or skipping some elements instead of searching for all elements. Implement this search and then analyze its running time.

23. The time complexity of jump search is between linear search ($O(n)$) and binary search ($O(Log\ n)$). True/False?

24. Best case of jump search will be when the first element of the array is the element that is being searched. In this case only one comparison will be required. Thus it will have a time complexity of $O(1)$. True/False?

25. The largest Japanese dictionary contains about 240,000 words. If you use binary search, in the worst case, how many words should be checked to find a word?

26. The binary search algorithm is a simple and fast algorithm. In fact, it is the most efficient algorithm and theoretically it can not be further improved. This algorithm is lower bound of searching. True/False?

27. Let A be an algorithm that searches by comparing a pair of items. If A can use fewer than $log\ n$ comparison operations, the search problem for n data cannot be solved by A. Prove this theorem.

8.2.4   Ternary Search

28. Unlike binary search, where a sorted array is divided into two parts, in ternary search the sorted array is divided into three parts, and then it is specified in which part the element is located.

(a) Give its algorithm.
(b) Give the recurrence relation of the search algorithm.
(c) Analyze its time complexity.

29. Consider the following sorted array containing 12 elements. Show the steps to find the number 5 using the ternary search.

$$A = [5,\ 23,\ 37,\ 40,\ 48,\ 54,\ 65,\ 75,\ 81,\ 96,\ 98,\ 100]$$

30. Consider the following array:

$$A = [-1,\ 0,\ 3,\ 5,\ 10,\ 23,\ 30,\ 45]$$

(a) How many comparisons does it take using a ternary search to find the elements $-1, 0, 5, 30$?
(b) In the array, find the average number of comparisons in both successful and unsuccessful search?

8.2.5 Binary Search Tree (BST)

31. Searching for a key in a binary search tree (BST) is very similar to searching in a sorted array. Recursion is more natural but for performance, a while-loop is preferred. True/False?

32. Given a BST T, write a recursive function that searches for key k, then write an iterative function.

33. Given a BST T and a key k, write a method that searches for the first entry larger than k.

8.2.6 Fibonacci Search

34. As a comparison-based algorithm, Fibonacci search uses Fibonacci numbers to search an element in a sorted array. So let us define the Fibonacci numbers first.

$$Fib(n) = Fib(n-1) + Fib(n-2), \quad Fib(0) = 0, \quad Fib(1) = 1$$

Implement this search.

35. Which algorithmic technique does Fibonacci search use?

  a. brute force technique
  b. divide and conquer
  c. greedy algorithm
  d. backtracking

36. Consider the following sorted array containing 12 elements. Show the steps to find the number 5 using the Fibonacci search.

$$A = [5, 23, 37, 40, 48, 54, 65, 75, 81, 96, 98, 100]$$

37. Consider the following array:

$$A = [-1, 0, 3, 5, 10, 23, 30, 45]$$

(a) How many comparisons does it take using a Fibonacci search to find the elements −1, 0, 5, 30?
(b) In the array, find the average number of comparisons in both successful and unsuccessful search?

38. Since Fibonacci search divides the array into two parts, although not equal, its time complexity is $O(log\ n)$, it is better than binary search in case of large arrays. True/False?

39. With each step of Fibonacci search, the search space is reduced by 1/3 on average, hence, the time complexity is $O(log_3\ n)$. True/False?

8.2.7   Exponential Search

40. Exponential Search—Explain how this search works? Analyze its running time and space complexity.

41. Exponential search requires the input array to be sorted. True/False?

42. In exponential search, we first find a range where the required elements should be present in the array. Then we apply binary search in this range. True/False?

43. In exponential search, we first find a range where the required elements should be present in the array. Then we apply binary search in this range. This takes $O(log\ n)$ time in the worst case. True/False?

44. Consider the following sorted array containing 12 elements. Show the steps to find the number 5 using the exponential search.

$$A = [5,\ 23,\ 37,\ 40,\ 48,\ 54,\ 65,\ 75,\ 81,\ 96,\ 98,\ 100]$$

45. Suppose we have the following array, and we want to find $X = 10$. Show the steps to find X using the exponential search.

$$A = [1,\ 2,\ 3,\ 4,\ 5,\ 6,\ 7,\ 8,\ 9,\ 10,\ 11]$$

46. Consider the following array:

$$A = [-1,\ 0,\ 3,\ 5,\ 10,\ 23,\ 30,\ 45]$$

(a) How many comparisons does it take using the exponential search to find the elements −1, 0, 5, 30?
(b) In array A, how many comparisons are made in successful and unsuccessful searches on average?

8.2.8 Interpolation Search

47. Interpolation search is an improvement over binary search for situations where values are uniformly distributed in a sorted array. Implement this type of search.

48. The time complexity of interpolation search algorithm is $O(log(log\ n))$. Justify needed.

49. Suppose we have the following array. As you can see in the array, the values vary from 1 to 600,000 while there are only eight values in the array. The values here are increasing almost exponentially. Explain why interpolation search on this array may not work well.

$$A = [1,\ 3,\ 15,\ 18,\ 90,\ 1500,\ 20000,\ 600000]$$

50. Do you think the following array could be a good input for using interpolation search?

$$A = [20000,\ 25000,\ 31000,\ 36000,\ 40000,\ 46000,\ 51000,\ 55000]$$

51. Interpolation search performs better than binary search for a sorted and uniformly distributed array. True/False?

52. Binary search performs better than interpolation search when the values are not distributed uniformly. True/False?

53. In case of non uniform distribution of values the time complexity of interpolation search is $O(n)$. True/False?

54. Which of the following searching algorithm is fastest when the input array is not sorted but has uniformly distributed values?

  a. jump search
  b. linear search
  c. binary search
  d. interpolation search

55. What is the formula used for calculating the position in interpolation search? (key = element being searched, A[] = input array, low and high are the leftmost and rightmost index of A[] respectively)

  a. ((key − A[low]) * (high − low))/(A[high] − A[low])
  b. high + ((key − A[low]) * (high − low))/(A[high] − A[low])
  c. low + ((key − A[low]) * (high − low))/(A[high] − A[low])
  d. key + ((key − A[low]) * (high − low))/(A[high] − A[low])

56. Show the steps to find the number 5 using the interpolation search on the following ascending sorted array.

$$A = [5, \ 23, \ 37, \ 40, \ 48, \ 54, \ 65, \ 75, \ 81, \ 96]$$

57. Suppose we have the following array, and we want to find $X = 10$. Show the steps to find X using the interpolation search.

$$A = [1, \ 2, \ 3, \ 4, \ 5, \ 6, \ 7, \ 8, \ 9, \ 10, \ 11]$$

8.2.9 Applications

58. Let A is a sorted array of integers and a key k is given. Write a function to return the index of first occurrence of k in A. Return 'No' if k does exist in A.

59. Provide an efficient algorithm that finds the index of the first occurrence an element larger than a specified key k. If any element is less than or equal to k, return -1.

60. Devise an algorithm to search a sorted array of distinct elements for finding an index that for $A[i] = i$ or indicating that no such index exists.

61. Find a 9-digit number that is not on file, according to the file, which contains approximately 300 million Social Security numbers (9-digit numbers). You have unlimited drive space but only 2 MB of RAM available.

62. Let A and B are two sorted arrays of size n and m, respectively. Considering the following conditions, give algorithms to return an array C, free of duplicates, containing the common elements A and B.

(a) $n \approx m$
(b) $n \ll m$

63. Assume in a very large stream sequence of strings separated by white space. The stream can be read twice by an algorithm. Develop an algorithm that uses only $O(k)$ memory to find all words that are in the stream for more $\lceil \frac{n}{k} \rceil$ times, where n is the length of the stream.

64. *Bisection algorithm*- Devise an algorithm to compute a root of an equation in one variable; namely, for what values of x does a continuous function $f(x) = 0$?

65. Explain block search and analyze its running time.

66. Let A and B are two sorted arrays of size n and m, respectively. Devise a $O(\log m + \log n)$ time algorithm for computing the k-*smallest* element in the union

of the two arrays (elements may be repeated).

67. Let M be a row-wise sorted matrix of size $r \times c$ that $r \times c$ is always odd. Devise an algorithm to find the median of the matrix.

Examples:
Input:
1 3 5
2 6 9
3 6 9
Output: Median is 5

68. Let A be an array of n positive integers. Find k largest elements from the array and print them in decreasing order.

Input:
N = 5, k = 2
Arr[] = {13, 6, 407, 2, 25}
Output: 407 25
Explanation: 1st largest element in the array is 407 and second largest is 25.

69. An element can be found in $O(log\ n)$ time in a sorted array through binary search. We rotate an ascending sorted array at some pivot is already unknown to you. For example, 3 4 5 6 7 might become 5 6 7 3 4. Explain how we can find an element in the rotated array in $O(log\ n)$ time.

70. A sorted array including distinct elements is rotated at some unknown point. Give an algorithm to find the minimum element in the array (Hint: use binary search).

71. Give an algorithm to find the kth smallest element in the given array including distinct elements.

72. Given an $n \times n$ matrix, where every row and column is sorted in non-decreasing order. Give an algorithm to find the kth smallest element in this matrix.

73. Suppose A is an array in which the elements are in any order. Using the following methods, give algorithms to find k largest elements in the given array.

(a) Use bubble sort k times
(b) Use temporary array
(c) Use merge sorting
(d) Use max heap
(e) Use order statistics
(f) Use min heap
(g) Using quick sort partitioning algorithm

74. Devise an algorithm to find the majority element of given array (if it exists), otherwise prints "No Majority Element". In an array of size n, an element is called a majority element if it appears more than $n/2$ times (and hence there is at most one such element)(Hint: use binary search tree).

Example:
Input: {4, 4, 6, 2, 6, 6, 2, 6, 6}
Output: 6

75. *Find a peak element.* Give an algorithm to find a peak element in the given array containing integers. An element is called a peak element if it is not smaller than its neighbours. Only one neighbor is considered for the corner elements. (Hint use binary search).

Example:
Input: array = {8, 13, 23, 18}
Output: 23
The element 23 has neighbours 13 and 18, both of them are less than 23.

76. Hashing is another approach to searching. Hashing is qualitatively different from binary search. Compare both of them.

8.3 Solutions

11. O(1)

12. $T(n) = T(n-1) + c$

14. (c)—binary search is not used for searching in an unordered list.

15. (d).

16. (b)—divide and conquer

18. True

20. When low and high are very large numbers, there can be an overflow if you use int mid = (high + low)/2. Instead, when you use int mid = low + (high–low)/2 or int mid = high – (high–low)/2 you still get the same result but keep the individual components small avoiding overflow.

21.

Algorithm:

1. We start from the middle of list to search data.
2. If matches exist, we return the item index and exit.
3. If there is not match, we check the situation.
4. We use the $mid = Low + ((High - Low) / (A[High] - A[Low])) * (X - A[Low])$
5. To divide the list and find new median.
6. If data larger than median, search the largest list.
7. If data smaller then median, search the smallest list.
8. These steps are repeated until a match is found.
9. End

22.

For example, suppose we have an ascending array [] of size n and a jump (step) of size s. Then we search in the indexes A[0], A[s], A[2s], ..., A[ks] and so on. When we find the range (A[ks] < x < A[(k + 1)s], a linear search operation is performed on the range found.

Time complexity:

In the worst case:
Number of jumps: n/s

Number of linear comparisons at worst: s–1

Total number of comparisons at worst: ((n/s) + s–1)

If: $s = \sqrt{n}$ the best case for the above function occurs then the time complexity is equal to: $O(\sqrt{n})$

Example:

A = [0, 6, 7, 10, 15, 18, 25, 38, 50]

x = 18 $\implies$ searching for x

s = 4

step 1: A[0] = 0 $\implies$ x > 0

step 2: A[4] = 15 $\Longrightarrow$ x > 15

step 3: A[8] = 50 $\Longrightarrow$ x < 50

So we do a linear search between steps two and three

Linear search has a time complexity of $O(n)$ and the time complexity of jump search is $O(\sqrt{n})$.

23. True—$(O\,(\log{(n)}) < O\left(\sqrt{n}\right) < O(n))$

25.
$$\log{(240000)} \approx 18$$

28. In binary search, the sorted array is divided into two parts while in ternary search, it is divided into three parts and then you determine in which part the element exists.

Steps:
$$first\_mid = low + \frac{high-low}{3}$$

$$second\_mid = high - \frac{high-low}{3}$$

Steps:

1. First compare the key with the *first_mid* . If they are the same return the *first_mid*.
2. If not:

   a. Compare the key with the *second_mid*. If they were same return *second_mid*.

3. If not:

   a. Check whether the key is less than the *first_mid*. If yes, then recur to the first part.

4. If not:

   a. Check whether the key is greater than the *second_mid*. If yes, then recur to the third part.

5. If not:

   a. Then we recur to the second (middle) part.

Algorithm:

1. Start.
2. Get array A with n elements from input and set low = 0 and high = n–1.

3. Then we should find *mid1* and *mid2* by the following equation:

$$mid1 = low + \frac{high - low}{3} \quad and \quad mid2 = high - \frac{high - low}{3}$$

4. If key is A[mid1] or A[mid2], go to the 8 and print the index of mid1 or mid2 as index of key we found, otherwise go to 5.

5. If key<A[mid1], on that case high = mid1−1 and go to 3, otherwise go to 6.

6. If key>A[mid2], on that case low = mid2+1 and go to 3, otherwise go to 7.

7. If is key>A[mid1] and key<A[mid2], on that case low = mid1+1 and high = mid2−1 and go to 3.

8. Print the index of as index of key we found otherwise

9. End.

(b) Recurrence relation for ternary search is $T(n) = T\left(\frac{n}{3}\right) + O(1)$

(c) The time complexity of ternary search is $O(\log_3 n)$

29.

mid1 = low + (high − low) /3

mid2 = high − (high − low) /3

Step1:

A = [5; 23; 37; 40; 48; 54; 65; 75; 81; 96]

A[mid1] = 40 and A[mid2] = 65

$\Longrightarrow 5 < 40 \Longrightarrow$ going to searching the 5 in first part

Step2:

A = [5; 23; 37; 40; 48; 54; 65; 75; 81; 96]

A[mid1] = 5 and A[mid2] = 37

$$\Longrightarrow 5 = 5$$

So we return mid1

30.

  (a)

$$mid1 = 0 + \frac{7-0}{3} = 2.3 = 2 \ \ \& \ \ mid2 = 7 - \frac{7-0}{3} = 4.6 = 4$$

$$A[2] = 3, \ A[4] = 10$$

Element −1:

$$-1 < 3 \ and \ -1 < 10 \rightarrow mid1 = 0 + \frac{1-0}{3} = 0.3 = 0 \rightarrow -1 = A[0] = -1$$

We came to a conclusion with 2 comparisons.

Element 0:

$$0 < 3 \ and \ 0 < 10 \rightarrow mid1 = 0 + \frac{1-0}{3} = 0.3 = 0$$

$$A[0] = -1 < 0 \rightarrow mid1 = 1 + \frac{1-1}{3} = 1 \rightarrow A[1] = 0 = 0$$

We came to a conclusion with 3 comparisons.

Element 5:

$$5 > 3 \ and \ 5 < 10 \rightarrow mid1 = 3 + \frac{3-3}{3} = 3 \rightarrow 5 = A[3] = 5$$

We came to a conclusion with 2 comparisons.

Element 30:

$$30 > 3 \ and \ 30 > 10 \rightarrow mid1 = 5 + \frac{7-5}{3} = 5.6 = 5 \rightarrow 30 > A[5] = 23$$

$$mid2 = 7 + \frac{7-5}{3} = 6.3 = 56 \rightarrow 30 = A[6] = 30$$

We came to a conclusion with 3 comparisons.

(a)

−1: 2 comparisons

0: 3 comparisons

5: 1 comparison

30: 1 comparison

(b)

number of successful search: $7 \implies 2 + 3 + 3 + 1 + 2 + 2 + 1 = 14$ comparisons

average: $\frac{14}{7} = 2$

34.

Algorithm:

1. **Start.**
2. We should find the smallest Fibonacci number greater than or equal to the size of the list in which we are searching for the key. We Suppose the size of the array is n and Fibonacci number is f_k. We must find a f_k that $f_k \geq n$.
3. Then we must compare the key with the element at Fibonacci number f_{k-1}.
4. If the key and array element at f_{k-1} are equal then the key is at $f_{k-1} + 1$
5. If the key is less than array element at f_{k-1}, then we search the left sub to f_{k-1}.
6. If the given key is greater than the array element at f_{k-1}, then we search the right sub to f_{k-1}.
7. If the key is not found, we repeat the step 2 to step 6 until $f_{k-1} = 0$, *Fibonacci number* $\geq n$. After each iteration the size of array(n) is reduced.
8. **End.**

36.

A = [5, 23, 37, 40, 48, 54, 65, 75, 81, 96]

Length of the array = 10

The smallest Fibonacci number which is equal to or greater than 10 is 13.

F(k) = 13, F(k–1) = 8, F(k–2) = 5

A[8] in the array is 81, which is greater than 5. So we remove 81 and above from the array.

F(k) = 8, F(k–1) = 5

A[5] in the array is 54, which is greater than 5. So, we remove 54 and above from the array.

F(k) = 5, F(k–1) = 3

A[3] in the array is 40, which is greater than 5. So, we remove 40 and above from the array.

F(k) = 3, F(k–1) = 2

A[2] in the array is 37, which is greater than 5. So, we remove 37 and above from the array.

F(k) = 2, F(k–1) = 1

A[1] in the array is 23, which is greater than 5. So, we remove 23 and above from the array.

F(k) = 1, F(k–1) = 0

Value of index 0 is equal to 5 so we print it.

40. Exponential search also known as finger search, searches for an element in a sorted array by jumping 2^i elements every iteration where i represents the value of loop control variable, and then verifying if the search element is present between last jump and the current jump.

How it works

1. Jump the array 2^i elements at a time searching for the condition $Array[2^{i-1}] <$ targetValue $< Array[2^i]$. If 2^i is greater than the length of array, then set the upper bound to the length of the array.
2. Do a binary search between $Array[2^{i-1}]$ and $Array[2^i]$

Time Complexity: O (Log n)

Space complexity: O(1)

41. True

Exponential search requires the input array to be sorted. The algorithm would fail to give the correct result if array is not sorted.

47.
In this search the position is computed as follows:

$$pos = low + \frac{key - arr\,[low] * (high - low)}{arr\,[high] - arr[low]}$$

```
def interpolation_search(arr, l, h, x):
    if (l <= h and x >= arr[l] and x <= arr[h]):
        pos = l + ((h – l) // (arr[h] – arr[l]) * (x – arr[l]))
        if arr[pos] == x:
            return pos
        if arr[pos] < x:
            return interpolation_search(arr, pos + 1, h, x)
        if arr[pos] > x:
            return interpolation_search(arr, l, pos – 1, x)
    return –1
```

49.

The interpolation search algorithm works best when the array elements are uniformly distributed, the reason for this is that interpolation search assumes linear growth of the elements in the array. But in this particular case as denoted in the question itself, growth of the array elements resembles exponential growth rather than a linear one.

51. True

55. c

60.

```
def binarySearch(arr, low, high):
    if high > = low:

        mid = low + (high – low)//2
        if mid == arr[mid]:
            return mid
        res = –1
        if mid + 1 <= arr[high]:
            res = binarySearch(arr, (mid + 1), high)
        if res !=–1:
            return res
        if mid-1 > = arr[low]:
            return binarySearch(arr, low, (mid –1))

    return –1
```

67.

Algorithm:

1. First, we find the minimum and maximum elements in the matrix. The minimum element can easily found by comparing the first element of each row, and also the maximum element can be found by comparing the last element of each row.
2. Second, we use binary search on range of numbers from minimum to maximum we find the mid of the min and max and get a count of numbers less than or equal to mid. And change the min or max.
3. For a number to be median, there should be $\frac{r*c}{2}$ numbers smaller than that number. So for every number, we get the count of numbers less than that by using upperbound() in each row of the matrix, if it is less than the count, the median must be greater than the selected number, else, the median must be less than or equal to the selected number.

72.

Algorithm:

1. We create a Min-heap to store the elements.
2. Traverse the first row from start to end and build a min-heap of elements from first row. A heap entry also stores column number and row number.
3. Then, run a loop k times to extract min element from heap in each iteration.
4. We get minimum element from Min-heap.
5. Then, find row number and column number of the minimum element.
6. Replace root with the next element from same column.
7. Finally, print the last extracted element, which is the kth minimum element.

76. Hash Table supports following operations in $\Theta(1)$ time.

(1) Search, (2) Insert, (3) Delete

The time complexity of above operations in a self-balancing Binary Search Tree (BST) (like Red-Black Tree, AVL Tree, Splay Tree, etc.) is O(Log n).

So, Hash Table seems to beating BST in all common operations. When should we prefer BST over Hash Tables, what are advantages? Following are some important points in favor of BSTs.

1. We can get all keys in sorted order by just doing Inorder Traversal of BST. This is not a natural operation in Hash Tables and requires extra efforts.
2. Doing order statistics, finding closest lower and greater elements, doing range queries are easy to do with BSTs. Like sorting, these operations are not a natural operation with Hash Tables.
3. BSTs are easy to implement compared to hashing, we can easily implement our own customized BST. To implement Hashing, we generally rely on libraries provided by programming languages.
4. With Self-Balancing BSTs, all operations are guaranteed to work in O(Log n) time. But with Hashing, $\Theta(1)$ is average time and some particular operations may be costly, especially when table resizing happens.

Chapter 9
Sorting

Abstract The purpose of sorting is to arrange the elements of a given list in a specific order (ascending or descending). Sorting algorithms are categorized according to the following–

- By number of comparisons
- By number of swaps
- By memory (space) usage
- By recursion
- By stability
- By adaptability

This chapter provides 431 exercises on the 35 sorting algorithms.

9.1 Lecture Notes

The purpose of sorting is to arrange the elements of a given list in a specific order (ascending or descending). Sorting algorithms are categorized according to the following–

By number of comparisons: Comparison-based sorting algorithms examine and sort list elements with key comparison operations, and for most inputs require at least an $O(n \log n)$ comparison. In this type of sorting, algorithms are classified based on the number of comparisons. For comparison-based sorting algorithms, the best case time complexity is $O(n \log n)$ and worst case time complexity is $O(n^2)$. Quicksort, insertion sort, counting sort, bucket sort, and radix sort are examples of comparison-based sorting algorithms.

By number of swaps: In this category, sorting algorithms are categorized based on the number of swaps between elements, also called inversion.

By memory (space) usage: Some sorting algorithms are "in place" and require $O(1)$ or $O(\log n)$ space to create auxiliary space for sorting the data.

By recursion: Some sorting algorithms perform sorting recursively (for example, quicksort) and some do non-recursive sorting (for example, selection sort and inser-

© The Author(s), under exclusive license to Springer Nature Switzerland AG 2022   289
H. Izadkhah, *Problems on Algorithms*,
https://doi.org/10.1007/978-3-031-17043-0_9

tion sort). Of course, there is a non-reversible version for some algorithms, for example, quicksort, and vice versa.

By stability: If two elements with equal values appear in the output in the same order as the input, the sorting algorithm is stable. In fact, stable sorting algorithms, unstable sorting algorithms, preserve the relative order of equal elements. Insertion sort, bubble sort, and radix sort are examples of stable sorting algorithms.

By adaptability: In some sorting algorithms, the time complexity of the algorithm varies depending on how it is input. For example, a pre-arranged array affects the execution time of an algorithm. Algorithms that consider this compatibility are known as adaptive algorithms. For example—insertion sort is an adaptive sorting algorithm because the time complexity of insertion sort depends on the initial order of input. If input is already sorted then time complexity becomes $O(n)$ and if input sequence is not sorted then time complexity becomes $O(n^2)$. In quicksort, on the other hand, the time complexity becomes $O(n^2)$ if the input is already sorted, and the time complexity becomes $O(n\log n)$ if the input sequence is not sorted. Selection sort, mergesort, and heapsort are examples of non-adaptive sorting algorithms.

Internal and external sorting: Sorting algorithms that perform the sorting of elements inside main memory are called internal sorting algorithms, and algorithms that also use external memory are called external sorting algorithms. Selection sort, insertion sort, and quicksort are examples of internal sorting and mergesort is example of external.

9.2 Exercises

9.2.1 Introduction

1. About sorting algorithm:

(a) What is in-place sorting?
(b) What are internal and external sorting?
(c) What is stable sorting?
(d) Which sorting algorithms are stable?
(e) Which sorting algorithms are unstable?
(f) Can we make any sorting algorithm stable?

2. *Count Inversions of an Array*—This count shows how far (or close) the array is from sorting. If the array is already sorted, the inversion count is 0, but if the array is sorted in the reverse order, the inversion count is the maximum.

Formally, two elements $a[i]$ and $a[j]$ (pair $(a[i], a[j])$) form an inversion if $a[i] > a[j]$ and $i < j$.

Example:
Input: arr[] = {10, 6, 4, 3}

Output: six inversions as follows
(10, 6), (6, 4), (10, 4), (10, 3), (6, 3), (4, 3).

(a) Give an algorithm to find the inversions and the total number of inversions.
(b) Analyze its time complexity.
(c) Analyze its space complexity.

3. To count the inversions in an array:

(a) Explain how we can use mergesort to count the inversions in an array.
(b) Use AVL tree as a self-balancing binary search tree to count the inversions in an array.
(c) Use Red-Black tree as a self-balancing binary search tree to count the inversions in an array.

4. Suppose an array of n distinct integers $a_1, a_2, \ldots, a_n$ and an integer k are given. Find out the number of sub-sequences of a such that $a_{i_1} > a_{i_2} > \ldots > a_{i_k}$, and $1 \le i_1 < i_2 < \ldots < i_k \le n$, indicating the total number of inversions of length k.

Example:
Input: arr[] = {11, 5, 8, 4, 3}, k = 3
Output: 7
The seven inversions are {11, 5, 4}, {11, 5, 3}, {11, 8, 4}, {11, 8, 3}, {11, 4, 3}, {5, 4, 3} and {8, 4, 3}.

9.2.2 Selection Sort

5. The selection sort algorithm sorts an array by repeatedly finding the minimum element (considering ascending order) from unsorted part placing it at the beginning of the unordered part.

(a) Give an algorithm to sort an array using selection sort.
(b) Analyze its time complexity.
(c) Analyze its space complexity (auxiliary space).

6. Give an algorithm to sort an array of strings using selection sort.

7. Selection sort is an in-place comparison-based sorting algorithm that requires no additional memory. True/False?

8. Selection sort only requires a constant amount of additional space (i.e., $O(1)$). True/False?

9. Selection sort is a stable algorithm. True/False?

10. In which of the following scenarios do you use selection sort?

a. The input array is already sorted.
b. A large file must be sorted.
c. Large values need to be sorted with small keys.
d. Small values need to be sorted with large keys.

11. Which of the following is an advantage of selection sort?

a. It requires no additional memory.
b. It is scalable.
c. It works best for inputs which are already sorted.
d. It is faster than any other sorting technique.

12. Selection sort is not a scalable algorithms as the input size increases, the performance of selection sort decreases. True/False?

13. Consider a situation where swap operation is very costly. In which of the following sorting algorithms is the number of swap operations generally minimal?

a. Bubble sort
b. Selection Sort
c. Insertion Sort
d. Mergesort

14. When all the input array elements are the same, which of the following sorting algorithms takes the least time?

a. Heapsort
b. Selection Sort
c. Insertion Sort
d. Mergesort

15. Which of the following sorting algorithms has the lowest worst-case complexity?

a. Bubble Sort
b. Selection Sort
c. Quicksort
d. Mergesort

16. Which of the following comparisons is correct for the following algorithms given their time complexity in the best case?

a. Mergesort > Quicksort > Insertion sort > Selection sort
b. Insertion sort < Quicksort < Mergesort < Selection sort
c. Mergesort > Selection sort > Quicksort > Insertion sort
d. Mergesort > Quicksort > Selection sort > Insertion sort

9.2.3 Bubble Sort

17. Bubble sort is the simplest sorting algorithm that sorts a given array by repeatedly swapping the adjacent elements if they are out of order.

(a) Explain how bubble sort works.
(b) Give an algorithm to sort an array using bubble sort.
(c) Analyze its time and auxiliary space complexity.

18. Worst and average case time complexity of bubble sort is: $O(n \times n)$. True/False?

19. In bubble sort, worst case occurs when array is reverse sorted. True/False?

20. Best case time complexity of bubble sort is: $O(n)$. Best case occurs when array is already sorted. True/False?

21. Auxiliary space of bubble sort is: $O(1)$. True/False?

22. Boundary Cases: bubble sort takes minimum time (Order of n) when elements are already sorted. True/False?

23. Sorting In Place: Yes. True/False?

24. Bubble sort is stable: Yes. True/False?

25. When does the best case of bubble sort to sort n distinct elements in ascending order occur?

  a. When elements are sorted in ascending order
  b. When elements are sorted in descending order
  c. When elements are not sorted by any order
  d. There is no best case for bubble sort. It always takes $O(n \times n)$ time

26. How would you optimise bubble sort?

9.2.4 Insertion Sort

27. About insertion sort algorithm:

(a) Explain how insertion sort works.
(b) Give an algorithm to sort an array using selection sort.
(c) Analyze its time and auxiliary space complexity.

28. Worst and average case time complexity: $O(n \times n)$. True/False?

29. Worst case occurs when array is reverse sorted. True/False?

30. Best Case Time Complexity: $O(n)$. Best case occurs when array is already sorted. True/False?

31. Auxiliary Space: $O(1)$. True/False?

32. Boundary Cases: Insertion sort takes minimum time (Order of n) when elements are already sorted. True/False?

33. Sorting In Place: Yes. True/False?

34. Stable: Yes. True/False?

35. What is binary insertion sort?

36. How to implement insertion sort for linked list?

37. How to sort the doubly linked list using the insertion sort technique?

38. Compare the best case, average case, worst case of selection sort, insertion sort, and bubble sort.

39. Compare the best case, average case, worst case of selection sort, insertion sort and selection sort.

40. True or False?

(a) Insertion Sort—Inserts the value in the presorted array to sort the set of values in the array.
(b) Selection Sort—Finds the minimum/maximum number from the list and sort it in ascending/descending order.

41. True or False?

(a) Insertion Sort—It is a stable sorting algorithm.
(b) Selection Sort—It is an unstable sorting algorithm.

42. True or False?

(a) Insertion Sort—The best-case time complexity is $O(n)$ when the array is already in ascending order.
(b) Selection Sort—There is no best case the time complexity is $O(n^2)$ in all cases.

43. True or False?

(a) Insertion Sort—The number of comparison operations performed in this sorting algorithm is less than the swapping performed.
(b) Selection Sort—The number of comparison operations performed in this sorting algorithm is more than the swapping performed.

44. True or False?

(a) Insertion Sort—It is more efficient than the Selection sort.
(b) Quicksort—It is less efficient than the Insertion sort.

45. True or False?

(a) Insertion Sort—Here the element is known beforehand, and we search for the correct position to place them.
(b) The location where to put the element is previously known we search for the element to insert at that position.

46. Insertion sort algorithm consist of ...... passes.

a. N
b. N-1
c. N+1
d. N^2

47. .... implementation is similar to the insertion sort.

a. Binary heap
b. Quicksort
c. Mergesort
d. Radix sort

48. Any sorting algorithm that sorts the elements by swapping adjacent elements require N^2 on average. True/False?

49. The average number of inversions in an array of n distinct numbers is ....

a. n(n-1)/4
b. n(n+1)/2
c. n(n-1)/2
d. n(n-1)/3

50. What is the time complexity of insertion sort algorithm for sorting an array of n elements when the input is pre-sorted?

a. n^2
b. O(n log n)
c. O(n)
d. O(log n)

51. In insertion sort, for the following numbers, how will the array elements look like after second pass?

$$35, \ 9, \ 65, \ 52, \ 33, \ 22$$

  a. 9, 22, 33, 35, 52, 65
  b. 9, 33, 35, 52, 65, 22
  c. 9, 35, 52, 65, 33, 22
  d. 9, 35, 65, 52, 33, 22

52. Binary search can be used in insertion sort algorithm to reduce the number of comparisons. True/False?

53. .... is the fastest sorting algorithm for sorting small arrays.

  a. Quicksort
  b. Insertion sort
  c. Shell sort
  d. Heapsort

54. For the best case input, the running time of insertion sort algorithm is?

  a. Linear
  b. Binary
  c. Quadratic
  d. Depends on the input

55. Which of the following cases represent the worst case input for insertion sort?

  a. array in sorted order
  b. array sorted in reverse order
  c. normal unsorted array
  d. large array

56. Which of the following is correct with regard to insertion sort?

  a. It is stable and in-place.
  b. It is unstable and in-place.
  c. It is stable and does not sort in-place.
  d. It is unstable and does not sort in-place.

57. Which of the following sorting algorithm is best suited when the elements are already sorted?

  a. Heapsort
  b. Quicksort
  c. Insertion Sort
  d. Mergesort

58. Let we use binary search to find the correct position for inserting element in insertion sort. What will be the worst case time complexity of insertion sort?

a. O(n log n)
b. $O(n^2)$
c. O(n)
d. O(log n)

59. Insertion sort is an example of an incremental algorithm. True/False?

60. .... is suitable sorting algorithm for sorting arrays that have less than 100 elements.

a. Quicksort
b. Selection Sort
c. Mergesort
d. Insertion Sort

61. Statement 1: In insertion sort, after n passes through the array, the first n elements will be in sorted order.

Statement 2: In insertion sort, after n passes through the array, the first n elements will be n smallest elements in the array.

a. Both the statements are true.
b. Statement 1 is true but statement 2 is false.
c. Statement 1 is false but statement 2 is true.
d. Both the statements are false.

9.2.5   Heapsort

62. Heapsort is a comparison-based sorting technique based on binary heap data structure.

(a) Explain how heapsort works.
(b) Give an algorithm to sort an array using heapsort.
(c) Analyze its time complexity.
(d) Analyze its space complexity (auxiliary space).

63. Sorting a nearly sorted array. Given an array of n elements, where each element is at most k away from its target position, devise an algorithm that sorts in $O(n \log k)$ time.

– use heapsort algorithm
– use insertion sort
– use a balanced binary search tree

64. Write a program for finding k largest elements in an array of any order.

– Use Bubble sort k times
– Use temporary array

 – Use Max Heap
 – Use Order Statistics
 – Use Min Heap
 – Using quicksort partitioning algorithm

65. Which of the following sorting algorithms performs best its typical implementation when applied to an array that is sorted or nearly sorted? (maximum 1 or two elements are misplaced).

  a. Quicksort
  b. Heapsort
  c. Mergesort
  d. Insertion Sort

66. Assume we have an array that every element in array is placed at most k distance from its position in sorted array where k is a positive integer smaller than size of array. Which sorting algorithm can be easily modified for sorting this array and what is the obtainable time complexity?

  a. Insertion sort with time complexity $O(kn)$
  b. Heapsort with time complexity $O(n \log k)$
  c. Quicksort with time complexity $O(k \log k)$
  d. Mergesort with time complexity $O(k \log k)$

67. You have to sort 2 GB of data with only 1000 MB of available main memory. Which sorting algorithm will be most appropriate?

  a. Heapsort
  b. Mergesort
  c. Quicksort
  d. Insertion sort

68. .... is most efficient sorting algorithm to sort string consisting of ASCII characters.

  a. Quicksort
  b. Heapsort
  c. Mergesort
  d. Counting sort

9.2.6 Shell Sort

69. About Shell sort

(a) Explain how shell sort works.
(b) Give an algorithm to sort an array using Shell Sort.
(c) Analyze its time complexity.

(d) Analyze its space complexity (auxiliary space).

70. Shell sort algorithm is the first algorithm to break the quadratic time barrier. True/False?

71. Shell sort algorithm is an example of?

a. External sorting
b. Internal sorting
c. In-place sorting
d. Bottom-up sorting

72. Which of the following sorting algorithms is closely related to shell sort?

a. Selection sort
b. Mergesort
c. Insertion sort
d. Bucket sort

73. What is the general form of Shell's increments?

a. $1, 2, 3, \ldots, n$
b. $1, 3, 7, \ldots, 2k-1$
c. $1, 3, 5, 7, \ldots, k-1$
d. $1, 5, 10, 15, \ldots, k-1$

74. Shell sort is also known as

a. diminishing decrement sort
b. diminishing increment sort
c. partition exchange sort
d. diminishing insertion sort

75. Statement 1: Shell sort is a stable sorting algorithm.
   Statement 2: Shell sort is an in-place sorting algorithm.

a. Both statements are true.
b. Statement 2 is true but statement 1 is false.
c. Statement 2 is false but statement 1 is true.
d. Both statements are false.

76. Shell sort is applied on the elements 28 60 50 38 16 91 82 40 and the chosen decreasing sequence of increments is (5, 3, 1). The result after the first iteration will be

a. 28 60 50 38 16 91 82 40
b. 28 60 38 50 16 91 82 40
c. 28 60 40 38 16 91 82 50
d. 16 60 50 38 28 91 82 40

77. Shell sort is an improvement on

a. Insertion sort
b. Selection sort
c. Binary tree sort
d. Quicksort

78. If Hibbard increments ($h_1 = 1, h_2 = 3, h_3 = 7, \ldots, h_k = 2^k - 1$) are used in a Shell sort implementation, then the best case time complexity will be

a. O(nlogn)
b. O(n)
c. O(n^2)
d. O(logn)

79. Shell sort is more efficient than insertion sort if the length of input arrays is small. True/False?

9.2.7 Introsort

80. Introsort is the in built sorting algorithm used by C++. It is an example of a hybrid sorting algorithm which means it uses more than one sorting algorithm as a routine. It may use quicksort, heapsort or insertion sort depending on the given situation. Explain how Introsort sort works.

81. Which of the following sorting algorithm is not stable?

a. Introsort
b. Brick sort
c. Bubble sort
d. Mergesort

82. Introsort sort is a comparison based sort. True/False?

83. .... is the best case time complexity of introsort.

a. O(n)
b. O(n log n)
c. O(n^2)
d. O(log n)

84. .... is the worst case time complexity of introsort.

a. O(n)
b. O(n log n)
c. O(n^2)

d. O(log n)

85. .... is the average time complexity of introsort.

a. O(n)
b. O(n log n)
c. $O(n^2)$
d. O(log n)

86. .... is the auxiliary space required in introsort.

a. O(n)
b. O(n log n)
c. $O(n^2)$
d. O(log n)

87. Heapsort preferred over mergesort for introsort implementation, because

a. heapsort is faster.
b. heapsort requires less space.
c. heapsort is easy to implement.
d. heapsort requires less time.

88. Insertion sort preferred over other sorting algorithms (such as selection sort and bubble sort) for introsort implementation, because

a. insertion sort is faster and adaptive.
b. insertion sort requires less space.
c. insertion sort is easy to implement.
d. insertion sort requires less time.

89. What is the cut-off for switching from quicksort to insertion sort in the implementation of introsort?

a. 4
b. 8
c. 16
d. 32

90. What is the cut-off for switching from quicksort to heapsort in the implementation of introsort?

a. 16
b. n^2
c. n log(n)
d. 2 log(n)

91. .... sorting algorithm will be preferred when the size of partition is between 16 and $2log(n)$ while implementing introsort.

a. Quicksort
b. Insertion sort
c. Heapsort
d. Mergesort

9.2.8   Tim Sort

92. Tim sort has been python's standard sorting algorithm since its version 2.3. It is an example of hybrid sorting algorithm which means it uses more than one sorting algorithm as a routine. It is derived from insertion sort and mergesort. Explain how Tim sort works.

93. Tim sort starts sorting a given array by using ....

   a. Selection sort
   b. Quicksort
   c. Insertion sort
   d. Mergesort

94. Tim sort is a comparison based sort. True/False?

95. .... is the best case time complexity of Tim sort.

   a. O(n)
   b. O(n log n)
   c. $O(n^2)$
   d. O(log n)

96. .... is the worst case time complexity of Tim sort.

   a. O(n)
   b. O(n log n)
   c. $O(n^2)$
   d. O(log n)

97. .... is the average time complexity of Tim sort.

   a. O(n)
   b. O(n log n)
   c. $O(n^2)$
   d. O(log n)

98. .... is the auxiliary space required in Tim sort.

   a. O(n)
   b. O(n log n)
   c. $O(n^2)$
   d. O(log n)

99. Which of the following algorithm is implemented internally in java when we use function *arrays.sort()*?

   a. Intro sort
   b. Quicksort
   c. Tim sort

d. Mergesort

100. In which case will tim sort will work as an insertion sort?

a. When number of elements are less than 64.
b. When number of elements are greater than 64.
c. When number of elements are less than size of run.
d. When number of elements are less than 32.

9.2.9   Binary Tree Sort

101. In binary tree sort, we first construct the BST and then we perform ....
traversal to get the sorted order.

a. inorder
b. postorder
c. preorder
d. level order

102. What is the worst case time complexity of the binary tree sort?

a. $O(n)$
b. $O(n \log n)$
c. $O(n^2)$
d. $O(\log n)$

103. What is the best case time complexity of the binary tree sort?

a. $O(n)$
b. $O(n \log n)$
c. $O(n^2)$
d. $O(\log n)$

104. Binary tree sort is an in-place sorting algorithm. True/False?

105. Which of the following is false about binary tree sort and quicksort algorithms?

a. They have same time complexity.
b. Binary tree sort used binary search tree as work area.
c. As the number of elements for sorting increases, binary tree sorting becomes more and more efficient.
d. Both are in place sorting algorithms.

106. Which of the following sorting algorithms can be considered as improvement to the binary tree sort?

a. Heapsort
b. Quicksort
c. Selection sort
d. Insertion sort

9.2.10   Counting Sort

107. Counting sort is a linear time sorting algorithm that sort in $O(n + k)$ time when elements are in the range from 1 to k.

(a) Explain how Counting sort works.
(b) Give an algorithm to sort an array using Counting Sort.
(c) Analyze its time complexity.
(d) Analyze its space complexity (auxiliary space).

708. Which of the following is not correct about comparison based sorting algorithms?

  a. For a random input array, the minimum possible time complexity is $O(n\log n)$.
  b. They can be made stable by using position as a criteria when two elements are compared.
  c. Counting sort is not a comparison based sorting algorithm.
  d. Heapsort is not a comparison based sorting algorithm.

109. How many comparisons will be made to sort the array arr $= \{1, 5, 3, 8, 2\}$ using counting sort?

  a. 5
  b. 7
  c. 9
  d. 0

110. Which of the following is not an example of non comparison sort?

  a. Bubble sort
  b. Counting sort
  c. Radix sort
  d. Bucket sort

111. Counting sort algorithm is efficient when range of data to be sorted is fixed. True/False?

112. If the elements are in the range from 1 to n^2, the counting sort can not be used because it will take $O(n^2)$ which is worse than comparison-based sorting algorithms. True/False?

113. Counting sort uses an extra constant space proportional to range of data. True/False?

114. Which sorting algorithms is most efficient to sort string consisting of ASCII characters?

  a. Quicksort
  b. Heapsort

c. Mergesort
d. Counting sort

115. Find duplicates in an array with values 1 to N using counting sort.

116. Time complexity of counting sort is given as $O(n + k)$ where n is the number of input elements and k is the range of input. So if range of input is not significantly larger than number of elements in the array then it proves to be very efficient. True/False?

117. Which of the following sorting techniques is most efficient if the range of input data is not significantly greater than a number of elements to be sorted?

a. Selection sort
b. Bubble sort
c. Counting sort
d. Insertion sort

118. What is the auxiliary space requirement of counting sort?

a. O(1)
b. O(n)
c. O(log n)
d. O(n+k), k = range of input

119. It is not possible to implement counting sort when any of the input element has negative value. True/False?

120. Which of the following uses the largest amount of auxiliary space for sorting?

a. Bubble sort
b. Counting sort
c. Quicksort
d. Heapsort

121. The time complexity of counting sort remains unvaried in all the three cases (best, average and worst case). It is given by $O(n + k)$. True/False?

122. Counting sort is used as a sub routine for radix sort as it is a stable and non comparison based sorting algorithm. True/False?

123. What is the advantage of counting sort over quicksort?

a. Counting sort has lesser time complexity when range is comparable to number of input elements.
b. Counting sort has lesser space complexity.
c. Counting sort is not a comparison based sorting technique.
d. It has no advantage.

124. Which of the following algorithm takes non linear time for sorting?

    a. Counting sort
    b. Quicksort
    c. Bucket sort
    d. Radix sort

125. Sort the following numbers in ascending order using Counting sort.

    (a) 5, 1, 10, 4, 30, 24, 400, 350, 6, 200
    (b) 1, 4, 5, 6, 10, 24, 30, 200, 350, 400
    (c) 400, 350, 200, 30, 24, 10, 6, 5, 3, 1

126. What are the characteristics of input array (best case input) so that Counting sorting has the best performance?

127. What are the characteristics of input array (worst case input) so that Counting sorting has the worst performance?

128. Analyze Counting sort time complexity (worst case, average case, and best case).

129. Analyze Counting sort space complexity (auxiliary space).

130. Counting sort is a stable sorting algorithm. True/False?

131. Counting sort is an in place sorting algorithm. True/False?

132. Counting sort is a comparison based sorting algorithm. True/False?

133. Implement the sorting algorithms shown below. Complete the following table. Implement the following sorting algorithms and then write the execution time of the algorithm on arrays of different sizes in the table below. In this table, A indicates that the input data is in ascending order, D indicates that the input data is in descending order, and R indicates that the input data is random.

Input size	n = 10			n = 32			n = 128			n = 1024			n = 10000		
Order of input	A	D	R	A	D	R	A	D	R	A	D	R	A	D	R
Insertion sort															
Quicksort															
Counting sort															

9.2.11   Radix Sort

134. Radix sort algorithm sorts the numbers digit by digit starting from least significant digit to most significant digit. Radix sort uses counting sort as a subroutine to sort.

(a) Explain how radix sort works.
(b) Give an algorithm to sort an array using radix sort.
(c) Analyze its time complexity.
(d) Analyze its space complexity (auxiliary space).

135. You are given an array containing numbers in range from 1 to n^4, which of the following algorithm can be used to sort these number in linear time?

  a. Not possible to sort in linear time
  b. Radix Sort
  c. Counting Sort
  d. Quicksort

136. Suppose each element is a five-digit octal number. .... is the maximum number of comparisons required to sort nine items using radix sort.

  a. 45
  b. 72
  c. 360
  d. 450

9.2.12   Mergesort

137. Mergesort uses .... technique to implement sorting.

  a. backtracking
  b. greedy algorithm
  c. divide and conquer
  d. dynamic programming

138. .... is the auxiliary space complexity of mergesort.

  a. O(1)
  b. O(log n)
  c. O(n)
  d. O(n log n)

139. .... method is used for sorting elements in mergesort.

  a. merging
  b. partitioning

c. selection

d. exchanging

140. Regarding mergesort, which statement is incorrect?

a. It is a comparison based sort.

b. It is an adaptive algorithm.

c. It is not an in place algorithm.

d. It is stable algorithm.

141. .... is not an in place sorting algorithm.

a. Mergesort

b. Quicksort

c. Heapsort

d. Insertion sort

142. Mergesort is preferred for arrays over linked lists. True/False?

143. Which of the following sorting algorithms has the lowest worst-case complexity?

a. Mergesort

b. Bubble Sort

c. Quicksort

d. Selection Sort

144. Which of the following statements is true about mergesort?

a. Mergesort performs better than quicksort when data is accessed from slow sequential memory.

b. Mergesort is stable sort by nature.

c. Mergesort outperforms heapsort in most of the practical situations.

d. All of the above.

145. Assume that a mergesort algorithm in the worst case takes 30 s for an input of size 64. Which of the following most closely approximates the maximum input size of a problem that can be solved in 6 min?

a. 256

b. 512

c. 1024

d. 2048

146. Compare the standard mergesort and in place mergesort.

147. Choose the incorrect statement about mergesort from the following?

a. Both standard mergesort and in-place mergesort are stable.

b. Standard mergesort has greater time complexity than in-place mergesort.

c. Standard mergesort has greater space complexity than in-place mergesort.

d. In place mergesort has O(log n) space complexity.

148. Compare the standard mergesort and bottom up mergesort.

149. The auxiliary space complexity of bottom up mergesort is same as standard mergesort as both uses the same algorithm for merging the sorted arrays which takes $O(n)$ space. But bottom up mergesort does not need to maintain a call stack. True/False?

150. Bottom up mergesort uses recursion. True/False?

151. Bottom up mergesort like standard mergesort is a stable sort. This implies that the relative position of equal valued elements in the input and sorted array remain same. True/False?

152. Choose the correct statement about bottom up mergesort from the following?

   a. Bottom up mergesort has greater time complexity than standard mergesort.
   b. Bottom up mergesort has lesser time complexity than standard mergesort.
   c. Bottom up mergesort saves auxiliary space required on call stack.
   d. Bottom up mergesort uses recursion.

9.2.13 QuickSort

153. Is quicksort stable? True/False?

154. The default implementation is not stable. However any sorting algorithm can be made stable by considering indexes as comparison parameter. True/False?

155. Is quicksort in-place? True/False?

156. What is 3-way quicksort?

157. Can we implement quicksort iteratively?

158. Can we apply quicksort on singly linked list.

159. Can we apply quicksort on doubly linked list.

160. Why mergesort is preferred over quicksort for linked lists?

161. How to optimize quicksort to take up $O(log\ n)$ extra space in worst case?

162. .... is recurrence relation indicating worst case of quicksort and .... is the worst case time complexity of quicksort.

  a. $T(n) = T(n-2) + O(n) - O(n^2)$
  b. $T(n) = T(n-1) + O(n) - O(n^2)$
  c. $T(n) = 2T(n/2) + O(n) - O(n \log n)$
  d. $T(n) = T(n/10) + T(9n/10) + O(n) - O(n \log n)$

163. Suppose a quicksort implementation uses an $O(n)$ time algorithm to find median of a given array and then using this median as pivot. What will be the time complexity of this version of quicksort at the worst case?

  a. $O(n^2 \log n)$
  b. $O(n^2)$
  c. $O(n \log n \log n)$
  d. $O(n \log n)$

164. .... has the best performance when applied on sorted or almost sorted arrays (maximum 1 or two elements are misplaced).

  a. Quicksort
  b. Heapsort
  c. Mergesort
  d. Insertion Sort

165. In a version of the quicksort algorithm, the pivot procedure splits the given array into two sub-lists each of which contains at least one-fifth of the elements. Let $T(n)$ be the number of comparisons required to sort n elements. Then

  a. $T(n) \leq 2T(n/5) + n$
  b. $T(n) \leq T(n/5) + T(4n/5) + n$
  c. $T(n) \leq 2T(4n/5) + n$
  d. $T(n) \leq 2T(n/2) + n$

166. Which of the following methods is the most effective for picking the pivot element?

  a. first element
  b. last element
  c. median-of-three partitioning
  d. random element

167. Compared to picking a first, last or random element as a pivot, median-of-three partitioning method is the best way to choose an appropriate pivot element. True/False?

168. To sort the sub arrays in quicksort, ... algorithm can be utilized.

  a. mergesort
  b. shell sort
  c. insertion sort
  d. bubble sort

169. Choosing the first element as pivot is the worst method because if the input is pre-sorted or in reverse order, then the pivot provides a poor partition. True/False?

170. Which among the following is the best cut-off range to perform insertion sort within a quicksort?

a. $N = 0–5$
b. $N = 5–20$
c. $N = 20–30$
d. $N > 30$

171. Consider the quicksort algorithm which sorts elements in ascending order using the first element as pivot. Then which of the following input sequence will require a maximum number of comparisons when this algorithm is applied on it?

a. 22 25 56 67 89
b. 52 25 76 67 89
c. 22 25 76 67 50
d. 52 25 89 67 76

172. A machine needs a minimum of 200 sec to sort 1000 elements by quicksort. The minimum time needed to sort 200 elements will be approximately

a. 60.2 s
b. 45.54 s
c. 31.11 s
d. 20 s

173. In a randomized quicksort,

a. the pivot is the leftmost element.
b. the pivot the rightmost element.
c. any element in the array is selected as the pivot.
d. a random number is generated which is used as the pivot.

174. Randomized quicksort helps in avoiding the worst case time complexity of $O(n^2)$ which occurs in case when the input array is already sorted. However the average case and best case time complexities remain unaltered. True/False?

175. Auxiliary space complexity of randomized quicksort is $O(log\ n)$ which is used for storing call stack formed due to recursion. True/False?

176. The average case time complexity of randomized quicksort is same as that of standard quicksort as randomized quicksort only helps in preventing the worst case. It is equal to $O(nlog\ n)$. True/False?

177. Randomized quicksort is an in place sort. True/False?

178. Randomized quicksort is a stable sort. True/False?

179. Which of the following is incorrect about randomized quicksort?

a. It has the same time complexity as standard quicksort.
b. It has the same space complexity as standard quicksort.
c. It is an in-place sorting algorithm.
d. It cannot have a time complexity of $O(n^2)$ in any case.

180. In the median of three technique, the median of first, last and middle element is chosen as the pivot. It is done so as to avoid the worst case of quicksort in which the time complexity shoots to $O(n^2)$. True/False?

9.2.14 Shell Sort

181. Shell Sort is mainly a variation of insertion sort.

(a) Explain how shell sort works.
(b) Give an algorithm to sort an array using Shell Sort.
(c) Analyze its time complexity.
(d) Analyze its space complexity (auxiliary space).

182. Shell sort algorithm is an example of?

a. External sorting
b. Internal sorting
c. In-place sorting
d. Bottom-up sorting

183. Shell sort is an example of internal sorting because sorting of elements is done internally using an array. True/False?

184. Given an array of the following elements 81, 94, 11, 96, 12, 35, 17, 95, 28, 58, 41, 75, 15. What will be the sorted order after 5-sort?

a. 11, 12, 15, 17, 28, 35, 41, 58, 75, 81, 94, 95, 96
b. 28, 12, 11, 35, 41, 58, 17, 94, 75, 81, 96, 95, 15
c. 35, 17, 11, 28, 12, 41, 75, 15, 96, 58, 81, 94, 95
d. 12, 11, 15, 17, 81, 94, 85, 96, 28, 35, 41, 58, 75

185. Mathematically, the lower bound analysis for shell sort using Hibbard's increments is $O(N^{3/2})$. True/False?

186. What is the general form of Shell's increments?

a. 1, 2, 3, ..., n
b. 1, 3, 7, ..., 2k-1

   c. 1, 3, 5, 7, ..., k-1
   d. 1 ,5, 10, 15, ..., k-1

187. What is the worst case analysis of shell sort using Shell's increments?

   a. O(N)
   b. $O(N^2)$
   c. $O(N^{1/2})$
   d. $O(N^{3/2})$

188. What is the worst case analysis of Shell sort using Sedgewick's increments?

   a. $O(N^2)$
   b. $O(N^{3/2})$
   c. $O(N^{4/3})$
   d. $O(N^{5/4})$

9.2.15 Cycle Sort

189. About Cycle Sort

(a) Explain how cycle sort works.
(b) Give an algorithm to sort an array using Cycle Sort.
(c) Analyze its time complexity.
(d) Analyze its space complexity (auxiliary space).

190. The time complexity of cycle sort is $O(n^2)$ in any case. True/False?

191. Cycle sort has an auxiliary space complexity of $O(1)$. So it qualifies to be an in-place sort. True/False?

192. Which of the following is an advantage of cycle sort?

   a. It can sort large arrays efficiently.
   b. It has a low time complexity.
   c. It requires minimal write operations.
   d. It is an adaptive sorting algorithm.

193. Cycle sort is a slow sorting algorithm but it requires a minimum number of write operations in order to sort a given array. So it is useful when the write/swap operation is expensive. True/False?

9.2.16 Library Sort

194. Library sort does not require the entire input data at the beginning itself in order to sort the array. It rather creates a partial solution in every step, so future elements are

not required to be considered. Hence it is an online sorting algorithm like insertion sort.

(a) Explain how Library Sort works.
(b) Give an algorithm to sort an array using Library Sort.
(c) Analyze its time complexity.
(d) Analyze its space complexity (auxiliary space).

195. Library sort is an online sorting algorithm. True/False?

196. Which of the following is a disadvantage of library sort when compared to insertion sort?

   a. Library sort has greater time complexity.
   b. Library sort has greater space complexity.
   c. Library sort makes more comparisons.
   d. It has no significant disadvantage.

197. Library sort requires the use of Insertion sort and binary search in its code. So it is a modified version of insertion sort. True/False?

198. Library sort is a stable algorithm. True/False?

199. Which of the following sorting algorithm requires the use of binary search in their implementation?

   a. Radix sort
   b. Library sort
   c. Odd-even sort
   d. Bead sort

200. Library sort is a comparison based sort. True/False?

201. Library sort uses binary search in order to insert elements in the sorted segment of the array which reduces its time complexity. So the average time complexity of library sort is $O(nlog\ n)$. True/False?

202. The best case time complexity of library sort is O(n). It occurs in the case when the input is already/almost sorted. True/False?

203. The worst case time complexity of library sort is the same as that of insertion sort. The worst case time complexity is $O(n^2)$. True/False?

204. Library sort has a better average time complexity as compared to insertion sort because it uses binary search for finding the index where the element has to be inserted in the sorted array. This makes the process faster. True/False?

205. Which of the following is an adaptive sorting algorithm?

a. Library sort
b. Mergesort
c. Heapsort
d. Selection sort

206. Library sort is an adaptive algorithm. It is because the time complexity of the algorithm improves when the input array is almost sorted. True/False?

9.2.17 Strand Sort

207. About Strand Sort

(a) Explain how strand sort works.
(b) Give an algorithm to sort an array using Strand Sort.
(c) Analyze its time complexity.
(d) Analyze its space complexity (auxiliary space).

208. Which one of the following sorting algorithm requires recursion?

a. Pigeonhole sort
b. Strand sort
c. Insertion sort
d. Counting sort

209. Strand sort requires the use of recursion for implementing its algorithm. True/False?

210. Strand sort is most efficient for data stored in?

a. linked list
b. arrays
c. trees
d. graphs

211. Strand sort is most efficient when data is stored in a linked list as it involves many insertions and deletions which is performed quite efficiently with the help of a linked list. True/False?

212. In which of the following case strand sort is most efficient?

  a. When input array is already sorted.
  b. When input array is reverse sorted.
  c. When input array is large.
  d. When input array is has randomly spread elements.

213. The best case of strand sort occurs when the input array is already sorted. In this case, it has linear time complexity. True/False?

214. The auxiliary space complexity of strand sort is $O(n)$. It is because a sub-list of size n has to be maintained. True/False?

215. Strand sort has an auxiliary space complexity of $O(n)$. So it is not an in place sorting algorithm. True/False?

216. Strand sort is a comparison based sorting algorithm. True/False?

217. Strand sort is a stable sorting algorithm. True/False?

218. Average case time complexity of strand sort is $O(n^2)$. So it is not as efficient as quicksort or mergesort. True/False?

219. Best case time complexity of strand sort is $O(n)$. It occurs in the case where the input array is already sorted. True/False?

220. Worst case time complexity of strand sort is $O(n^2)$. It occurs in the case where the input array is in reverse sorted order. True/False?

221. Strand sort is an adaptive sorting algorithm. This is because it gives a better performance when the input list is almost sorted. True/False?

9.2.18   Cocktail Sort

222. About Cocktail Sort

  (a) Explain how Cocktail Sort works.
  (b) Give an algorithm to sort an array using Cocktail Sort.
  (c) Analyze its time complexity.
  (d) Analyze its space complexity (auxiliary space).

223. Cocktail sort is also known by the name of ripple sort. It is also known by other names like—bidirectional bubble sort, cocktail shaker sort, shuttle sort, and shuffle sort. True/False?

224. Cocktail sort is very similar to bubble sort. It works by traversing an array in both directions alternatively. It compares the adjacent elements in each iteration and swaps the ones which are out of order. True/False?

225. In cocktail sort manipulation is done on the input array itself. So no extra space is required to perform sorting. Thus it requires constant auxiliary space. True/False?

226. Cocktail sort is a stable sorting algorithm. True/False?

227. Cocktail sort is an in place sorting technique as it only requires constant auxiliary space for manipulating the input array. Rest all other options are not in place. True/False?

228. Cocktail sort is a comparison based sort. True/False?

229. Worst case complexity is observed when the input array is reverse sorted. This is the same as the worst case complexity of bubble sort. True/False?

230. Best case complexity is observed when the input array is already sorted. This is the same as the best case complexity of bubble sort. True/False?

231. Cocktail sort takes $O(n^2)$ time on average as it keeps on applying bubble sort on the elements in two phases until they are sorted. This is the same as the average time complexity of bubble sort. True/False?

232. Bubble sort performs better as compared to cocktail sort. True/False?

233. Both bubble sort and cocktail sort has the same time complexities. But cocktail sort has a comparatively better performance. True/False?

9.2.19 Comb Sort

234. About Comb Sort

(a) Explain how comb sort works.
(b) Give an algorithm to sort an array using Comb Sort.
(c) Analyze its time complexity.
(d) Analyze its space complexity (auxiliary space).

235. Comb sort is an improved version of

a. Selection sort
b. Bubble sort
c. Insertion sort

d. Mergesort

236. Worst case complexity is observed when the input array is reverse sorted. This is same as the worst case complexity of bubble sort. True/False?

237. Auxiliary space used by comb sort is $O(1)$ as it does not use any extra space for manipulating the input. True/False?

238. Comb sort is a stable sorting algorithm. True/False?

239. Best case complexity for comb sort and bubble sort is $O(nlog\ n)$ and $O(n)$ respectively. It occurs when the input array is already sorted. True/False?

240. What is the advantage of comb sort over mergesort?

   a. Comb sort is an in place sorting algorithm.
   b. Comb sort is a stable sorting algorithm.
   c. Comb sort is more efficient.
   d. It has no advantage.

9.2.20   Gnome Sort

241. Explain how Gnome sort works.

242. Give an algorithm to sort an array using Gnome sort.

243. Sort the following numbers in ascending order using Gnome sort.

   (a) 5, 1, 10, 4, 30, 24, 400, 350, 6, 200
   (b) 1, 4, 5, 6, 10, 24, 30, 200, 350, 400
   (c) 400, 350, 200, 30, 24, 10, 6, 5, 3, 1

244. What are the characteristics of input array (best case input) so that Gnome sorting has the best performance?

245. What are the characteristics of input array (worst case input) so that Gnome sorting has the worst performance?

246. Analyze Gnome sort time complexity (worst case, average case, and best case).

247. Analyze Gnome sort space complexity (auxiliary space).

248. Gnome sort is a stable sorting algorithm. True/False?

249. Gnome sort is an in place sorting algorithm. True/False?

250. Gnome sort is a comparison based sorting algorithm. True/False?

251. Implement the sorting algorithms shown below. Complete the following table. Implement the following sorting algorithms and then write the execution time of the algorithm on arrays of different sizes in the table below. In this table, A indicates that the input data is in ascending order, D indicates that the input data is in descending order, and R indicates that the input data is random.

Input size	n = 10			n = 32			n = 128			n = 1024			n = 10000		
Order of input	A	D	R	A	D	R	A	D	R	A	D	R	A	D	R
Insertion sort															
Quicksort															
Gnome sort															

9.2.21 Bogo Sort

252. Explain how Bogo sort works.

253. Give an algorithm to sort an array using Bogo sort.

254. Sort the following numbers in ascending order using Bogo sort.

(a) 5, 1, 10, 4, 30, 24, 400, 350, 6, 200
(b) 1, 4, 5, 6, 10, 24, 30, 200, 350, 400
(c) 400, 350, 200, 30, 24, 10, 6, 5, 3, 1

255. What are the characteristics of input array (best case input) so that Bogo sorting has the best performance?

256. What are the characteristics of input array (worst case input) so that Bogo sorting has the worst performance?

257. Analyze Bogo sort time complexity (worst case, average case, and best case).

258. Analyze Bogo sort space complexity (auxiliary space).

259. Bogo sort is a stable sorting algorithm. True/False?

260. Bogo sort is an in place sorting algorithm. True/False?

261. Bogo sort is a comparison based sorting algorithm. True/False?

262. Implement the sorting algorithms shown below. Complete the following table. Implement the following sorting algorithms and then write the execution time of the algorithm on arrays of different sizes in the table below. In this table, A indicates that the input data is in ascending order, D indicates that the input data is in descending order, and R indicates that the input data is random.

Input size	n = 10			n = 32			n = 128			n = 1024			n = 10000		
Order of input	A	D	R	A	D	R	A	D	R	A	D	R	A	D	R
Insertion sort															
Quicksort															
Bogo sort															

9.2.22   Sleep Sort

263. Explain how Sleep sort works.

264. Give an algorithm to sort an array using Sleep sort.

265. Sort the following numbers in ascending order using Sleep sort.

(a) 5, 1, 10, 4, 30, 24, 400, 350, 6, 200
(b) 1, 4, 5, 6, 10, 24, 30, 200, 350, 400
(c) 400, 350, 200, 30, 24, 10, 6, 5, 3, 1

266. What are the characteristics of input array (best case input) so that Sleep sorting has the best performance?

267. What are the characteristics of input array (worst case input) so that Sleep sorting has the worst performance?

268. Analyze Sleep sort time complexity (worst case, average case, and best case).

269. Analyze Sleep sort space complexity (auxiliary space).

270. Sleep sort is a stable sorting algorithm. True/False?

271. Sleep sort is an in place sorting algorithm. True/False?

272. Sleep sort is a comparison based sorting algorithm. True/False?

273. Implement the sorting algorithms shown below. Complete the following table. Implement the following sorting algorithms and then write the execution time of the algorithm on arrays of different sizes in the table below. In this table, A indicates that the input data is in ascending order, D indicates that the input data is in descending order, and R indicates that the input data is random.

Input size	n = 10			n = 32			n = 128			n = 1024			n = 10000		
Order of input	A	D	R	A	D	R	A	D	R	A	D	R	A	D	R
Insertion sort															
Quicksort															
Sleep sort															

9.2.23 Pigeonhole Sort

274. Explain how Pigeonhole sort works.

275. Give an algorithm to sort an array using Pigeonhole sort.

276. Sort the following numbers in ascending order using Pigeonhole sort.

(a) 5, 1, 10, 4, 30, 24, 400, 350, 6, 200
(b) 1, 4, 5, 6, 10, 24, 30, 200, 350, 400
(c) 400, 350, 200, 30, 24, 10, 6, 5, 3, 1

277. What are the characteristics of input array (best case input) so that Pigeonhole sorting has the best performance?

278. What are the characteristics of input array (worst case input) so that Pigeonhole sorting has the worst performance?

279. Analyze Pigeonhole sort time complexity (worst case, average case, and best case).

280. Analyze Pigeonhole sort space complexity (auxiliary space).

281. Pigeonhole sort is a stable sorting algorithm. True/False?

282. Pigeonhole sort is an in place sorting algorithm. True/False?

283. Pigeonhole sort is a comparison based sorting algorithm. True/False

284. Implement the sorting algorithms shown below. Complete the following table. Implement the following sorting algorithms and then write the execution time of the algorithm on arrays of different sizes in the table below. In this table, A indicates that the input data is in ascending order, D indicates that the input data is in descending order, and R indicates that the input data is random.

Input size	$n = 10$			$n = 32$			$n = 128$			$n = 1024$			$n = 10000$		
Order of input	A	D	R	A	D	R	A	D	R	A	D	R	A	D	R
Insertion sort															
Quicksort															
Pigeonhole sort															

9.2.24 Bucket Sort (Uniform Keys)

285. Explain how Bucket sort works.

286. Give an algorithm to sort an array using Bucket sort.

287. Sort the following numbers in ascending order using Bucket sort.

(a) 5, 1, 10, 4, 30, 24, 400, 350, 6, 200
(b) 1, 4, 5, 6, 10, 24, 30, 200, 350, 400
(c) 400, 350, 200, 30, 24, 10, 6, 5, 3, 1

288. What are the characteristics of input array (best case input) so that Bucket sorting has the best performance?

289. What are the characteristics of input array (worst case input) so that Bucket sorting has the worst performance?

290. Analyze Bucket sort time complexity (worst case, average case, and best case).

291. Analyze Bucket sort space complexity (auxiliary space).

292. Bucket sort is a stable sorting algorithm. True/False?

293. Bucket sort is an in place sorting algorithm. True/False?

294. Bucket sort is a comparison based sorting algorithm. True/False?

295. Implement the sorting algorithms shown below. Complete the following table. Implement the following sorting algorithms and then write the execution time of the algorithm on arrays of different sizes in the table below. In this table, A indicates that the input data is in ascending order, D indicates that the input data is in descending order, and R indicates that the input data is random.

Input size	n = 10			n = 32			n = 128			n = 1024			n = 10000		
Order of input	A	D	R	A	D	R	A	D	R	A	D	R	A	D	R
Insertion sort															
Quicksort															
Bucket sort															

9.2.25 Bead Sort

296. Explain how Bead sort works.

297. Give an algorithm to sort an array using Bead sort.

298. Sort the following numbers in ascending order using Bead sort.

(a) 5, 1, 10, 4, 30, 24, 400, 350, 6, 200
(b) 1, 4, 5, 6, 10, 24, 30, 200, 350, 400
(c) 400, 350, 200, 30, 24, 10, 6, 5, 3, 1

299. What are the characteristics of input array (best case input) so that Bead sorting has the best performance?

300. What are the characteristics of input array (worst case input) so that Bead sorting has the worst performance?

301. Analyze Bead sort time complexity (worst case, average case, and best case).

302. Analyze Bead sort space complexity (auxiliary space).

303. Bead sort is a stable sorting algorithm. True/False?

304. Bead sort is an in place sorting algorithm. True/False?

305. Bead sort is a comparison based sorting algorithm. True/False?

306. Implement the sorting algorithms shown below. Complete the following table. Implement the following sorting algorithms and then write the execution time of the algorithm on arrays of different sizes in the table below. In this table, A indicates that the input data is in ascending order, D indicates that the input data is in descending order, and R indicates that the input data is random.

Input size	n = 10			n = 32			n = 128			n = 1024			n = 10000		
Order of input	A	D	R	A	D	R	A	D	R	A	D	R	A	D	R
Insertion sort															
Quicksort															
Bead sort															

9.2.26   Pancake Sort

307. Explain how Pancake sort works.

308. Give an algorithm to sort an array using Pancake sort.

309. Sort the following numbers in ascending order using Pancake sort.

(a) 5, 1, 10, 4, 30, 24, 400, 350, 6, 200
(b) 1, 4, 5, 6, 10, 24, 30, 200, 350, 400
(c) 400, 350, 200, 30, 24, 10, 6, 5, 3, 1

310. What are the characteristics of input array (best case input) so that Pancake sorting has the best performance?

311. What are the characteristics of input array (worst case input) so that Pancake sorting has the worst performance?

312. Analyze Pancake sort time complexity (worst case, average case, and best case).

313. Analyze Pancake sort space complexity (auxiliary space).

314. Pancake sort is a stable sorting algorithm. True/False?

315. Pancake sort is an in place sorting algorithm. True/False?

316. Pancake sort is a comparison based sorting algorithm. True/False?

317. Implement the sorting algorithms shown below. Complete the following table. Implement the following sorting algorithms and then write the execution time of the algorithm on arrays of different sizes in the table below. In this table, A indicates that the input data is in ascending order, D indicates that the input data is in descending order, and R indicates that the input data is random.

Input size	n = 10			n = 32			n = 128			n = 1024			n = 10000		
Order of input	A	D	R	A	D	R	A	D	R	A	D	R	A	D	R
Insertion sort															
Quicksort															
Pancake sort															

9.2.27 Odd-Even Sort

318. Explain how Odd-Even sort works.

319. Give an algorithm to sort an array using Odd-Even sort.

320. Sort the following numbers in ascending order using Odd-Even sort.

(a) 5, 1, 10, 4, 30, 24, 400, 350, 6, 200
(b) 1, 4, 5, 6, 10, 24, 30, 200, 350, 400
(c) 400, 350, 200, 30, 24, 10, 6, 5, 3, 1

321. What are the characteristics of input array (best case input) so that Odd-Even sorting has the best performance?

322. What are the characteristics of input array (worst case input) so that Odd-Even sorting has the worst performance?

323. Analyze Odd-Even sort time complexity (worst case, average case, and best case).

324. Analyze Odd-Even sort space complexity (auxiliary space).

325. Odd-Even sort is a stable sorting algorithm. True/False?

326. Odd-Even sort is an in place sorting algorithm. True/False?

327. Odd-Even sort is a comparison based sorting algorithm. True/False?

328. Implement the sorting algorithms shown below. Complete the following table. Implement the following sorting algorithms and then write the execution time of the algorithm on arrays of different sizes in the table below. In this table, A indicates that the input data is in ascending order, D indicates that the input data is in descending order, and R indicates that the input data is random.

Input size	n = 10			n = 32			n = 128			n = 1024			n = 10000		
Order of input	A	D	R	A	D	R	A	D	R	A	D	R	A	D	R
Insertion sort															
Quicksort															
Odd-even sort															

9.2.28 Stooge Sort

329. Explain how Stooge sort works.

330. Give an algorithm to sort an array using Stooge sort.

331. Sort the following numbers in ascending order using Stooge sort.

(a) 5, 1, 10, 4, 30, 24, 400, 350, 6, 200
(b) 1, 4, 5, 6, 10, 24, 30, 200, 350, 400
(c) 400, 350, 200, 30, 24, 10, 6, 5, 3, 1

332. What are the characteristics of input array (best case input) so that Stooge sorting has the best performance?

333. What are the characteristics of input array (worst case input) so that Stooge sorting has the worst performance?

334. Analyze Stooge sort time complexity (worst case, average case, and best case).

335. Analyze Stooge sort space complexity (auxiliary space).

336. Stooge sort is a stable sorting algorithm. True/False?

337. Stooge sort is an in place sorting algorithm. True/False?

338. Stooge sort is a comparison based sorting algorithm. True/False?

339. Implement the sorting algorithms shown below. Complete the following table. Implement the following sorting algorithms and then write the execution time of the algorithm on arrays of different sizes in the table below. In this table, A indicates that the input data is in ascending order, D indicates that the input data is in descending order, and R indicates that the input data is random.

Input size	$n = 10$			$n = 32$			$n = 128$			$n = 1024$			$n = 10000$		
Order of input	A	D	R	A	D	R	A	D	R	A	D	R	A	D	R
Insertion sort															
Quicksort															
Stooge sort															

9.2.29 Permutation Sort

340. Explain how Permutation sort works.

341. Give an algorithm to sort an array using Permutation sort.

342. Sort the following numbers in ascending order using Permutation sort.

(a) 5, 1, 10, 4, 30, 24, 400, 350, 6, 200
(b) 1, 4, 5, 6, 10, 24, 30, 200, 350, 400
(c) 400, 350, 200, 30, 24, 10, 6, 5, 3, 1

343. What are the characteristics of input array (best case input) so that Permutation sorting has the best performance?

344. What are the characteristics of input array (worst case input) so that Permutation sorting has the worst performance?

345. Analyze Permutation sort time complexity (worst case, average case, and best case).

346. Analyze Permutation sort space complexity (auxiliary space).

347. Permutation sort is a stable sorting algorithm. True/False?

348. Permutation sort is an in place sorting algorithm. True/False?

349. Permutation sort is a comparison based sorting algorithm. True/False?

350. Implement the sorting algorithms shown below. Complete the following table. Implement the following sorting algorithms and then write the execution time of the algorithm on arrays of different sizes in the table below. In this table, A indicates that the input data is in ascending order, D indicates that the input data is in descending order, and R indicates that the input data is random.

Input size	n = 10			n = 32			n = 128			n = 1024			n = 10000		
Order of input	A	D	R	A	D	R	A	D	R	A	D	R	A	D	R
Insertion sort															
Quicksort															
Permutation sort															

9.2.30   Recursive Bubble Sort

351. Explain how Recursive Bubble sort works.

352. Give an algorithm to sort an array using Recursive Bubble sort.

353. Sort the following numbers in ascending order using Recursive Bubble sort.

(a) 5, 1, 10, 4, 30, 24, 400, 350, 6, 200
(b) 1, 4, 5, 6, 10, 24, 30, 200, 350, 400
(c) 400, 350, 200, 30, 24, 10, 6, 5, 3, 1

354. What are the characteristics of input array (best case input) so that Recursive Bubble sorting has the best performance?

355. What are the characteristics of input array (worst case input) so that Recursive Bubble sorting has the worst performance?

356. Analyze Recursive Bubble sort time complexity (worst case, average case, and best case).

357. Analyze Recursive Bubble sort space complexity (auxiliary space).

358. Recursive Bubble sort is a stable sorting algorithm. True/False?

359. Recursive Bubble sort is an in place sorting algorithm. True/False?

360. Recursive Bubble sort is a comparison based sorting algorithm. True/False?

361. Implement the sorting algorithms shown below. Complete the following table. Implement the following sorting algorithms and then write the execution time of the algorithm on arrays of different sizes in the table below. In this table, A indicates that the input data is in ascending order, D indicates that the input data is in descending order, and R indicates that the input data is random.

Input size	n = 10			n = 32			n = 128			n = 1024			n = 10000		
Order of input	A	D	R	A	D	R	A	D	R	A	D	R	A	D	R
Insertion sort															
Quicksort															
Recursive bubble sort															

9.2.31 Binary Insertion Sort

362. Explain how binary insertion sort works.

363. Give an algorithm to sort an array using binary insertion sort.

364. Sort the following numbers in ascending order using binary insertion sort.

(a) 5, 1, 10, 4, 30, 24, 400, 350, 6, 200
(b) 1, 4, 5, 6, 10, 24, 30, 200, 350, 400
(c) 400, 350, 200, 30, 24, 10, 6, 5, 3, 1

365. What are the characteristics of input array (best case input) so that binary insertion sorting has the best performance?

366. What are the characteristics of input array (worst case input) so that binary insertion sorting has the worst performance?

367. Analyze binary insertion sort time complexity (worst case, average case, and best case).

368. Analyze binary insertion sort space complexity (auxiliary space).

369. Binary insertion sort is a stable sorting algorithm. True/False?

370. Binary insertion sort is an in place sorting algorithm. True/False?

371. Binary insertion sort is a comparison based sorting algorithm. True/False?

372. Implement the sorting algorithms shown below. Complete the following table. Implement the following sorting algorithms and then write the execution time of the algorithm on arrays of different sizes in the table below. In this table, A indicates that the input data is in ascending order, D indicates that the input data is in descending order, and R indicates that the input data is random.

Input size	n = 10			n = 32			n = 128			n = 1024			n = 10000		
Order of input	A	D	R	A	D	R	A	D	R	A	D	R	A	D	R
Insertion sort															
Quicksort															
Binary insertion sort															

9.2.32 Recursive Insertion Sort

373. Explain how recursive insertion sort works.

374. Give an algorithm to sort an array using recursive insertion sort.

375. Sort the following numbers in ascending order using recursive insertion sort.

(a) 5, 1, 10, 4, 30, 24, 400, 350, 6, 200
(b) 1, 4, 5, 6, 10, 24, 30, 200, 350, 400
(c) 400, 350, 200, 30, 24, 10, 6, 5, 3, 1

376. What are the characteristics of input array (best case input) so that recursive insertion sorting has the best performance?

377. What are the characteristics of input array (worst case input) so that recursive insertion sorting has the worst performance?

378. Analyze recursive insertion sort time complexity (worst case, average case, and best case).

379. Analyze recursive insertion sort space complexity (auxiliary space).

380. Recursive insertion sort is a stable sorting algorithm. True/False?

381. Recursive insertion sort is an in place sorting algorithm. True/False?

382. Recursive insertion sort is a comparison based sorting algorithm. True/False?

383. Implement the sorting algorithms shown below. Complete the following table. Implement the following sorting algorithms and then write the execution time of the algorithm on arrays of different sizes in the table below. In this table, A indicates that the input data is in ascending order, D indicates that the input data is in descending order, and R indicates that the input data is random.

Input size	n = 10			n = 32			n = 128			n = 1024			n = 10000		
Order of input	A	D	R	A	D	R	A	D	R	A	D	R	A	D	R
Insertion sort															
Quicksort															
Recursive insertion sort															

9.2.33 Tree Sort

384. Explain how tree sort works.

385. Give an algorithm to sort an array using tree sort.

386. Sort the following numbers in ascending order using tree sort.

(a) 5, 1, 10, 4, 30, 24, 400, 350, 6, 200
(b) 1, 4, 5, 6, 10, 24, 30, 200, 350, 400
(c) 400, 350, 200, 30, 24, 10, 6, 5, 3, 1

387. What are the characteristics of input array (best case input) so that tree sorting has the best performance?

388. What are the characteristics of input array (worst case input) so that tree sorting has the worst performance?

389. Analyze tree sort time complexity (worst case, average case, and best case).

390. Analyze tree sort space complexity (auxiliary space).

391. Tree sort is a stable sorting algorithm. True/False?

392. Tree sort is an in place sorting algorithm. True/False?

393. Tree sort is a comparison based sorting algorithm. True/False?

394. Implement the sorting algorithms shown below. Complete the following table. Implement the following sorting algorithms and then write the execution time of the algorithm on arrays of different sizes in the table below. In this table, A indicates that the input data is in ascending order, D indicates that the input data is in descending order, and R indicates that the input data is random.

Input size	n = 10			n = 32			n = 128			n = 1024			n = 10000		
Order of input	A	D	R	A	D	R	A	D	R	A	D	R	A	D	R
Insertion sort															
Quicksort															
Tree sort															

9.2.34 Cartesian Tree Sorting

395. Explain how cartesian tree sort works.

396. Give an algorithm to sort an array using cartesian tree sort.

397. Sort the following numbers in ascending order using cartesian tree sort.

(a) 5, 1, 10, 4, 30, 24, 400, 350, 6, 200
(b) 1, 4, 5, 6, 10, 24, 30, 200, 350, 400
(c) 400, 350, 200, 30, 24, 10, 6, 5, 3, 1

398. What are the characteristics of input array (best case input) so that cartesian tree sorting has the best performance?

399. What are the characteristics of input array (worst case input) so that cartesian tree sorting has the worst performance?

400. Analyze cartesian tree sort time complexity (worst case, average case, and best case).

401. Analyze cartesian tree sort space complexity (auxiliary space).

402. Cartesian tree sort is a stable sorting algorithm. True/False?

403. Cartesian tree sort is an in place sorting algorithm. True/False?

404. Cartesian tree sort is a comparison based sorting algorithm. True/False?

405. Implement the sorting algorithms shown below. Complete the following table. Implement the following sorting algorithms and then write the execution time of the algorithm on arrays of different sizes in the table below. In this table, A indicates that the input data is in ascending order, D indicates that the input data is in descending order, and R indicates that the input data is random.

Input size	n = 10			n = 32			n = 128			n = 1024			n = 10000		
Order of input	A	D	R	A	D	R	A	D	R	A	D	R	A	D	R
Insertion sort															
Quicksort															
Cartesian tree sort															

9.2.35   3-Way Quicksort

406. Explain how 3-way quicksort sort works.

407. Give an algorithm to sort an array using 3-way quicksort sort.

408. Sort the following numbers in ascending order using 3-way quicksort sort.

(a) 5, 1, 10, 4, 30, 24, 400, 350, 6, 200
(b) 1, 4, 5, 6, 10, 24, 30, 200, 350, 400
(c) 400, 350, 200, 30, 24, 10, 6, 5, 3, 1

409. What are the characteristics of input array (best case input) so that 3-way quick-sort sorting has the best performance?

410. What are the characteristics of input array (worst case input) so that 3-way quicksort sorting has the worst performance?

411. Analyze 3-way quicksort sort time complexity (worst case, average case, and best case).

412. Analyze 3-way quicksort sort space complexity (auxiliary space).

413. 3-way quicksort sort is a stable sorting algorithm. True/False?

414. 3-way quicksort sort is an in place sorting algorithm. True/False?

415. 3-way quicksort sort is a comparison based sorting algorithm. True/False?

416. Implement the sorting algorithms shown below. Complete the following table. Implement the following sorting algorithms and then write the execution time of the algorithm on arrays of different sizes in the table below. In this table, A indicates that the input data is in ascending order, D indicates that the input data is in descending order, and R indicates that the input data is random.

Input size	n = 10			n = 32			n = 128			n = 1024			n = 10000		
Order of input	A	D	R	A	D	R	A	D	R	A	D	R	A	D	R
Insertion sort															
Quicksort															
3-Way quicksort															

9.2.36 3-Way Mergesort

417. Explain how 3-way mergesort works.

418. Give an algorithm to sort an array using 3-way mergesort.

419. Sort the following numbers in ascending order using 3-way mergesort.

(a) 5, 1, 10, 4, 30, 24, 400, 350, 6, 200
(b) 1, 4, 5, 6, 10, 24, 30, 200, 350, 400
(c) 400, 350, 200, 30, 24, 10, 6, 5, 3, 1

420. What are the characteristics of input array (best case input) so that 3-way merge-sort has the best performance?

421. What are the characteristics of input array (worst case input) so that 3-way mergesort has the worst performance?

422. Analyze 3-way mergesort time complexity (worst case, average case, and best case).

423. Analyze 3-way mergesort space complexity (auxiliary space).

424. 3-way mergesort is a stable sorting algorithm. True/False?

425. 3-way mergesort is an in place sorting algorithm. True/False?

426. 3-way mergesort is a comparison based sorting algorithm. True/False?

427. Implement the sorting algorithms shown below. Complete the following table. Implement the following sorting algorithms and then write the execution time of the algorithm on arrays of different sizes in the table below. In this table, A indicates that the input data is in ascending order, D indicates that the input data is in descending order, and R indicates that the input data is random.

Input size	n = 10			n = 32			n = 128			n = 1024			n = 10000		
Order of input	A	D	R	A	D	R	A	D	R	A	D	R	A	D	R
Insertion sort															
Quicksort															
3-way mergesort															

428. What is the most efficient sorting algorithm for each of the following situations:

(a) A small array of integers.
(b) A large array whose entries are random numbers.
(c) A large array of integers that is already almost sorted.
(d) A large collection of integers that are drawn from a very small range.
(e) A large collection of numbers most of which are duplicates.
(f) Stability is required i.e., the relative order of two records that have the same sorting key should not be changed.

429. Sort a file containing 10^{12} 100 byte strings.

430. Most sorting algorithms rely on a basic swap step. When records are of different lengths, the swap step becomes nontrivial. Sort lines of a text file that has a million lines such that the average length of a line is 100 characters but the longest line is one million characters long.

431. When a comparison operation is used as a basic operation, no algorithm can solve the sorting problem with fewer than $\Omega(n \log n)$ operations.

9.3 Solutions

11. a
12. True
17.
Bubble Sort is a simple algorithm which is used to sort a given set of n elements in an array with n number of elements. Bubble Sort compares all the element one by one and sort them based on their values.

If the given array has to be sorted in ascending order, then bubble sort will start by comparing the first element of the array with the second element, if the first element is bigger than the second element, it will swap both the elements, and then move on to compare the second and the third element, and so on.

Algorithm

1. Start.
2. Starting with the first element (index = 0), compare the current element with the next element of the array.
3. If the current element is greater than the next element of the array, swap them.
4. If the current element is less than the next element, move to the next element. Repeat Step 2.
5. End.

Worst Case Time Complexity: O(n*n). Worst case happens when array is reverse sorted.

Average Time Complexity: O(n*n).

Best Case Time Complexity: O(n). Best case happens when array is already sorted.

Space complexity: O(1)

18. True

19. True

20 .True

21. True

22. True

23. True

24. True

25. a

26. Bubble sort can be optimized by using a *flag=swapped* variable that exits the loop once swapping is done. The best complexity of a bubble sort can be O(n). O(n) is only possible if the array is sorted. The value of swapped is set true if there occurs swapping of elements. Otherwise, it is set false. After an iteration, if there is no swapping, the value of swapped will be false. This means elements are already sorted and there is no need to perform more iterations.

27. Insertion sort is a sorting algorithm that works similar to the way we sort playing cards in our hands. The array is virtually split into a sorted and unsorted part. values from the unsorted part are picked and placed at the correct position in the sorted part.

Time complexity $=O\left(n^2\right)$, Auxiliary space $=O\left(1\right)$

28. True, worst and average case time complexity of insertion sort is $O(n^2)$.

29. True

30. True

31. True

32. True

33. True

34. True

35. Binary insertion sort uses binary search to find the proper location to insert the selected item at iteration. We can reduce the sorting time in normal insertion to $O(log\ i)$ by using binary search.

38.

	Best case	Average case	Worst case	Space worst case
Selection sort	$O(n^2)$	$O(n^2)$	$O(n^2)$	$O(1)$
Insertion sort	$O(n)$	$O(n^2)$	$O(n^2)$	$O(1)$
Bubble sort	$O(n)$	$O(n^2)$	$O(n^2)$	$O(1)$

40.
a. True
b. True

41. a. True
b, True

42. (a) True
(b) True

46. (b) $N - 1$
An insertion algorithm consist of $N - 1$ passes when an array of N elements is given.

48. True

49. (a) N(N-1)/4

The total number of pairs is N(N-1)/2. thus an average list has half this amount.

50. (c) O(N)

51.
(d) 8, 34, 64, 51, 32, 21

52. True

53. (b) insertion sort
insertion sort runs even faster than quicksort.

54. (a) Linear

55. (b) array sorted in reverse order

56. (a) insertion sort is stable and it sorts in-place
during insertion sort, the relative order of elements is not changed and requires only
O(1) of additional memory space.

57. (c) insertion sort
The best case running time of the insertion sort is O(n).

58. (b) $O(n^2)$
because of the series of swapping operations required for each insertion.

59. True

60. (d) insertion sort

61. (b) statement 1 is True but statement 2 is False.

65.
    (d) Insertion sort

66.
    (b) Heapsort with time complexity $O(n\log k)$

67.
    (b) Mergesort

68.
    (c) Counting sort

81. (a) Introsort

82.
    TRUE

83.
    (b) O(n log n)

84.
    (b) O(n log n)

85. O(n log n)

86. O(log n)

87. (b) Because heapsort requires less space

89.
(c) 16

92. We divide the array into blocks known as Run. We sort those runs using insertion sort one by one and then merge those runs using the combine function used in mergesort. If the size of the array is less than run, then array gets sorted just by using insertion sort. The size of the run may vary from 32 to 64 depending upon the size of the array. Note that the merge function performs well when size subarrays are powers of 2. The idea is based on the fact that insertion sort performs well for small arrays.

93. (c) insertion sort

94. True

95. (a) O(n)
Best case time complexity occurs when the input array is already sorted.

96. (b) $O(n\ logn)$

97. (b) $O(n\ logn)$

98. (a) $O(n)$

99. (c) Tim sort

100. (c) When no. of elements are less than size of run.

101. (a) inorder

102. (c) $O(n^2)$. For the binary tree sort the worst case when the BST constructed is unbalanced. BST gets unbalanced when the elements are already sorted. So, in the worst case, O(n*n) time is required to build the BST and O(n) time to traverse the tree. so, the worst-case time complexity is O (n*n).

103. (b) O(n log n).
The best case happens when the BST is balanced. So, when tree is balanced, we require O(nlogn) time to build the tree and O(n) time to traverse the tree. So, the best-case time complexity of the binary tree sort is O(nlogn).

104. False

105. (d) Both quicksort and binary tree are in place sorting algorithms.

106. (a) Heapsort

134.
(a) Radix sort is a non-comparative sorting algorithm. It avoids comparison by and distributing elements into buckets according to their radix. For elements with more than one significant digit, this bucketing process is repeated for each digit, while preserving the ordering of the prior step, until all digits have been considered.

(b) algorithm:

1. Define 10 queues each representing a bucket for each digit from 0 to 9.
2. Consider the least significant digit of each number in the list which is to be sorted.
3. Insert each number into their respective queue based on the least significant digit.
4. Group all the numbers from queue 0 to queue 9 in the order they have inserted into their respective queues.
5. Repeat from step 3 based on the next least significant digit.
6. Repeat from step 2 until all the numbers are grouped based on the most significant digit.

(c) Time complexity:

worst-case time complexity: $O(nk)$

average-case time complexity: $\Theta(nk)$

best-case time complexity: $\Omega(nk)$

(d) Space complexity:

worst-case space complexity: $O(n+k)$

135.
(b) Radix Sort

136.

1. We have 8 buckets, numbered as b0, b1, b2, ..., b7.
2. For a number, on first iteration, it takes maximum 8 (due to 8 buckets only)
3. Given that we have 9 numbers, so on first iteration total comparisons

$8*9 = 72$
Totally we have five iterations
Total comparisons $72 * 5 = 360$.

(c) 360

137.
The correct answer is c. Mergesort uses divide and conquer in order to sort a given array. This is because, the mergesort process is it divides the array into two halves and applies mergesort algorithm to each half individually after which the two sorted halves are merged together.

138.
The correct answer is c. An additional space of O(n) is required in order to merge two sorted arrays. Thus mergesort is not an in place sorting algorithm.

139.
The correct answer is a. Mergesort algorithm divides the array into two halves and applies mergesort algorithm to each half individually after which the two sorted halves are merged together. Thus its method of sorting is called merging.

140. b. Mergesort is not an adaptive sorting algorithm. This is because it takes O(n log n) time complexity irrespective of any case.

141.
a. Quicksort, Heapsort, and insertion sort are in-place sorting algorithms, whereas an additional space of O(n) is required in order to merge two sorted arrays. Even though we have a variation of mergesort (to do in-place sorting), it is not the default option. So, among the given choices, mergesort is the most appropriate answer.

142.
False. Mergesort is preferred for linked list over arrays. It is because in a linked list the insert operation takes only O(1) time and space which implies that we can implement merge operation in constant time.

143.
The correct answer is a.

144.
The correct answer is d.

145.
Time complexity of mergesort is O(nlogn).

$$c * 64\log 64 = 30$$

$$c * 64 * 6 = 30$$

$$c = \frac{5}{64}$$

$$c * n\log n = 6 * 60$$

$$\frac{5}{64} * n\log n = 6 * 60$$

$$n\log n = 512 * 9$$

$$n = 512$$

The correct answer is b.

147.
The correct answer is b.

149.
False. Bottom up mergesort unlike standard mergesort uses an iterative algorithm for sorting. Thus, it saves auxiliary space required by the call stack.

150.
False. Bottom up mergesort uses the iterative method in order to implement sorting. It begins by merging a pair of adjacent array of size 1 each and then merge arrays of size 2 each in the next step and so on.

151.
True.

152.
c. Bottom up mergesort unlike standard mergesort uses an iterative algorithm for sorting. Thus, it saves auxiliary space required by the call stack.

153. False

154. True

155. True

156.

1. arr[l] to arr[i] elements less than pivot.
2. arr[i+1] to arr[j-1] elements equal to pivot.
3. arr[j] to arr [r] elements greater than pivot.

158. Yes

159. Yes

160. In the case of linked lists, the nodes may not be present at adjacent memory locations, therefore mergesort is used.

162. (b) Recurrence is T(n) = T(n-1) + O(n) and time complexity is $O(n^2)$

163. (d) O(nLogn)

164. (d) Insertion Sort

165. (b) $T(n) \leq T(n/5) + T(4n/5) + n$

166. (c) median-of-three partitioning

167. True

168. (c) Insertion sort

169. True

170. (b) N = 5–20

171. (a) 22 25 56 67 89

173. (d) A random number is generated which is used as the pivot

174. True

175. True

176. True

177. True

178. False

180. True

252. It works by recursively calling itself with smaller and smaller copies of the beginning of the list to see if they are sorted.

253.

```
def bogoSort(a):
    n = len(a)
    while (is_sorted(a)== False):
```

```
      shuffle(a)
def is_sorted(a):
   n = len(a)
   for i in range(0, n-1):
      if (a[i] > a[i+1] ):
         return False
   return True
def shuffle(a):
   n = len(a)
   for i in range (0,n):
      r = random.randint(0,n-1)
      a[i], a[r] = a[r], a[i]
```

255. When array given is already sorted.

256. This algorithm has no upper bound.

257. worst case= $O(\infty)$

average case= $O(n*n!)$

best case= $O(n)$

258. $O(1)$

259. False

260. True

329. Stooge Sort is a recursive sorting algorithm. It is inefficient but interesting sorting algorithm. It divides the array into two overlapping parts (2/3 each). Then it performs sorting in first 2/3 part and then it performs sorting in last 2/3 part. After that, sorting is done on first 2/3 part to ensure the array is sorted.

330.

1. If the value at the start is larger than the value at the end, swap them.
2. Recursively sort the first 2/3 part of the array.
3. Recursively sort the last 2/3 part of the array.
4. Again, sort the first 2/3 part of the array.
5. The array becomes finally sorted.

335. space complexity - $O(n)$

336. False

337. False

338. True

340.
It successively generates permutations of its input until it finds one that is sorted.

362. In binary insertion sort, we divide the array into two sub-arrays sorted and unsorted. The first element of the array is in the sorted sub-array, and the rest of the elements are in the unsorted one. We then iterate from the second element to the last element.

363.

Algorithm:

1. **Start**
2. **Iterate the array from the second element to the last element.**
3. **Store the current element $A[i]$ in a variable key.**
4. **Find the position of the element just greater than $A[i]$ in the sub-array from $A[0]$ to $A[i-1]$ using binary search.**
5. **End**

365. The best-case input of binary insertion sort will be when the element is already in its sorted position. In this case, we don't have to shift any of the elements, we can insert the element in $O(1)$.

367. Worst case: $O(nlogn)$ because the time complexity does not change when we use binary insertion sort in place of standard insertion sort.

Average case: $O(n^2)$

Best case: $O(n)$ It occurs in the case when the input is already sorted position.

368. The auxiliary space of binary insertion sort is $O(1)$. So it qualifies as an in-place sorting algorithm.

369. The answer is True because the elements with identical values appear in the same order in the output array as they were in the input array, so binary insertion sort is the only algorithm which is stable.

Fig. 9.1 Tree for question
397(a)

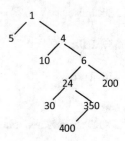

370. The answer is True because the auxiliary space required by binary insertion sort
is $O(1)$.

371. The answer is True because in insertion sort we need to compare elements in
order to find the minimum element in each iteration. So, we can say that it uses
comparisons in order to sort the array. Therefore, it qualifies as a comparison-based
sort.

395.
Cartesian Sort is an adaptive sorting as it sorts the data faster if data is partially sorted.
In fact, there are very few sorting algorithms that make use of this fact.
This kind of sorting tree is obeying in the min (or max) heap property (each node is
less or greater than its children).
A symmetric (in-order) traversal of the tree results in the original sequence.

396.

1. Build a (min-heap) Cartesian Tree from the array
2. Pop the node at the top of the priority queue and add it to a list.
3. Push left child of the popped node first if exists.
4. Push right child of the popped node next if exists.

397.
(a) 5, 1, 10, 4, 30, 24, 400, 350, 6, 200

See Fig. 9.1

1. 1 is pushed initially.
2. pop 1 and push 5 and 4
3. pop 4 and 5 then push 10 and 6
4. pop 6&10 then push 200 and 24
5. pop 24 and push 350 and 30
6. pop 30 and 200 and 350 then push 400
7. pop 400

Fig. 9.2 Tree for question 397(b)

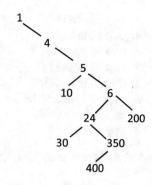

Fig. 9.3 Tree for question 397(c)

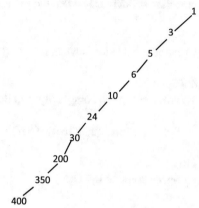

And the sorted array is : 1,4,5,6,10,24,30,200,350,400

(b) 1, 4, 5, 6, 10, 24, 30, 200, 350, 400
See Fig. 9.2

1. 1 is pushed initially
2. pop 1 and push 4
3. pop 4 and push 5
4. pop 5 and push 6 and 10
5. pop 6 and 10 push 24 and 200
6. pop 24 and push 30 and 350
7. pop 30 and 200 and 350 then push 400
8. pop 400

and the sorted array is: 1,4,5,6,10,24,30,200,350,400
(c) 400, 350, 200, 30, 24, 10, 6, 5, 3, 1
See Fig. 9.3

1. pop 1 push 3
2. pop 3 push5

3. pop 5 push 6
4. pop 6 push 10
5. pop 10 push 24
6. pop 24 push30
7. pop 30 push 200
8. pop 200 push 350
9. pop 350 push 400
10. pop 400

And the sorted array is: 1,3,5,6,10,24,30,200,350,400
398.
best-case behaviour is when the input data is partially sorted.

399.
worst-case behavior is when the input data is not partially sorted.

400.
   Worst case time complexity: $\Theta(NlogN)$

   Best case time complexity: $O(N)$

401.
   Space complexity: $O(N)$

402. False

403. False

Chapter 10
Divide and Conquer

Abstract The divide and conquer method is used for solving problems. In this type of method, the main problem is divided into sub-problems that are exactly similar to the main problem but smaller in size. The first phase of this method, as its name suggests, is to break or divide the problem into sub-problems, and the second phase is to solve smaller problems and then integrate the answers aiming to find the answer to the main problem. The divide and conquer method is not a general solution to all problems and can only be used for problems that are inherently divisible into smaller problems. We use this method when the number of data is large and also the problem can be divided into k sub-problem. This chapter provides 163 exercises for addressing different aspects of the divide and conquer method. To this end, this chapter provides exercises on binary search, finding minimum and maximum, greatest common divisor (gcd), mergesort, quicksort, finding the median, integer multiplication, matrix multiplication, and several other applications.

10.1 Lecture Notes

The first method we present for solving problems in this book is the divide and conquer method. In this type of algorithm, the main problem is divided into sub-problems that are exactly similar to the main problem but smaller in size.

The first phase of this method, as its name suggests, is to break or divide the problem into sub-problems, and the second phase is to solve smaller problems and then integrate the answers aiming to find the answer to the main problem. The divide and conquer method is not a general solution to all problems and can only be used for problems that are inherently divisible into smaller problems. We use this method when the number of data is large and also the problem can be divided into k sub-problem.

In this case, k is a number between 1 and n. It is necessary, first, to solve the k sub-problem. Finally, we must have a way of combining these sub-problems so that we can solve the main problem. If, after dividing the problem, the sub-problems are still large and unsolvable, we divide them again into several other sub-problems. So problems have to be small enough to be solved by successive divisions.

© The Author(s), under exclusive license to Springer Nature Switzerland AG 2022
H. Izadkhah, *Problems on Algorithms*,
https://doi.org/10.1007/978-3-031-17043-0_10

Many algorithms have a recursive structure. These algorithms need to call them-selves one or more to solve problems. By generating smaller sub-problems similar to the main problem and solving them, the answer to the main problem is obtained. In this case, we say that these algorithms use the divide and conquer method.

In summary, the divide and conquer algorithms follow the following three steps.

1. Division: divides the problem into sub-problems.
2. Conquer: Solves sub-problems recursively. If the sub problems are small enough, solves them only directly (not recursively).
3. Combination: integrates the solved sub-problems to solve the main problem.

The divide and conquer method is a top-down method. That is, solving a high-level sample of the problem is achieved by going down and solving smaller samples.

The recurrence relation of most algorithms designed by divide and conquer method is as follows:

$$T(n) = \begin{cases} T(1) & n = 1 \\ aT(\frac{n}{b}) + f(n) & n > 1 \end{cases}$$

where a is the number of sub-problems that must be solved recursively. That is, the number of divisions of the main problem into sub-problems. n is the input size and b is the number of divisions of the input size, and $f(n)$ is the time required to combine.

Example 10.1 Suppose algorithm A solves a sample problem by dividing it into four sub-problems of half the size, recursively solves each of these four sub-problems, and then combines their solutions to form the solution of the input problem in O(n log n) time. The recurrence for the algorithm is as follows:

$$a = 4, \quad b = 2, \quad f(n) = n\log n$$

$$T(n) = 4T\left(\frac{n}{2}\right) + O(n\log n)$$

10.2   Exercises

10.2.1   Preliminary

1. If $T(n) = aT(n/b) + O(n^d)$ for some constants $a > 0, b > 1$, and $d \geq 0$, prove

$$T(n) = \begin{cases} O(n^d) & \text{if } d > \log_b a \\ O(n^d \log n) & \text{if } d = \log_b a \\ O(n^{\log_b a}) & \text{if } d < \log_b a \end{cases}$$

2. Suppose algorithm A solves a sample problem by dividing it into four sub-problems of half the size, recursively solves each of these four sub-problems, and then combines their solutions to form the solution of the input problem in $O(n)$ time. The recurrence for the algorithm is as follows:

$$T(n) = 4T\left(\frac{n}{2}\right) + O(n)$$

$$T(1) = 1$$

(a) Solve this recurrence using the master theorem.
(b) Solve this recurrence using the recursion tree method.
(c) Solve this recurrence using the substitution method.

3. Suppose algorithm A solves a sample problem of size n by dividing it into two sub-problems of size $n - 1$, recursively solves each of these two sub-problems, and then combines their solutions to form the solution of the input problem in constant time. The recurrence for the algorithm is as follows:

$$T(n) = 2T(n - 1) + c$$

$$T(1) = 1$$

(a) Solve this recurrence using the master theorem.
(b) Solve this recurrence using the recursion tree method.
(c) Solve this recurrence using the substitution method.

4. Suppose the following three algorithms can solve a specific problem of size n. Which one of the three algorithms would you choose to solve the problem? Justify your answer.

– *M1* solves problem by dividing it into five subproblems of half the size, recursively solves each of these five sub-problems, and then combines their solutions to form the solution of the input problem in $O(n)$ time (linear time).
– Algorithm *M2* solves problem by dividing it into two subproblems of size $n-1$, recursively solves each of these two sub-problems, and then combines their solutions to form the solution of the input problem in constant time.
– Algorithm *M3* solves problem by dividing it into nine subproblems of size $\frac{n}{3}$, recursively solves each of these nine sub-problems, and then combines their solutions to form the solution of the input problem in quadratic time (n^2).

10.2.2 Binary Search

5. *Binary Search.* In this search, we divide our array of size n into two parts (by taking a mid) and discard half of our search space at each iteration.

(a) Describe an iterative algorithm for binary search and then calculate its time complexity.
(b) Describe a recursive algorithm for binary search and then calculate its time complexity.
(c) Implement both algorithms in a programming language and compare the speed of both algorithms on arrays of different sizes (e.g., n = 32, 64, 128, 512, 1024, ...).

6. Consider the following array:

$$A = [85\ 80\ 75\ 40\ 27\ 23\ 21\ 20\ 18\ 12\ 11]$$

For x = 18, x = 40, x = 84, apply binary search algorithm. In a table shows low, high, mid, and A[mid] in each iterative of the algorithm.

7. Consider the following array:

$$A = [-1\ 0\ 3\ 5\ 10\ 23\ 30\ 45]$$

(a) How many comparisons does it take using a binary search to find the elements −1, 0, 5, 30?
(b) In the array, find the average number of comparisons in both successful and unsuccessful search?

8. For binary search problem, analyze the average number of comparisons when

(a) all elements in the array have an equal probability of being searched for.
(b) each element in the array have different probability of being searched for.

9. Let n is not necessarily a power of 2. Show that the number of comparisons used by the binary search algorithm is $\lceil \log n \rceil$?

10. Does the following binary search algorithm work properly?

```
Function SEARCH(A, x, low, high)
    comment find x in A[low..high]
    if (low = high) then return(low)
    else
        m := ⌊(low + high)/2⌋
        if (x ≤ A[m])
            then return(SEARCH(A, x, low, m))
            else return(SEARCH(A, x, m, high))
```

11. Is the following algorithm for binary search correct? If so, prove it. If not, give an example on which it fails.

> Function SEARCH(A, x, low, high)
>   comment find x in A[low..high]
>   if (low = high) then return(low)
>   else
>   m := ⌊low + (high − low + 1)/2⌋
>   if (x ≤ A[m])
>   then return(SEARCH(A, x, low, m))
>   else return(SEARCH(A, x, m + 1, high))

12. Modify the binary search algorithm so that it splits the input not into two sets of almost-equal sizes, but into two sets of sizes approximately one-third and two-thirds.

13. Find number of rotations in a circularly sorted array. Given a circularly sorted array of integers, find the number of times the array is rotated. Assume there are no duplicates in the array and the rotation is in anti-clockwise direction.

For example:
Input: arr = [8, 9, 10, 2 5, 6]
Output: The array is rotated 3 times. We can see that the number of rotations is equal to number of elements before minimum element of the array or index of the minimum element.

(a) Describe a linear search algorithm and then calculate its time complexity.
(b) Modify binary search algorithm to solve this problem and then calculate its time complexity.

14. Find first or last occurrence of a given number is sorted array. Given a sorted array of integers, find index of first or last occurrence of a given number. If the element is not found in the array, report that as well.

For example:
Input: arr = [2, 5, 5, 5, 6, 6, 8, 9, 9, 9]
Target = 5
Output: First occurrence of element 5 is found at index 1.
Last occurrence of element 5 is found at index 3.

(a) Describe a linear search algorithm and then calculate its time complexity.
(b) Modify binary search algorithm to solve this problem and then calculate its time complexity.

15. Count occurrence of a number in sorted array with duplicates. Given a sorted array of integers containing duplicates, count occurrence of a number provided. If the element is not found in the array, report that as well.

For example:
Input: arr = [2, 5, 5, 5, 6, 6, 8, 9, 9, 9]
Target = 5
Output: Element 5 occurs three times.

(a) Describe a linear search algorithm and then calculate its time complexity.
(b) Modify binary search algorithm to solve this problem and then calculate its time complexity.

16. Find smallest missing element from a sorted array. Given a sorted array of distinct non-negative integers, find smallest missing in it.

For example:
Input: arr = [0, 1, 2, 6, 9, 11, 15]
Output: The smallest missing element is 3.

(a) Describe a linear search algorithm and then calculate its time complexity.
(b) Modify binary search algorithm to solve this problem and then calculate its time complexity.

17. *Find peak element in an array.* Given an array, find an element in it that is greater than its neighbors.

(a) Describe a linear search algorithm and then calculate its time complexity.
(b) Modify binary search algorithm to solve this problem and then calculate its time complexity.

18. *Ternary Search.* In this search, we divide our array into three parts (by taking two mid) and discard two-third of our search space at each iteration.

(a) Describe an iterative algorithm for ternary search and then calculate its time complexity.
(b) Describe a recursive algorithm for ternary search and then calculate its time complexity.
(c) Describe a recurrence relation for this algorithm and solve this recurrence relation.
(d) Compare the efficiency of ternary and binary search algorithms in terms of time complexity.

19. Given a monotonically increasing function $f(x)$ on the non-negative integers, find the value of x where $f(x)$ becomes positives for the first time. In other words, find a positive integer x such that $f(x - 1), f(x - 2), \ldots$ are negative and $f(x + 1), f(x + 2), \ldots.$ are positive. (Hint: use binary search algorithm.)

20. If an array is sorted from 1 to 1024 integers. How many comparisons are needed to find number 4 by binary search algorithm?

21. Consider the following array:

$$[-1, 0, 1, 2, 3, 4, 5, 6, 7]$$

What is the average number of comparisons for a successful search?

22. In an array, n numbers are in descending order. If we use binary search to find the number, what is the maximum number of comparisons?

23. Suppose a 200-element array is arranged. What is the worst-case comparison number to find the known element x in array A using binary search?

24. Prove in the binary search the following relations hold.

$$S(n) = (I + n)/n$$

$$U(n) = E/(n + 1)$$

where
$S(n)$ = number of comparisons in case of successful search
$U(n)$ = number of comparisons in case of unsuccessful search
I = internal path length of the binary tree, and
E = external path length of the binary tree.
n is the number of nodes in the binary tree.

25. Considering above question, show the following relations hold.

$$S(n) = \Theta(n)$$

$$S(n) = \left(1 + \frac{1}{n}\right) U(n) - 1$$

26. Let T(n) denote the run time of the binary search algorithm on an array of size n. Show the recurrence relation for this algorithm is as follows:

$$T(n) = T(n/2) + 1$$

$$T(1) = 1$$

27. Use the binary search technique to devise an algorithm for the problem of finding square-roots of natural numbers: Given an n-bit natural number N, compute $\lceil \sqrt{n} \rceil$ using only $O(n)$ additions and shifts.

10.2.3 Finding Minimum and Maximum

28. The following algorithm computes the maximum value in an array A[1..n].

> Function MAXIMUM(n)
>   Comment return max of A[1..n].
> 1.   if n≤1 then return (A[1])
> 2.   else return (MAX (MAXIMUM (n-1), A[n]))
>
> Function MAX (x, y)
>   Comment return largest of x and y.
>   if x ≥ y then return (x) else return (y)

(a) Prove that this recursive algorithm for computing the maximum value in an array
    A[1..n] is correct.
(b) Analyze this algorithm. How many comparisons does it use (that is, how many
    times is line 2 executed) in the worst case?

29. Consider the following procedure. How many the number of comparisons
required in line 5 in terms of n?

> FINDMAX (A, i, j)
> comment finding maximum on array A[i..j]
> 1. n ← j - i + 1
> 2. if (n=1) then return A[i]
> 3. m1 ← FINDMAX(A, i, i+[n/2]-1)
> 4. m2 ← FINDMAX(A, i+[n/2], j)
> 5. if m1 < m2
> 6.   then return m2
> 7.   else return m1

a. $n - 1$
b. n
c. $n - (\lceil \frac{n}{2} \rceil - \lfloor \frac{n}{2} \rfloor)$
d. $\lceil logn \rceil$

30. *Finding minimum and maximum.* Considering the problem of finding minimum
and maximum in an array of size n, answer the following questions.

(a) Give a naive algorithm to find the maximum element in the array.
(b) Give a naive algorithm to find the minimum element in the array.
(c) Give a naive algorithm to simultaneously find both the minimum and the maxi-
    mum in the array.

31. True or False, and Justify. Given an array of integers of size n, using a naive
algorithm, the number of comparisons are required to determine the maximum element in the array is $n - 1$; and the number of comparisons are required to determine

the minimum element in the array is $n - 1$. Therefore, the number of comparisons required to simultaneously find the smallest and largest elements is $2n - 2$.

32. *Simultaneous finding minimum and maximum.* Design a divide and conquer algorithm to simultaneous find the minimum and the maximum in the array of size n. An application of this problem is in a graphics program. A graphics program may need to scale a set of(x, y) data to fit onto a rectangular display screen or other graphical output device. To do so, the program must first determine the minimum and maximum of each coordinate.

33. The following recursive algorithm is designed to find the maximum element in array A of size n (n is a power of 2).

> Function MAXIMUM(x, y)
> > comment return maximum in A[x..y]
> > if $y - x \le 1$ then return(max(A[x], A[y]))
> > else
> > > max1:=MAXIMUM(x, $\lfloor (x + y)/2 \rfloor$)
> > > max2:=MAXIMUM($\lfloor (x + y)/2 + 1 \rfloor$, y)
> > return(max(max1, max2))

(a) Prove the correctness of the algorithm.
(b) Derive a recurrence relation for the worst-case number of key comparisons used by this algorithm. Solve this recurrence relation. Explain your answer.
(c) Find the time complexity of this algorithm.

34. Consider the problem of finding the minimum number of comparisons required to determine the minimum and the maximum of an integer array with n numbers using divide and conquer approach.

(a) Draw recursion tree for following list.

$$[12, 13, 0, -3, 15, 43, 57, 31, 47]$$

(b) The minimum number of comparisons required to find the *min* and the *max* in this array is ....

35. Given an array of integers of size n, if n is even, how many the minimum number of comparisons required to find the minimum and the maximum of the array?

36. Given an array of integers of size n, if n is odd, how many the minimum number of comparisons required to find the minimum and the maximum of the array?

37. If T(n) is the number of comparisons, prove the recurrence relation for finding the minimum number of comparisons required to find the minimum and the maximum of an integer array with n numbers is

$$T(n) = \begin{cases} T(\lceil \frac{n}{2} \rceil) + T(\lfloor \frac{n}{2} \rfloor) + 2 & n > 2 \\ 1 & n = 2 \\ 0 & n = 1 \end{cases}$$

38. The minimum number of comparisons required to find the minimum and the maximum of an array of size 200 is ....

39. The following algorithm is designed to find maximum and minimum of the array A[1..n].

> Procedure MAXMIN(A)
>> comment computes maximum and minimum of A[1..n]
>> if n is odd then max := A[n]; min := A[n];
>>> else max := $-\infty$; min := ∞;
>> for i := 1 to $\lfloor n/2 \rfloor$ do
>>> if A[2i - 1] $\leq$ A[2i]
>>>> then small := S[2i - 1]; large := S[2i];
>>>> else small := A[2i]; large := A[2i - 1];
>>>> if small < min then min:=small;
>>>> if large > max then min:=small;

(a) Apply this algorithm to the array below.

$$[12, 13, 0, -3, 15, 43, 57, 31, 47]$$

(b) In this algorithm, how many comparisons are required to find the minimum and the maximum of the array.
(c) Is it likely to be faster or slower than the divide-and-conquer algorithm in practice?

Consider the maximum partial sum problem (MPS). This problem is defined as follows. Given an array A[1..n] of integers, find values of i and j with $1 \leq i \leq j \leq n$ such that

$$\sum_{k=i}^{j} A[k]$$

is maximized.

   Example: For the array $[4, -5, 6, 7, 8, -10, 5]$, the solution to MPS is $i = 3$ and $j = 5$ (sum 21).

   To help us design an efficient algorithm for the maximum partial sum problem, we consider the *left position l maximal partial sum* problem ($LMPS_l$). This problem consists of finding value j with $l \leq j \leq n$ such that

$$\sum_{k=l}^{j} A[k]$$

is maximized. Similarly, the *right position r maximal partial sum* problem ($RMPS_r$), consists of finding value i with $1 \leq i \leq r$ such that

$$\sum_{k=i}^{r} A[k]$$

is maximized.

Example: For the array $[4, -5, 6, 7, 8, -10, 5]$, the solution to e.g. $LMPS_4$ is j $= 5$ (sum 15) and the solution to $RMPS_7$ is i $= 3$ (sum 16).

Answer the following questions.

40. Describe $O(n)$ time algorithms for solving $LMPS_l$ and $RMPS_r$ for given l and r.

41. Using an $O(n)$ time algorithm for $LMPS_l$, describe a simple $O(n^2)$ algorithm for solving MPS.

42. Using an $O(n)$ time algorithms for $LMPS_l$ and $RMPS_r$, describe an $O(n \log n)$ divide-and-conquer algorithm for solving MPS.

43. Design a divide-and-conquer algorithm to compute k^n for $k > 0$ and integer $n \geq 0$.

44. *K'th smallest element in union of two sorted arrays.* Let A and B are two sorted lists of size n and m. We need to find the kth smallest element in the union of the two sorted lists. Design an algorithm with $O(\log n + \log m)$ time.

45. *K'th smallest element in unsorted array.* Given a collection of data, the kth order statistic is the kth smallest value in the data set. Given an array and a number k where k is smaller than size of array, we need to find the kth smallest element in the given array. Answer the following questions.

(a) One way to solve this problem is to sort and then output the kth element. Explain why this simple algorithm is $O(n \log n)$.
(b) Design an algorithm which uses min-heap to find kth smallest element, and then compute its time complexity.
(c) Design an algorithm which uses max-heap to find kth smallest element, and then compute its time complexity.
(d) Design an algorithm which uses median-of-medians algorithm to find kth smallest element, and then compute its time complexity.

46. Assume that quicksort is used as the sorting algorithm in the first step in the selecting the smallest kth element. Note that do not complete the quick sort, but stop at the point where the pivot itself is kth smallest element. Also, do not repeat the choosing pivot for both left and right, but repeat for one of them according to the

position of the pivot. Show the worst case time complexity of this method is $O(n^2)$, but it works in $O(n)$ on average.

47. Let A denote an array of size n and integer $k \leq n$. Also, let T(n, k) denote the expected time to find the kth smallest in the array A, and let $T(n) = \max_k T(n, k)$. Show that $T(n) < 4n$.

48. Given array A of size n and integer $k \leq n$, consider the following algorithm to find kth smallest element in an unsorted array.

1. Divide the array into $n/5$ parts of size 5 and find the median of each part.
2. Recursively, find the true median of the medians. Call this m.
3. Use m as a pivot to split the array into subarrays LESS and GREATER.
4. Recurse on the appropriate piece.

(a) Apply this algorithm to the array below.

$$[12, 13, 0, -3, 15, 43, 57, 31, 47, 20, 1, -5, 42, 14, 15, 8, 100, 32, 85, 41]$$

(b) Show this algorithm makes $O(n)$ comparisons to find the kth smallest in an array of size n.

49. Let T(n) be maximum number of comparisons to find the kth smallest in an array of size n. Show

$$T(n) \leq cn + T(n/5) + T(7n/10)$$

50. Show $n + \lceil \log n \rceil - 2$ comparisons are sufficient to find the second-largest of n elements.

51. Prove the lower bound for finding kth smallest element in an array of size n $(n > 1)$ is

$$n + (k - 1) \left\lceil \log\left(\frac{n}{k-1}\right) \right\rceil - k$$

10.2.4 Greatest Common Divisor (gcd)

52. Let a and b be two positive integers. The *greatest common divisor* (gcd) of a and b is the largest integer which divides them both.

(a) Check whether the following recursive relation can calculate the gcd correctly?

$$gcd(a, b) = \begin{cases} 2gcd(a/2, b/2) & \text{if } a, b \text{ are even} \\ gcd(a, b/2) & \text{if } a \text{ is odd, } b \text{ is even} \\ gcd((a - b)/2, b) & \text{if } a, b \text{ are odd} \end{cases}$$

(b) Describe a recursive algorithm for *gcd(a, b)*.

(c) Let *a* and *b* are *n*-bit integers. Compare the algorithm described in (b) to Euclid's algorithm.

53. Explain how we can use the *gcd* function described in exercise 52 for more than two arguments. Give a recursive algorithm for this problem. (Hint: $gcd(a_0, a_1, \ldots, a_n) = gcd(a_0, gcd(a_1, a_2, \ldots, a_n)))$

54. Prove for any nonnegative integer *a* and any positive integer *b*,

$$gcd(a, b) = gcd(b, a \bmod b).$$

55. Let *lcm* denotes the least common multiple of the *n* integers. Explain how we can use *gcd* as a subroutine to compute $lcm(a_1, a_2, \ldots, a_n)$.

10.2.5 Mergesort

56. Let A and B are two sorted arrays of size *n* and *m*, respectively. The following algorithm is used to merge these to arrays and form a sorted array C of size *p* (where $p = n + m$).

```
void MERGE (int A[ ], int n, int B[ ] , int m , int C[ ], int p)
    {
    int i=0, j=0, k;
    for (k=0; i<n && j<m; k++)
        if (A[i]< B[j])
            C[k]=A[i++];
        else
            C[k]=B[j++];
        while (i<n)
            C[k++]=A[i++];
        while (j<m)
            C[k++]=B[j++];
    }
```

(a) Using the merge algorithm, merge the following arrays.

$$A = [3, 5, 7, 8, 10], \quad B = [1, 2, 6, 12, 20]$$

(b) Provide two sorted arrays A and B that require a maximum number of comparisons to merge (worst case happens).

(c) Provide two sorted arrays A and B that require a minimum number of comparisons to merge (best case happens).
(d) How many comparisons would be required to merge two sorted arrays of lengths
m and n in the worst case and the best case?

57. Suppose we have several sorted arrays. Give two multi-merge algorithms for
merging them.

58. Let A and B are two sorted arrays so that the number of elements of array A is
much less than the elements of array B. Provide an efficient algorithm for merging
the two arrays. (Hint: use binary search algorithm.)

59. For the mergesort algorithm:

(a) Describe an iterative algorithm for mergesort.
(b) Describe a recursive algorithm for mergesort.
(c) Implement both algorithms in a programming language and compare the speed
of both algorithms on arrays of different sizes (e.g., n = {32, 64, 128, 512, 1024,
...}).
(d) When does the best-case of mergesort occur?
(e) When does the worst-case of mergesort occur?

60. Consider the following mergesort algorithm.

> MERGESORT(low, high, A[])
> Comment A[low..high] is an array
> if (low < high)
>   mid = $\lfloor (low + high)/2 \rfloor$
>   MERGESORT(low, mid, A)
>   MERGESORT(mid + 1, high, A)
>   merge(low, mid, high, A)

Merge function combines the two sorted subarrays A[low..mid] and A[mid +
1..high] into a sorted sequence. For the following array

$$310, 285, 179, 625, 351, 423, 861, 254, 450, 520$$

(a) How many calls (mergeSort) are made?
(b) How many calls are made for merge function?
(c) What is the maximum number of comparisons between array elements?

61. Draw the recurrence tree for the following array.

$$20, 10, 7, 15$$

62. Let T(n) denotes the run time of mergesort on an array of size n. Show the
recurrence relation of this algorithm is as follows:

$$T(n) = T\left(\left\lfloor \frac{n}{2} \right\rfloor\right) + T\left(\left\lceil \frac{n}{2} \right\rceil\right) + n - 1$$

63. Show the exact solution for the recurrence relation for mergesort is as follows:

$$T(n) = n \log n - (n - 1)$$

64. Solve the recurrence relation for mergesort using the master theorem, recursion tree method, and the substitution method.

65. For the mergesort algorithm, justify:

(a) Let T(n) denotes the number of recursive calls. We have

$$T(n) = T\left(\left\lfloor \frac{n}{2} \right\rfloor\right) + T\left(\left\lceil \frac{n}{2} \right\rceil\right) + 1, \quad T(1) = 1$$

(b) Let T(n) denotes the number of merge function calls in the mergesort algorithm. We have

$$T(n) = T\left(\left\lfloor \frac{n}{2} \right\rfloor\right) + T\left(\left\lceil \frac{n}{2} \right\rceil\right) + 1, \quad T(1) = 0$$

(c) Let T(n) denotes the number of comparisons in the merge function. We have

$$T(n) = T\left(\left\lfloor \frac{n}{2} \right\rfloor\right) + T\left(\left\lceil \frac{n}{2} \right\rceil\right) + n - 1, \quad T(1) = 0$$

66. We use the mergesort algorithm for sorting an array of size 10:

(a) Get the number of recursive calls.
(b) Get the number of merge function calls in the mergesort algorithm.
(c) Get the number of comparisons in the merge function.

67. Apply mergesort to sort the list G, F, O, B, D, A, E in alphabetical order. Draw the tree of the recursive calls made.

68. Explain why for the following description, among sorting algorithms mergesort algorithm is preferred.

You have to sort 1 GB of data with only 100 MB of available main memory.

69. The algorithm Merge-1 is presented for merging the two arrays A and B with n and m elements, respectively. The result is stored in sorted array C of length $m + n$. The array index starts from 1. For which of the following statements does this algorithm work correctly instead of *conditional statement*?

Merge-1 (A, B, n, m)
i ← 1; j ← 1; k ← 0

while *conditional statement* **do**
  k ← k + 1
  if (j > m) or ((j ≤ n) and (A[i] ≤ B[j]))
  then C[k] ← A[i]
    i ← i + 1
    else C[k] ← B[j]
     j ← j + 1

a. (i < n and j < m)
b. (i ≤ n and j ≤ m)
c. (i ≤ n or j ≤ m)
d. (i < n or j < m)

70. The algorithm Merge-2 is presented for merging the two arrays A and B with n and m elements, respectively. The result is stored in array C of length $m + n$. The array index starts from 1. For which of the following statements does this algorithm work correctly instead of *conditional statement*?

Merge-2 (A, B, n, m)
i ← 1; j ← 1; k ← 0
while (i ≤ n) or (j ≤ m) **do**
  k ← k + 1
  if *conditional statement*
    then if (A[i] ≤ B[j])
      then C[k] ← A[i]
       i ← i + 1
      else C[k] ← B[j]
       j ← j + 1
    else if (i > n)
      then C[k] ← A[i]
       i ← i + 1
      else C[k] ← B[j]
       j ← j + 1

a. (i < n and j < m)
b. (i ≤ n and j ≤ m)
c. (i ≤ n or j ≤ m)
d. (i < n or j < m)

71. The algorithm Merge-3 is presented for merging the two arrays A and B with n and m elements, respectively. The result is stored in array C of length $m + n$. The array index starts from 1. When does this algorithm work?

Merge-3 (A, B, n, m)
i ← 1; j ← 1; k ← 0
while (i ≤ n) or (j ≤ m) **do**

$$\mathbf{if}\ (j > m)\ \text{or}\ (A[i] \le B[j])$$
$$\mathbf{then}\ C[k] \leftarrow A[j]$$
$$i \leftarrow i + 1$$
$$\mathbf{else}\ C[k] \leftarrow B[j]$$
$$j \leftarrow j + 1$$
$$k \leftarrow k + 1$$

a. It always works right.
b. It never works properly.
c. $B[m] < A[n]$
d. $\forall i\ A[i] < B[1]$

72. To sort array S of length n containing real numbers, the following algorithm is suggested:

(a) If $n \le 2$, sort it in a simple way (by comparison).
(b) If $n \ge 3$, divide S (left and right) by sub-arrays A, B, and C so that the sizes A and B are $\lceil \frac{n}{3} \rceil$ and $\lfloor \frac{n}{3} \rfloor$, respectively.
(c) Recursively sort the sub-arrays A and B.
(d) Recursively sort the sub-arrays B and C.
(e) Recursively sort the sub-arrays A and B.

Does this algorithm always work correctly?

73. Mergesort outperforms heapsort in most of the practical situations. True or False?

74. A mergesort algorithm takes 30 s to sort an input of size 64 in the worst case. Suppose a problem is solved by this algorithm in 6 min. The input size of this problem is closer to which of the following options.

a. 256
b. 512
c. 1024
d. 2048

75. Which sorting algorithm is the best sorting method to use for an array with more than ten million elements? Mergesort or quicksort?

76. Write is the auxiliary space complexity of mergesort.

77. Suppose that in the mergesort, the list is divided into four equal parts at each step, and then in the merge step, these four lists are merged. What is the complexity of the algorithm?

78. Suppose that in the mergesort, we divide the list by 9 to 1 at each step. That is, one part is 9 times the other, and then in the merge phase, the two lists are merged.

What is the complexity of the algorithm?

79. If in mergesort, insertion sort is used to sort lists with fewer than 20 elements, what will be the time complexity of the algorithm?

80. *A k-way merge operation.* Assume we have k sorted arrays of size n, and our aim is to merge them into a single sorted array of size kn.

(a) Using the merge procedure, merge the first two arrays, then merge in the third, then merge in the fourth, and so on. What is the time complexity of this algorithm, in terms of k and n?
(b) Using divide and conquer technique, devise an efficient algorithm to this problem.

81. Let $A[1..n]$ be an array of n distinct numbers. If $i < j$ and $A[i] > A[j]$, then the pair (i, j) is called an inversion of A.

(a) List the five inversions of the array $< 3, 5, 10, 7, 1 >$.
(b) Write an array of the set $\{1, 2, \ldots, n\}$ that has the most inversion.
(c) What is the relationship between the time complexity of insertion sort and the number of inversions in the input array? Justify your answer.
(d) Give an algorithm that determines the number of inversions in any permutation on n elements in $\Theta(n \log n)$ worst-case time. (Hint: modify mergesort.)

10.2.6   Quicksort

82. Quicksort is the widely used sorting algorithm. In quicksort, first pick a pivot element and then partition or rearrange the array into two sub-arrays such that each element in the left sub-array is less than or equal to the pivot element and each element in the right sub-array is larger than the pivot element. Picking a good pivot is necessary for the fast working of quicksort. Some of the ways of choosing a pivot are as follows:

– Pivot can either be the first element or the last element of the given array.
– Pivot can be random, i.e. selecting the random pivot from the given array.
– Use the "median-of-three" rule to pick the pivot: selecting median of leftmost, middle and rightmost element of the array.

In the following function, the first element is selected as pivot.

```
PARTITION (A[1..n], int low, int high, int & pivot)
    right = high
    left = low + 1
    pivot = A[low]
```

```
pivotIndex=low
while(left<right)
{
    while(A[left] < pivot)
        left = left + 1
    while (A[right] ≥ pivot)
        right = right -1
    if (left < right)
        swap(A[left], A[right])
}
swap(A[pivotIndex], A[right])
```

Partition the following arrays using the PARTITION function.

(a) 55, 11, 66, 10, 99, 22, 88, 33, 77, 44
(b) 1, 2, 3, 4, 5, 6, 7, 8, 9, 10
(c) 10, 9, 8, 7, 6, 5, 4, 3, 2, 1

83. One way to improve quicksort is to focus on the pivot element being selected more carefully than selecting a random element from the subarray. One common approach for computing *pivot* is called median-of-three: where *pivot* is chosen as the median of the first, last, and middle elements of the array A; i.e. median(A[0], A[(n - 1)/2], A[n - 1]). Modify the PARTITION function to support the median-of-three.

Perform the partitioning step of quicksort on the following arrays, where the pivot is chosen using the median-of-three heuristic.

(a) 55, 11, 66, 10, 99, 22, 88, 33, 77, 44
(b) 1, 2, 3, 4, 5, 6, 7, 8, 9, 10
(c) 10, 9, 8, 7, 6, 5, 4, 3, 2, 1

84. Consider the following quicksort algorithm.

```
QUICKSORT(A[], low, high)
    if low < high
        q = PARTITION(A, low, high)
        QUICKSORT(A, low, q-1)
        QUICKSORT(A, q+1, high)
```

Sort the following arrays using quicksort algorithm (show the steps):

(a) 55, 11, 66, 10, 99, 22, 88, 33, 77, 44
(b) 1, 2, 3, 4, 5, 6, 7, 8, 9, 10
(c) 10, 9, 8, 7, 6, 5, 4, 3, 2, 1

85. For quicksort algorithm:

(a) The best general-purpose internal sorting algorithm is quicksort, although it requires tuning effort to achieve maximum performance. True/False?

(b) Let the first element is selected as pivot. Are arrays made up of all equal elements the worst-case input, the best- case input, or neither?

(c) Let the first element is selected as pivot. Are strictly decreasing arrays the worst-case input, the best-case input, or neither?

(d) For quicksort with the median-of-three pivot selection, are strictly increasing arrays the worst-case input, the best-case input, or neither?

(e) For quicksort with the median-of-three pivot selection, are strictly decreasing arrays the worst-case input, the best-case input, or neither?

(f) On the average, quicksort makes only 39% more comparisons than in the best case. True/False?

(g) When does the best-case of quicksort occur? Show that its best-case running time on an array of size n is $\Theta(n\log n)$.

(h) When does the worst-case of quicksort occur? Show that its worst-case running time on an array of size n is $\Theta(n^2)$.

(i) Quicksort usually runs faster than mergesort and heapsort on randomly ordered arrays of nontrivial sizes. True/False?

(j) Show that its expected running time satisfies the recurrence relation

$$T(n) \leq O(n) + \frac{1}{n} \sum_{i=1}^{n-1} (T(i) + T(n-i))$$

Then, show that the solution to this recurrence is $O(n \log n)$.

86. Perform quicksort to sort the list S, A, Y, M, Y, L, S in alphabetical order. Draw the recursive calls tree made.

87. True or False and justify?

(a) The running time of the quicksort algorithm depends on how the *partition()* method will split the input array.

(b) The quicksort algorithm performs the worst when the pivot is the largest or smallest value in the array.

88. In a new variant of quicksort, to select the pivot from n elements, we select the first $2\sqrt{n} + 1$ elements of the array and sort them with a simple algorithm, e.g., insertion sort. Pivot is the middle element of these sorted elements. The rest of the algorithm is done using quicksort. Which of the following recurrence relations shows the worst-case performance of this algorithm?

a. $T(n) \leq T(n - \sqrt{n}) + O(\sqrt{n})$
b. $T(n) \leq T(2\sqrt{n}) + T(n - 2\sqrt{n}) + O(n)$
c. $T(n) \leq T(\sqrt{n}) + T(n - \sqrt{n}) + O(n)$
d. $T(n) \leq T(2\sqrt{n}) + T(n - 2\sqrt{n}) + O(n)$

89. For every n > 1, there are n-element arrays that are sorted faster by insertion sort than by quicksort. True/False?

90. Consider the following variant of quicksort. The values to be sorted are in an array A[1..n].

> Procedure QUICKSORT(low, high)
> 1. comment sort A[low...high]
> 2. i := low; j := high;
> 3. a := some element from A[low..high];
> 4. repeat
> 5.   while (A[i] < a) do i := i + 1;
> 6.   while (A[j] > a) do j := j - 1;
> 7.   if (i ≤ j) then
> 8.   swap A[i] and A[j];
> 9.   i := i + 1; j := j - 1;
> 10. until i > j;
> 11. if (low < j) then QUICKSORT(low, j);
> 12. if (i < high) then QUICKSORT(i, high);

(a) Prove that this variant of quicksort is correct.
(b) Analyze that variant of quicksort. How many comparisons does it use (that is, how many times are lines 5 and 6 executed) in the worst case?

91. What is the running time of quicksort when all elements of array A have the same value?

92. Assume that array A contains distinct elements and is sorted in descending order. Show that the quicksort time complexity on this array is $\Theta(n^2)$.

93. There are several schemes that can be used in quicksort to partition the array. Consider Hoare's and Lomuto partition schemes:

(a) Using an example, explain how Hoare's and Lomuto's partition schemes work?
(b) Write high-level version of Hoare's quicksort algorithm.
(c) Write high-level version of Lomuto's quicksort algorithm.
(d) Implement Hoare's and Lomuto partition schemes in quicksort and compare them performance.

94. Banks often record transactions on an account in order of the times of the transactions, but many people like to receive their bank statements with checks listed in order by check number. People usually write checks in order by check number, and merchants usually cash them with reasonable dispatch. The problem of converting time-of-transaction ordering to check-number ordering is therefore the problem of sorting almost-sorted input. Argue that the procedure INSERTION-SORT would tend to beat the procedure quicksort on this problem.

95. True or False and justify? For a randomized algorithm, we analyze the average-case, not worst-case.

96. Given an array of positive and negative numbers, we would like to rearrange them such that all negative numbers appear before all the positive numbers in the array.

(a) Use an auxiliary array to rearrange the numbers in the array.
(b) Use partition process of quicksort to rearrange the numbers.

97. Let T(n) be the worst-case time for the procedure quicksort on an input of size n. Show

$$T(n) = \max_{0 \le q \le n-1} (T(q) + T(n - q - 1)) + \Theta(n)$$

where the parameter q ranges from 0 to $n - 1$ because the procedure PARTITION produces two subproblems with total size $n - 1$.

98. The running time of quicksort can be improved in practice by taking advantage of the fast running time of insertion sort when its input is "nearly" sorted. When quicksort is called on a subarray with fewer than k elements, let it simply return without sorting the subarray. After the top-level call to quicksort returns, run insertion sort on the entire array to finish the sorting process. Argue that this sorting algorithm runs in $O(nk + n\log(n/k))$ expected time. How should k be picked, both in theory and in practice?

99. Suppose quicksort always selects $\lceil \frac{n}{3} \rceil$th smallest value as the pivot element. How many comparisons would be made then in the worst case?

100. Consider the following implementation of quicksort on an array A[1..n].

```
Procedure QUICKSORT(low, high)
1. comment sort A[low...high]
2. i := low; j := high;
3. a := some element from A[low..high];
4. repeat
5.     while (A[i] < a) do i := i + 1;
6.     while (A[j] > a) do j := j - 1;
7.     if (i ≤ j) then
8.         swap A[i] and A[j];
9.         i := i + 1; j := j - 1;
10. until i > j;
11. if (low < j) then QUICKSORT(low, j);
12. if (i < high) then QUICKSORT(i, high);
```

(a) How many number of comparisons needed to sort n values in the worst case?
(b) How many number of comparisons needed to sort n values on average?
(c) Suppose the preceding algorithm pivots on the middle value, that is, line 3 is replaced by a:= $S[\lfloor (low + high)/2 \rfloor]$. Give an input of 8 values for which quicksort exhibits its worst-case behavior.
(d) Suppose the above algorithm pivots on the middle value, that is, line 3 is replaced by a:= $S[\lfloor (low + high)/2 \rfloor]$. Give an input of n values for which quicksort exhibits its worst-case behavior.

101. For the quicksort:

(a) Estimate how many times faster quicksort will sort an array of one million random numbers than insertion sort.
(b) There are arrays that are sorted faster by insertion sort than by quicksort. True or False and Explain?
(c) To improve the quicksort performance, we can use the following rule:
"switching to insertion sort on small subarrays (between 5 and 15 elements)"

102. Show that any array of integers x[1..n] can be sorted in $O(n + M)$ time, where

$$M = \max_i x_i - \min_i x_i$$

For small M, this is linear time: why doesn't the $\Omega(n \log n)$ lower bound apply in this case?

103. *Randomized-Select, or QuickSelect for finding kth Smallest Element in Unsorted Array.* We want to select kth smallest element using Quicksort. To this end, after the partitioning step, we choose the subarray that contains Kth smallest element just by looking at their sizes. This is repeated recursively to find the desired element. Let "LESS" and "GREATER" subarrys indicate the elements less than the pivot and elements greater than the pivot, respectively. The following algorithm is used for this aim.

QUICKSELECT: Given array A[1..n] and integer $m \le n$,

1. Choose a random pivot p from A.
2. Use partition procedure in quicksort to divide array A into subarrays LESS and GREATER by comparing each element to pivot p. Simultaneously count the number of elements (Q) entered into LESS.
3. **a.** If $Q = m - 1$, then output p
   b. If $Q > m - 1$, output QUICKSELECT(LESS, m).
   c. If $Q < m - 1$, output QUICKSELECT(GREATER, m - Q - 1)

(a) Implement this algorithm and apply it on an array of size 547 to find 281'th smallest element.
(b) Show the number of comparisons for QUICKSELECT is $O(n)$.

104. In the finding the k smallest element of the array A[1..n], we first divide all the elements into s-parts (except possibly the last part). We get the median of each part and then we find the median of medians recursively. Select this element as the pivot element and apply the partition algorithm to the array elements, then run the same algorithm recursively (and for another k) on one of the parts to find the element. Write the recurrence relation for algorithm.

105. Quicksort requires a stack to store parameters of subarrays that are yet to be sorted. The stack depth required for quicksort will be $\Theta(n)$ in the worst case. If the size of the input array is large, the required stack depth can be an obstacle to its efficient implementation. To reduce the maximum stack space of recursive calls, it is necessary to decrease the number of recursive calls.

The algorithm QUICKSORT 1 removes the second recursive call from the body of the algorithm, and the algorithm QUICKSORT 2 first solves the smaller subarray (of the two subarrays obtained from the partition) by recursive calling. Specify the required stack depth in both algorithms in worst case.

```
Algorithm QUICKSORT 1(A[low..high])
    while low<high do
        s = Partition(A[low..high])    //s is split position
        QUICKSORT 1(A[low..s-1])
    low = s + 1
```

```
Algorithm QUICKSORT 2(A[low..high])
    while low<high do
        s = Partition(A[low..high])    //s is split position
        if s-low < high-s //first subarray is smaller
            QUICKSORT 2(A[low..s-1])
            low = s + 1
        else
            QUICKSORT 2(A[s+1..high])
            high = s - 1
```

106. Which of the following heuristics improves the performance of the quicksort algorithm? Justify your answer.

– Use randomization—Randomly permuting the keys before sorting
– Median of three elements as pivot
– Terminating the quicksort recursion and switching to insertion sort when the sub-arrays get small, for example less than 20 elements.
– Sort the smaller partition first

10.2.7 Finding the Median

107. Let A denote an array of size n containing distinct numbers and k is a positive integer $k \leq n$. Give an algorithm to find the k numbers in A that are closest to the median of A.

108. Let $A[1..n]$ and $B[1..n]$ are two sorted arrays. Give an algorithm to find the median of all $2n$ elements in arrays A and B. Analyze its running time.

109. The median of an array of size n is the $\lceil \frac{n}{2} \rceil$th smallest value. Suppose quicksort always selects the median element as the pivot element. How many comparisons would be made then in the worst case?

10.2.8 Integer Multiplication

110. Assume n is a power of 2. We partition two numbers u and v into lower and upper digits as $u = x \times 10^{n/2} + y$ and $v = w \times 10^{n/2} + z$. Thus, the product becomes $A \times 10^n + (B + C) \times 10^{n/2} + D$, where $A = x \times w$, $B = y \times w$, $C = x \times z$ and $D = y \times z$. Consider the following recursive algorithm to multiply two numbers u and v.

MULTIPLY(u, v):

(1) Assume n = length(u) = length(v), can pad 0's for shorter number
(2) if $length(u)$ and $length(v) = 1$ then return $u \times v$
(3) Partition u,v into $u = x \times 10^{n/2} + y$ and $v = w \times 10^{n/2} + z$
(4) A = MULTIPLY(x, w)
(5) B = MULTIPLY(y, w)
(6) C = MULTIPLY(x, z)
(7) D = MULTIPLY(y, z)
(8) Return $A \times 10^n + (B + C) \times 10^{n/2} + D$

Answer the following questions.

(a) Let T(n) be the running time to multiply two n-digit numbers, u and v. Show

$$T(n) = 4T\left(\frac{n}{2}\right) + O(n)$$

(b) Analyze the MULTIPLY(u, v) algorithm running time.
(c) Calculate the multiplication of two numbers 4301×2021 using the above algorithm.

(d) Suppose we have an abstract machine that works on the basis of 10 and can only multiply one-digit numbers, but there is no limit to adding or subtracting two integers, and the digits are multiplied by 10^x easily through shifts. What is the minimum number of multiplication used by this machine to multiply 1234 and 5618?

111. Consider the following recursive algorithm to multiply two numbers u and v.

MULTIPLY(u, v):

(1) Assume n = length(u) = length(v), can pad 0's for shorter number
(2) if $length(u) \geq 1$ then return $u \times v$
(3) Partition u, v into $u = x \times 10^{n/2} + y$ and $v = w \times 10^{n/2} + z$
(4) A = MULTIPLY(x, w)
(5) B = MULTIPLY(y, w)
(6) C = MULTIPLY(x, z)
(7) D = MULTIPLY(y, z)
(8) Return $A \times 10^n + (B + C) \times 10^{n/2} + D$

Multiply 2345 with 678 using the above algorithm.

112. Consider the following recursive algorithm to multiply two numbers u and v.

MULTIPLY(u, v):

(1) Assume n = length(u) = length(v), can pad 0's for shorter number
(2) if $length(u) \geq 1$ then return $u \times v$
(3) Partition u, v into $u = x \times 10^{n/2} + y$ and $v = w \times 10^{n/2} + z$
(4) A = MULTIPLY(x, w)
(5) B = MULTIPLY(y, z)
(6) C = MULTIPLY(x + y, z + w)
(7) Return $A \times 10^n + (C - A - B) \times 10^{n/2} + B$

Let T(n) be the running time to multiply two n-digit numbers, u and v. Show

$$T(n) = 3T\left(\frac{n}{2}\right) + O(n)$$

113. Similar to the standard integer multiplication using divide and conquer technique, we would like to design an algorithm for performing integer multiplication. Let x and y are two integer numbers. Instead of breaking the inputs x and y into two parts, we break them into three parts. Suppose x and y have n bits, where n is a power of 3. Break x into three parts, a, b, c, each with $\frac{n}{3}$ bits. Break y into three parts, d, e, f, each with $\frac{n}{3}$ bits. Then,

$$xy = ad2^{4n/3} + (ae + bd)2^n + (af + cd + be)2^{2n/3} + (bf + ce)2^{n/3} + cf$$

We have:
$$r_1 := ad$$
$$r_2 := (a + b)(d + e)$$
$$r_3 := be$$
$$r_4 := (a + c)(d + f)$$
$$r_5 := cf$$
$$r_6 := (b + c)(e + f)$$

$$z := r_1 2^{4n/3} + (r_2 - r_1 - r_3)2^n + (r_3 + r_4 - r_1 - r_5)2^{2n/3} + (r_6 - r_3 - r_5)2^{n/3} + r_5$$

(a) Show that $z = xy$.
(b) Compute the multiplication of two numbers 4301×2021 by this divide and conquer algorithm.
(c) Show that the running time of this algorithm is $O(n^{1.63})$.

114. In the large-integer multiplication algorithm, explain why we did not include multiplications by 10^n in the multiplication count.

115. Let $A(x) = a_0 + a_1 x + a_2 x^2 + \cdots + a_n x^n$ and $B(x) = b_0 + b_1 x + b_2 x^2 + \cdots + b_n x^n$ are two polynomials. To compute the multiplication $C(x) = A(x) \times B(x)$:

(a) Write a naive algorithm to compute this multiplication.
(b) Give two divide and conquer algorithms to compute this multiplication.

116. Let $A = a^x + b$ and $B = c^x + d$ are two polynomials. Explain how we can multiply $A \times B$ using only three multiplications. (Hint: one of multiplications is $(a + b)(c + d)$.)

117.

(a) Let A and B are two polynomials of degree-bound n. Describe a divide and conquer algorithm that run in time $\Theta(n^{\log 3})$ so that it divide the input polynomial coefficients into a high half and a low half.
(b) Let A and B are two polynomials of degree-bound n. Describe a divide and conquer algorithm that run in time $\Theta(n^{\log 3})$ so that it divide them according to whether their index is even or odd.

118. To multiply two numbers A and B by n and $2n$ digits, respectively, for $5000 \le n \le 10000$:

a. It is better to use the usual and classic method for multiplying them.
b. It is better to use the divide and conquer algorithm and it does not matter how.
c. After dividing B into 4 parts, it is better to multiply each of these parts by A by divide and conquer algorithm.

d. After adding 3 zeros to the left of A, it is better to multiply the result by B with the help of divide and conquer algorithm.

119. Let A and B be two n-bit integers:

(a) Explain how we can use divide and conquer integer multiplication algorithm to multiply A and B.
(b) Show that two n-bit integers can be multiplied in $O(n^{\log 3})$ steps, where each step operates on at most a constant number of 1-bit values.
(c) Multiply the two binary integers 11110011 and 10001010 using the algorithm described in (a).

10.2.9   Matrix Multiplication

120. Consider the following matrix multiplication algorithm.

```
Procedure MATRIXMULTIPLY(X, Y, n);
        Comment multiplies n×n matrices X and Y
1.      for i:=1 to n do
2.          for j:=1 to n do
3.              Z[i, j] :=0;
4.              for k :=1 to n do
5.                  Z[i, j] := Z[i, j]+ X[i, k].Y[k, j];
6. return (Z)
```

(a) Prove that this matrix multiplication algorithm is correct.
(b) Analyze the matrix multiplication algorithm. How many multiplications does it use (that is, how many times is line 5 executed) in the worst case?

121. Let us assume that n is a power of 2 in each of the $n \times n$ matrices for A and B. This simplifying assumption allows us to break a big $n \times n$ matrix into smaller blocks or quadrants of size $n/2 \times n/2$. Let A and B be two square $n \times n$ matrices, where n is even. Let A_{11}, A_{12}, A_{21}, and A_{22} represent the four $n/2 \times n/2$ submatrices of A that correspond to its four quadrants. For example, A_{11} consists of rows 1 through $n/2$ of A whose entries are restricted to columns 1 through $n/2$. Similarly, A_{12} consists of rows 1 through $n/2$ of A whose entries are restricted to columns $n/2 + 1$ through n. Finally, A_{21} and A_{22} represent the bottom half of A. Thus, A can be written as

$$A = \begin{pmatrix} A_{11} & A_{12} \\ A_{21} & A_{22} \end{pmatrix} \quad B = \begin{pmatrix} B_{11} & B_{12} \\ B_{21} & B_{22} \end{pmatrix} \quad C = A \times B = \begin{pmatrix} C_{11} & C_{12} \\ C_{21} & C_{22} \end{pmatrix}$$

The following divide and conquer algorithm is used to multiply the two matrices A and B.

MMULT(A, B, n):

(1) If n = 1 Output A × B
(2) Else
(3) Compute A11, B11, ..., A22, B22
(4) X1 ← MMULT(A11, B11, n/2)
(5) X2 ← MMULT(A12, B21, n/2)
(6) X3 ← MMULT(A11, B12, n/2)
(7) X4 ← MMULT(A12, B22, n/2)
(8) X5 ← MMULT(A21, B11, n/2)
(9) X6 ← MMULT(A22, B21, n/2)
(10) X7 ← MMULT(A21, B12, n/2)
(11) X8 ← MMULT(A22, B22, n/2)
(12) C11 ← X1 + X2
(13) C12 ← X3 + X4
(14) C21 ← X5 + X6
(15) C22 ← X7 + X8
(16) Output C
(17) End If

Multiply the following matrices using the above algorithm.

$$A = \begin{pmatrix} 1 & 2 & 3 & 4 \\ 5 & 6 & 7 & 8 \\ 1 & 2 & 3 & 4 \\ 5 & 6 & 7 & 8 \end{pmatrix} \quad B = \begin{pmatrix} 9 & 8 & 7 & 5 \\ 2 & 3 & 1 & 5 \\ 6 & 6 & 3 & 4 \\ 7 & 9 & 1 & 2 \end{pmatrix}$$

122. Let T(n) be the total number of mathematical operations performed by MMULT (A, B, n), show

$$T(n) = 8T\left(\frac{n}{2}\right) + \Theta(n^2)$$

123. Strassen's algorithm to multiply the two matrices is based on the following observation:
C11 = P5 + P4 − P2 + P6
C12 = P1 + P2
C21 = P3 + P4
C22 = P1 + P5 − P3 − P7
where
P1 = A11(B12 − B22)
P2 = (A11 + A12)B22
P3 = (A21 + A22)B11
P4 = A22(B21 − B11)
P5 = (A11 + A22)(B11 + B22)
P6 = (A12 − A22)(B21 + B22)
P7 = (A11 − A21)(B11 + B12)

The above formulas can be used to compute A × B recursively as follows:

STRASSEN(A, B):

(1) If n = 1 Output A × B
(2) Else
(3) Compute A11, B11, ..., A22, B22
(4) P1 ← STRASSEN(A11, B12 - B22)
(5) P2 ← STRASSEN(A11 + A12, B22)
(6) P3 ← STRASSEN(A21 + A22, B11)
(7) P4 ← STRASSEN(A22, B21 - B11)
(8) P5 ← STRASSEN(A11 + A22, B11 + B22)
(9) P6 ← STRASSEN(A12 - A22, B21 + B22)
(10) P7 ← STRASSEN(A21 - A11, B11 + B12)
(11) C11 ← P5 + P4 - P2 + P6
(12) C12 ← P1 + P2
(13) C21 ← P3 + P4
(14) C22 ← P1 + P5 - P3 + P7
(15) Output C
(16) End If

Multiply the following matrices using the above algorithm.

$$A = \begin{pmatrix} 0 & 1 & 2 & 3 \\ 1 & 2 & 3 & 0 \\ 2 & 3 & 0 & 1 \\ 3 & 0 & 1 & 2 \end{pmatrix} \quad B = \begin{pmatrix} 3 & 2 & 1 & 0 \\ 0 & 2 & 3 & 1 \\ 2 & 2 & 1 & 0 \\ 1 & 1 & 0 & 3 \end{pmatrix}$$

124. Regarding Strassen's algorithm

(a) Let T(n) be the total number of mathematical operations performed by STRASSEN (A, B), show

$$T(n) = 7T\left(\frac{n}{2}\right) + \Theta(n^2)$$

(b) What is the running time of Strassen's algorithm for matrix multiplication?

125. In multiplication of matrices by Strassen's algorithm, if the small problem is multiplication of 2 × 2 matrices, how many numerical multiplications are done for multiplication of two 8 × 8 matrices?

126. Use Strassen's algorithm to multiply the two matrices

$$\begin{pmatrix} 0 & 2 \\ 1 & 5 \end{pmatrix} \cdot \begin{pmatrix} 7 & 2 \\ 0 & 5 \end{pmatrix}$$

127. Use Strassen's algorithm to multiply the two matrices

$$\begin{pmatrix} 0 & 0 & 3 & 1 \\ 3 & 1 & 2 & 1 \\ 1 & 0 & 4 & 1 \\ 5 & 0 & 2 & 1 \end{pmatrix} \cdot \begin{pmatrix} 1 & 0 & 0 & 5 \\ 3 & 0 & 0 & 1 \\ 2 & 0 & 1 & 1 \\ 0 & 2 & 4 & 0 \end{pmatrix}$$

the multiplication ends when n = 2, i.e., use the naive algorithm to compute the products of 2 × 2 matrices.

128. Suppose for a specific crossover point, Strassen's algorithm use the brute force method after matrix sizes become smaller than the crossover point. Implement this version of the Strassen's algorithm and find the appropriate crossover point on your computer system.

129. Explain how Strassen's algorithm can be applied to multiply $n \times n$ matrices in which n is not a power of 2?

130. Give a divide and conquer algorithm to multiply $n \times n$ matrices in time $O(n^{\log 7})$.

131. V. Pan has discovered a way of multiplying 68×68 matrices using 132,464 multiplications, a way of multiplying 70×70 matrices using 143,640 multiplications, and a way of multiplying 72×72 matrices using 155,424 multiplications. Which method yields the best asymptotic running time when used in a divide and conquer matrix multiplication algorithm? How does it compare to Strassen's algorithm?

132. Let X and Y are two matrix. Given the program of naive matrix multiplication algorithm.

```
for i=1 to n do
    for j=1 to n do
        Z[i][j] = 0;
        for k=1 to n do
            ..........
```

Fill in the blanks with appropriate formula.

a. $Z[i][j] = Z[i][j] + X[i][k] \times Y[k][j]$
b. $Z[i][j] = Z[i][j] + X[i][k] + Y[k][j]$
c. $Z[i][j] = Z[i][j] \times X[i][k] \times Y[k][j]$
d. $Z[i][j] = Z[i][j] \times X[i][k] + Y[k][j]$

133. What is the recurrence relation used in Strassen's algorithm?

a. $7T(n/2) + \Theta(n^2)$
b. $8T(n/2) + \Theta(n^2)$
c. $7T(n/2) + O(n^2)$

Fig. 10.1 Another version $s_1 = G + H$ $m_1 = s_2 s_6$ $t_1 = m_1 + m_2$
of Strassen's algorithm $s_2 = s_1 - E$ $m_2 = EI$ $t_2 = t_1 + m_4$
                              $s_3 = E - G$ $m_3 = FK$
                              $s_4 = F - s_2$ $m_4 = s_3 s_7$
                              $s_5 = J - I$ $m_5 = s_1 s_5$
                              $s_6 = L - s_5$ $m_6 = s_4 L$
                              $s_7 = L - J$ $m_7 = H s_8$
                              $s_8 = s_6 - K.$

d. $8T(n/2) + O(n^2)$

134. The square of a matrix A is its product with itself, AA.

(a) Show that five multiplications are sufficient to compute the square of a 2×2
    matrix.
(b) Can you give a more efficient algorithm than the Strassen's algorithm for this
    problem?
(c) True or False, and Justify. Squaring a matrix is easier than matrix multiplication.

135. Suppose we want to evaluate the polynomial $p(x) = a_0 + a_1 x + a_2 x^2 + \cdots +$
$a_n x^n$ at point x.

(a) Show that the following simple function, known as Horner's rule, does the task
    and provides the answer in z.

    z = a_n
    for i = n - 1 downto 0
        z = zx + a_i

(b) How many additions and multiplications does this function use? Write as a
    function of n. Can you find a polynomial for which an alternative method is
    substantially better?

136. The standard description of Strassen's algorithm assumes that n is a power of
2. Devise an algorithm that runs in time $O(n^{2.81})$ when n is not necessarily a power
of 2.

137. Another version of Strassen's algorithm uses the following identities. To compute

$$\begin{bmatrix} A & B \\ C & D \end{bmatrix} = \begin{bmatrix} E & F \\ G & H \end{bmatrix} \times \begin{bmatrix} I & J \\ K & L \end{bmatrix}$$

first compute the values shown in Fig. 10.1, then

$$A = m_2 + m_3$$

$$B = t_1 + m_5 + m_6$$

$$C = t_2 - m_7$$

$$D = t_2 + m_5$$

(a) Prove that this algorithm is correct.
(b) Analyze its running time. Is it likely to be faster or slower than the standard version of Strassen's algorithm in practice?

138. Suppose we were to come up with a variant of Strassen's algorithm based on the fact that 3×3 matrices can be multiplied in only m multiplications instead of the normal 27. How small would m have to be for this algorithm to be faster than Strassen's algorithm for large enough n?

139. Using Strassen's algorithm, give a divide and conquer algorithm to multiply a $kn \times n$ matrix by an $n \times kn$ matrix.

10.2.10 Application

140. Let A be a sorted array of n distinct integers. Describe a divide and conquer algorithm in time $O(logn)$ that finds an index i such that $A[i] = i$ (if one exists).

141. *Finding the missing number.* Suppose you are given an unsorted array A of all integers in the range 0 to n except for one integer, denoted the *missing number*. Assume $n = 2^k - 1$. Answer the following questions.

(a) Design a Divide and Conquer algorithm to find the missing number.
(b) Analyze its running time
(d) Show that the recursive relation for this algorithm is: $T(n) = T(\frac{n}{2}) + n$.

142. *Fast Exponentiation.* Given a number a and given an integer n, compute a to the power n. The simplest algorithm performs $n - 1$ multiplications, by computing $a \times a \times \cdots \times a$. The running time of this simple algorithm is $O(n)$. However, we can do better by observing that $n = \lceil \frac{n}{2} \rceil + \lfloor \frac{n}{2} \rfloor$. The idea to calculate a^n is to express a^n as:

$$a^n = \begin{cases} a^{n/2}.a^{n/2} & \text{if } n \text{ is even;} \\ a^{(n-1)/2}.a^{(n-1)/2}.a & \text{if } n \text{ is odd;} \end{cases}$$

(a) Give a divide and conquer algorithm for this problem.

(b) Show that the recursive relation for this algorithm is:

$$T(n) = T(n/2) + 1$$

143. Assume A be an array of size n and a number p is given. We need to find a pair of elements in the array whose sum is exactly p. If there is a such pair, print the answer "yes," and otherwise "no." For example, given the array [2, 5, 3] and $p = 5$, the answer is "yes," but given $p = 4$ the answer is "no."

(a) Use mergesort and binary search algorithms to solve this problem. Analyse its running time.
(b) Give a $\Theta(n \log n)$-time algorithm (different above) to this problem.

144. Let $S[1..n]$ is an array of size n. For some $m \leq n$, a subsequence is called ascending if $A[i_1] \leq A[i_2] \leq \cdots \leq A[i_m]$. Using divide and conquer algorithm, describe an algorithm to find the longest ascending subsequence (i.e., the largest m).

145. What is the time complexity of computing the sum of $\sqrt{n}$ smallest elements in an unsorted array of length n?

146. Let A is an array containing n elements. The problem is to build a heap tree H on this array using divide and conquer algorithm.

Recall that a heap is an (almost) perfectly balanced binary tree where key$(v) \geq$ key(parent(v)) for all nodes v. We assume $n = 2^h - 1$ for some constant h, such that H is perfectly balanced (leaf level is "full"). The following, SLOWHEAP$(1, n)$, is an algorithm for this aim. The algorithm, first constructs (a pointer to) H by finding the minimal element x in A, making x the root in H, and recursively constructing the two sub-heaps below x (each of size approximately $\frac{n-1}{2}$).

SLOWHEAP(i, j)

> If $i = j$ then return pointer to heap consisting of node containing $A[i]$
> Find $i \leq l \leq j$ such that $x = A[l]$ is the minimum element in $A[i \ldots j]$
> Exchange $A[l]$ and $A[j]$
> $Ptr_{left} = $ SLOWHEAP$(i, \lfloor \frac{i+j-1}{2} \rfloor)$
> $Ptr_{right} = $ SLOWHEAP$(\lfloor \frac{i+j-1}{2} \rfloor + 1, j - 1)$
> Return pointer to heap consisting of root r containing x with child pointers Ptr_{left} and Ptr_{right}

End

The following algorithm, FASTHEAP$(1, n)$, is another divide and conquer algorithm to build a heap tree H on an array. This algorithm constructs (a pointer to) H by placing an arbitrary element x from A (the last one) in the root of H, recursively constructing the two sub-heaps below x, and finally performing a DOWN- HEAPIFY operation on x to make H a heap.

FASTHEAP(i, j)

$Ptr_{\text{left}} = \text{FASTHEAP}(i, \lfloor \frac{i+j-1}{2} \rfloor)$

$Ptr_{\text{right}} = \text{FASTHEAP}(\lfloor \frac{i+j-1}{2} \rfloor + 1, j - 1)$

Let Ptr be pointer to tree consisting of root r containing $x = A[j]$ with child pointers Ptr_{left} and Ptr_{right}

Perform DOWN- HEAPIFY on Ptr

Return Ptr

End

(a) Define and solve a recurrence equation for the running time of SLOWHEAP.
(b) Define and solve a recurrence equation for the running time of FASTHEAP.

147. *2-D maxima finding problem.* A point (x_1, y_1) dominates (x_2, y_2) if $x_1 > x_2$ and $y_1 > y_2$. A point is called a maximum if no other point dominates it. Straightforward method for this problem is to compare every pair of points. In this case time complexity is $O(n^2)$.

(a) Devise a divide-and-conquer algorithm for 2-D maxima finding problem.
(b) Write down its recursive relation and then computes its time complexity.

148. *2-D closest pair problem.* Given a set of points $\{p_1, \ldots, p_n\}$, find a pair of points, $\{p_i, p_j\}$, which are closest together.

Input: A set of points in the plane, $p_1 = (x_1, y_1)$, $p_2 = (x_2, y_2), \ldots, p_n = (x_n, y_n)$

Output: The closest pair of points: that is, the pair $p_i \neq p_j$ for which the distance between p_i and p_j, that is,

$$\sqrt{(x_i - x_j)^2 + (y_i - y_j)^2}$$

is minimized.

Here's a high-level overview of the divide-and-conquer algorithm:

(1) Split the points with line L so that half the points are on each side.
(2) Recursively find the pair of points closest in each half.

(a) Devise a divide-and-conquer algorithm for closest pair problem.
(b) Write down its recursive relation and then computes its time complexity.

149. 1-dimension problem is a special case of 2-D closest pair problem to find a pair of points, $\{p_i, p_j\}$, which are closest together.

(a) Using sorting, devise an algorithm for 1-dimension closest pair problem.
(b) Devise a divide-and-conquer algorithm for 1-dimension closest pair problem.

150. Practice with the Fast Fourier transform (FFT).

(a) What is the FFT of $(1, 0, 0, 0)$? What is the appropriate value of ω in this case? And of which sequence is $(1, 0, 0, 0)$ the FFT?
(b) Repeat for $(1, 0, 1, -1)$.

151. Consider Fast Fourier Transform (FFT) method:

(a) Explain how to multiply two polynomials using the FFT.
(b) Multiply the two polynomials $2x + 3$ and $3x^2 + 4$ using the FFT.
(c) Multiply the two polynomials $2 + 2x + 3x^2$ and $x^2 + 1$ using the FFT.

152. An array $A[1..n]$ is said to have a majority element if more than half of its entries are the same. Given an array, the task is to design an efficient algorithm to tell whether the array has a majority element, and, if so, to find that element. The elements of the array are not necessarily from some ordered domain like the integers, and so there can be no comparisons of the form "is $A[i] > A[j]$?". (Think of the array elements as GIF files, say.) However you can answer questions of the form: "is $A[i] = A[j]$?" in constant time.

(a) Show how to solve this problem in $O(n \log n)$ time. (Hint: Split the array A into two arrays A_1 and A_2 of half the size.) Does knowing the majority elements of A_1 and A_2 help you figure out the majority element of A? If so, you can use a divide-and-conquer approach.)
(b) Can you give a linear-time algorithm? (Hint: Here's another divide-and-conquer approach:

   • Pair up the elements of A arbitrarily, to get $n/2$ pairs
   • Look at each pair: if the two elements are different, discard both of them; if they are the same, keep just one of them
     Show that after this procedure there are at most n/2 elements left, and that they have a majority element if and only if A does.)

153. Can you design a more efficient algorithm than the one based on the bruteforce strategy to solve the closest-pair problem for n points $x_1, x_2, \ldots, x_n$ on the real line?

154. Let $x_1 < x_2 < \cdots < x_n$ be real numbers representing coordinates of n villages located along a straight road. A post office needs to be built in one of these villages.

(a) Design an efficient algorithm to find the post-office location minimizing the average distance between the villages and the post office.
(b) Design an efficient algorithm to find the post-office location minimizing the maximum distance from a village to the post office.

155. Find the convex hulls of the following sets and identify their extreme points (if they have any):

- a line segment
- a square

– the boundary of a square
– a straight line

156. Design a linear-time algorithm to determine two extreme points of the convex hull of a given set of $n > 1$ points in the plane.

157. What modification needs to be made in the brute-force algorithm for the convex-hull problem to handle more than two points on the same straight line?

158. Write a program implementing the brute-force algorithm for the convex-hull problem.

159. Let A denote an array of size n:

(a) Using divide-and-conquer technique, give an algorithm to find the position of the largest element in the array A.
(b) Apply the algorithm on arrays with several elements of the largest value and print the output.
(c) Provide a recurrence relation for the number of key comparisons performed by the algorithm.
(d) Compare the designed algorithm with the brute-force algorithm for this problem.

160. *Tromino puzzle*. Atromino (more accurately, a right tromino) is an L-shaped tile formed by three 1×1 squares. The problem is to cover any $2^n \times 2^n$ chessboard with a missing square with trominoes. Trominoes can be oriented in an arbitrary way, but they should cover all the squares of the board except the missing one exactly and with no overlaps (See Fig. 10.2).

Fig. 10.2 Tromino puzzle

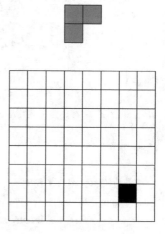

Design a divide-and-conquer algorithm for this problem.

161. *Nuts and bolts.* Suppose there are a collection of n bolts of different widths and n corresponding nuts. You can try a nut and bolt together, from which you can determine whether the nut is larger than the bolt, smaller than the bolt, or matches the bolt exactly. However, there is no way to compare two nuts together or two bolts together. The problem is to match each bolt to its nut. Design an algorithm for this problem with average-case efficiency in $\Theta(n \log n)$.

162. We can compute the number of levels in a binary tree by the dividing the tree into left and right sub-trees and computing the height of each sub trees recursively. Give a divide and conquer algorithm for computing the number of levels in a binary tree. What is the running time of your algorithm?

163. Let $A[1..n]$ denotes an array of size n. An element is referred a majority element of A if it occur in more than $n/2$ positions. Assume that elements cannot be ordered or sorted, but can be compared for equality. Give a divide and conquer algorithm to find a majority element in A (or determine that no majority element exists).

10.3   Solutions

1.

$$T(n) = 1n^d + a\left(\frac{n}{b}\right)^d + a^2\left(\frac{n}{b^2}\right)^d + \cdots + a^L\left(\frac{n}{b^L}\right)^d$$

$$= n^d[1 + a\left(\frac{1}{b}\right)^d + a^2\left(\frac{1}{b^2}\right)^d + \cdots + a^L\left(\frac{1}{b^L}\right)^d$$

$$= n^d[1 + \frac{a}{b^d} + \left(\frac{a}{b^d}\right)^2 + \cdots + \left(\frac{a}{b^d}\right)^L]$$

Case 1:

$$if\ d > log_b a\ ,\ then\quad \frac{a}{b^d} < 1\quad \rightarrow\quad T(n) = O\left(n^d - 1\right) = O(n^d)$$

Case 2:

$$if\ d = log_b a\ ,\ then\quad \frac{a}{b^d} = 1\quad \rightarrow\quad T(n) = O\left(n^d(L+1)\right) = O\left(n^d.L\right)$$

$$we\ know\ n = b^L\ so\ L = log_b n = \log n :$$

$$T(n) = O(n^d.\log n)$$

Case 3:

$$if \ d < \log_b a \ , \ then \ \frac{a}{b^d} > 1 \ \rightarrow \ T(n) = O\left(n^d \left(\frac{a}{b^d}\right)^L\right) = O(n^{\log_b a})$$

2.

a. $a = 4, \ b = 2, \ f(n) = n \ \Rightarrow \ n^{\log_b a} = n^2 > O(n) \ \Rightarrow T(n) = O\left(n^{\log_b a}\right) = O\left(n^2\right)$

b.

$$T(n) = n + 4\left(\frac{n}{2}\right) + 8\left(\frac{n}{4}\right) + \ldots \overset{height = \log_2 n}{\Longrightarrow} T(n) = n + 2n + 4n + \cdots + 2^{\log_2 n} n$$

$$= n\left(2^0 + 2^1 + 2^2 + \cdots + 2^{\log_2 n}\right) = n\frac{2^{\log_2 n} - 1}{2 - 1} = n\left(2^{\log_2 n} - 1\right) = n\left(n^{\log_2 2} - 1\right)$$

$$= n(n-1) = n^2 - n \Rightarrow T(n) = O(n^2)$$

Substitution Method: We make a guess for the solution and then we use mathematical induction to prove the guess is correct or incorrect.

I guess the solution as $T(n) = O(n^2)$. Now I use induction to prove my guess, I need to prove that $T(n) \le cn^2$

$$T(n) = 4T\left(\frac{n}{2}\right) + n \le 4c\left(\frac{n}{2}\right)^2 + n$$

$$= cn^2 + n \le n^2 \Rightarrow T(n) = O(n^2)$$

3.

(a) Can't solve with master theorem: because $T(n) = 2T(n-1) + c$ don't have $(b > 1)$ condition.

(b) See Fig. 10.3.
We know $T(1) = 1$

$$T(n) = C + 2C + 4C + \cdots + \left(2^{n-1}\right)C = \left(2^n - 1\right)C$$

$$T(n) = O(2^n)$$

(c)
Guess: $T(n) = O(2^n) \quad or \quad T(n) \le k2^n + d$

$$T(n) = 2T(n-1) + c$$

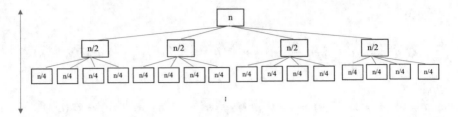

Fig. 10.3 Recursive tree for question 3

$$T(n) \leq 2\left(k2^{n-1} + d\right) + c$$

$$T(n) \leq k2^n + 2d + c$$

$$T(n) \leq k2^n + d + (d + c)$$

$$T(n) \leq k2^n + d$$

$$T(n) = O(2^n)$$

4.

$$\begin{cases} M_1 \Rightarrow T(n) = 5T\left(\frac{n}{2}\right) + O(n) \Rightarrow O(n^{2.33}) \\ M_2 \Rightarrow T(n) = 2T(n-1) + c \Rightarrow \quad O\left(2^n\right) \\ M_3 \Rightarrow T(n) = 9T\left(\frac{n}{3}\right) + n^2 \Rightarrow O\left(n^2 \log n\right) \end{cases} \Rightarrow \boldsymbol{M_3 \textit{ is the best algorithm here}}$$

We have used master theorem.

5.
(a)
```
int Search(int[] array, int x)
   {
     int l = 0, r = array.Length - 1;
     while (l <= r) {
       int m = l + (r - l) / 2;
       if (array[m] == x)
         return m;
       if (array[m] < x)
         l = m + 1;
       else
         r = m - 1;

   }
```

(b)

```
int Search(int[] array, int l, int r, int x) {
    if (r >= l) {
    int m = l + (r - l) / 2;
    if (array[m] == x)
      return m;
    if (array[m] > x)
      return Search(array, l, m - 1, x);
      return Search(array, m + 1, r, x);
    }
    return -1;
}
```

Both of them has $O (\log n)$ time complexity, but iterative approach has $O(1)$ space complexity, it means iterative algorithm has constant space complexity. The space complexity of recursive algorithm is $O (\log n)$

7.
(a)

$-1 : 3$ $0 : 2$ $5 : 1$ $30 : 3$

(b)

$$\frac{19}{8} = 2/3$$

16. (a) Linear search algorithm

1. If array[0] is not 0, return 0.
2. Otherwise traverse the input array starting from index 0, and for each pair of elements array[i] and array[i+1], find the difference between them.
3. If the difference is greater than 1 then array[i]+1 is the missing number.

Time Complexity: O(n)

17.
(a) Linear search algorithm

1. If in the array, the first element is greater than the second or the last element is greater than the second last, print the respective element and terminate the program.
2. Else traverse the array from the second index to the second last index
3. If for an element array[i], it is greater than both its neighbors, (array[i] > array[i-1] and array[i] > array[i+1]), then print that element and terminate.

Time complexity: O(n)

(b) Modified binary search

1. Create two variables, low and high, initialize $low = 0$ and $high = n - 1$
2. Iterate the steps below till $low <= high$:
3. Compare middle element with its neighbors (if neighbors exist), and check if the mid value or index $mid = low + (high - low)/2$, is the peak element or not, if yes then print the element and terminate.
4. If middle element is not peak and its left neighbors is greater than it, then left half must have a peak element. So, if the element on the left side of the middle element is greater, then check for peak element on the left side.
5. If middle element is not peak and its right neighbors is greater than it, then right half must have a peak element. So, if the element on the right side of the middle element is greater, then check for peak element on the right side.

Time Complexity: O(log n)

21.

Comparisons for each element: $\{-1 \Rightarrow 3, \quad 0 \Rightarrow 2, \quad 1 \Rightarrow 3, \quad 2 \Rightarrow 4, \quad 3 \Rightarrow 1, \quad 4 \Rightarrow 3, \quad 5 \Rightarrow 3, \quad 6 \Rightarrow 3, \quad 7 \Rightarrow 4\}$
So, average of this comparisons equal to: 3.17

34.
(b) The minimum number of comparisons required to find the min and the max in this array is 8 (see Fig. 10.4).

52.
(a) No.

Example   (1) $gcd\,(7, 24) = gcd\,(7, 12) = gcd\,(7, 6) = gcd\,(7, 3) = gcd\,(2, 3) =?$

Example (2) $gcd\,(24, 8) = 2gcd\,(12, 4) = 4gcd\,(6, 2) = 8gcd\,(3, 1) = 8gcd\,(1, 1) = 8gcd\,(0, 1) =?$

There is no relation when a is even and b is odd, $(Ex1)$, $(Ex2) \Rightarrow$ Algorithm (a) is not working, no exit condition.

The correct algorithm is as follows:

$$gcd\,(a, b) = a \qquad a \neq 0,\ b = 0$$

$$gcd\,(a, b) = b \qquad a = 0, b \neq 0$$

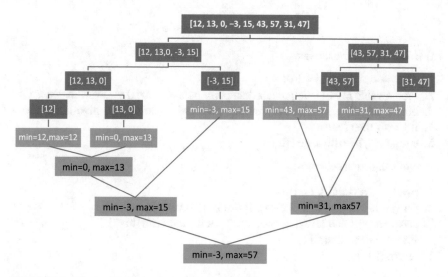

Fig. 10.4 Tree for question 34

$$gcd\,(a, b) = 2gcd\,(a/2, b/2) \qquad a\ is\ even,\ b\ is\ even$$

$$gcd\,(a, b) = gcd\,(a, b/2) \qquad a\ is\ odd,\ b\ is\ even$$

$$gcd\,(a, b) = gcd\,(a/2, b) \qquad a\ is\ even,\ b\ is\ odd$$

$$gcd\,(a, b) = gcd\left(\frac{a - b}{2}, b\right) \qquad a\ is\ odd,\ b\ is\ odd \quad a \ge b$$

$$gcd\,(a, b) = gcd\left(a, \frac{b - a}{2}\right) \qquad a\ is\ odd,\ b\ is\ odd \quad b \ge a$$

(b)

GCD (a, b)
    if b = 0 then
       return a;
    else
      GCD (b, a% b);

(c) The time complexity for the above algorithm is O (log(max (a, b))), but time complexity for the Euclidean Algorithm is O(log(min (a, b)))

93.

(a) Hoare's partition schemes:

1. pivot = array[low], i = low − 1, j = high + 1
2. Find a value in left side greater than pivot: while array[i] < pivot: increase i
3. Find a value in right side smaller than pivot: while (array[j] > pivot): decrease j
4. if i >= j then return j
5. swap array[i] with array[j]

Lomuto's partition schemes:

1. pivot = array[high], i = (low − 1)
2. for (j = low; j <= high- 1; j++): if (array[j] < pivot): increase i
3. swap array[i] and array[j], swap array[i + 1] and array[high]
4. if i >= j then return j
5. return (i + 1)

(d) Hoare's scheme does three times fewer swaps. So, it is more efficient than Lomuto's partition scheme. Both of them has time complexity of $O(n^2)$.

107.

1. Calculate median of array A and store it in X.
2. Then we will search for the crossover point.
   The point before which elements are smaller than or equal to X and after which elements are greater.

   Consider the following pseudo-code for finding crossover point:

   ```
   def crossOver(arr, l, h, x):
       if (arr[h] <= x) then return h
       if (arr[l] > x) then return l
       mid = (l + h) / 2
       if (arr[mid] <= x and arr[mid + 1] > x) then return mid
       if(arr[mid] < x) then return crossOver(arr, mid + 1, h, x)
       return crossOver(arr, l, mid - 1, x)
   ```
   Note: This step takes O(n) time.
3. Now we can compare elements on both sides of crossover point to print k closest elements as the answer.

Note: This step takes O(k) time.
Time complexity is O(n).

108.
Algorithm:

1. Declare three pointer i, j, k

2. Set i=1, j=1, k=1 (first elements of A and B)
3. Initialize array C for merge
4. Do:

   a. Compare A[i],B[j]
   b. If A[i] <= B[j] : C[k++] = A[i++]
   c. Else if A[i] > B[j] : C[k++] = B[j++]
   d. Repeat until (i == n or j == n)

5. Copy remaining elements from A or B to rest of C
6. Calculate sum of C[n/2] and C[n/2 + 1], and half the result
7. Result of step 6 is answer.

Time complexity: this algorithm simulates the "merge" part of mergesort, and this function takes $O(n)$

110(c).

$$4301 = 43 \times 10^2 + 1$$

$$2021 = 20 \times 10^2 + 21$$

$$uv = xw \times 10^{\frac{n}{2}} + (xz + wy) \times 10^{\frac{n}{2}} + yz$$

$$4301 \times 2021 = (43 \times 20) \times 10^2 + (43 \times 21 + 1 \times 20) \times 10^2 + 1 \times 21$$

$$43 = 4 \times 10 + 3$$

$$20 = 2 \times 10 + 0$$

$$43 \times 20 = 4 \times 2 \times 10 + (4 \times 0 + 2 \times 3) \times 10 + 3 \times 0 = 860$$

$$43 = 4 \times 10 + 3$$

$$21 = 2 \times 10 + 1$$

$$43 \times 21 = 4 \times 2 \times 10 + (4 \times 1 + 2 \times 3) \times 10 + 3 \times 1 = 903$$

$$1 = 0 \times 1 + 1$$

$$20 = 2 \times 10 + 0$$

$$1 \times 20 = 0 \times 2 \times 10 + (0 \times 0 + 2 \times 1) \times 10 + 1 \times 0 = 20$$

$$1 = 0 \times 1 + 1$$

$$21 = 2 \times 10 + 1$$

$$1 \times 20 = 0 \times 2 \times 10 + (0 \times 1 + 2 \times 1) \times 10 + 1 \times 1 = 21$$

$$4301 \times 2021 = 860 \times 10^2 + (903 + 20) \times 10^2 + 21 = 8692321$$

111.

$$u = 2345$$

$$v = 678$$

$$u = x \times 10^2 + y = 2345 \rightarrow x = 23, \ y = 45$$

$$v = w \times 10^1 + z = 678 \rightarrow w = 6, \ z = 78$$

$$A = Multiply \ (x, w) = Multiply \ (23, 6) \rightarrow 2 \times 10 + 3 \ , \ 6 \times 10^0 + 0$$

$$\rightarrow A_1 = 2 \times 6 = 12, \ B_1 = 3 \times 6 = 18, \ C_1 = 2 \times 0 = 0, \ D_1 = 3 \times 0 = 0$$

$$\rightarrow A_1 \times 10^n + (B_1 + C_1) \times 10^{\frac{n}{2}} + D \rightarrow 12 \times 10 + (18 + 0) \times 1 + 0 = 138$$

$$B = Multiply \ (y, w) = Multiply \ (45, 6) \rightarrow 4 \times 10 + 5 \ , \ 6 \times 10^0 + 0$$

$$\rightarrow A_1 = 4 \times 6 = 24 \ , \ B_1 = 5 \times 6 = 30, \ C_1 = 4 \times 0 = 0, \ D_1 = 5 \times 0 = 0$$

$$\rightarrow 24 \times 10 + (30 + 0) \times 1 + 0 = 270$$

$$C = Multiply \ (x, z) = Multiply \ (23, 78) \rightarrow 2 \times 10 + 3 \ , \ 7 \times 10 + 8$$

$$\rightarrow A_1 = 2 \times 7 = 14 \ , \ B_1 = 3 \times 7 = 21, \ C_1 = 2 \times 8 = 16, \ D_1 = 3 \times 8 = 24$$

$$\rightarrow 14 \times 100 + (21 + 16) \times 10 + 24 = 1794$$

$$D = Multiply \ (y, z) = Multiply \ (45, 78) \rightarrow 4 \times 10 + 5 \ , \ 7 \times 10 + 8$$

$$\rightarrow A_1 = 4 \times 7 = 28 \ , \ B_1 = 5 \times 7 = 35, \ C_1 = 4 \times 8 = 32, \ D_1 = 5 \times 8 = 40$$

$$\rightarrow 28 \times 100 + (35 + 32) \times 10 + 40 = 3510$$

$$u \times v = A \times 10^n + (B + C) \times 10^{\frac{n}{2}} + D \rightarrow 138 \times 100 + (270 + 1794) \times 10 + 3510 = 1589910$$

112.

In this algorithm, we divide each number to 2 parts,

And next step is to calculate multiplication of x, w, y, z, and x + y, z + w (as formula),

so, we need to call this procedure recursively 3 times. After that, the part of merging result from each call is takes O(n) time.

as the master theorem:

$$T(n) = aT\left(\frac{n}{2}\right) + f(n)$$

That **a** is number of recursive calls inside algorithm, **b** is number of parts that input splits, and **f(n)** is the callback actions runtime.

In this algorithm, we assume a = 3 (three recursive call), b = 2 (each step, the input is halved) and f(n) = O(n).

So, we have:

$$T(n) = 3T\left(\frac{n}{2}\right) + O(n)$$

115(a).
    Simple solution is to one by one consider each term of A(x) and multiply it with terms of B(x) by the help of for loop.
Time complexity: $O(n^2)$

Consider the following pseudo-code:

Note: array A represents coefficients of A(x) and array B in our pseudo-code, represents coefficients of B(x).


```
def multiply(A, B, a_len, b_len):
        product is an array with len: a_len – b_len + 1
        for i from 0 to a_len – 1:
         for j from 0 to b_len – 1:
          product [i + j] += A[i] * B[j];
         return product
```
121.

$$A_{11} = \begin{bmatrix} 1 & 2 \\ 5 & 6 \end{bmatrix} \quad A_{12} = \begin{bmatrix} 3 & 4 \\ 7 & 8 \end{bmatrix} \quad A_{21} = \begin{bmatrix} 1 & 2 \\ 5 & 6 \end{bmatrix} \quad A_{22} = \begin{bmatrix} 3 & 4 \\ 7 & 8 \end{bmatrix}$$

$$B_{12} = \begin{bmatrix} 9 & 8 \\ 2 & 3 \end{bmatrix} \quad B_{12} = \begin{bmatrix} 7 & 5 \\ 1 & 5 \end{bmatrix} \quad B_{21} = \begin{bmatrix} 6 & 6 \\ 7 & 9 \end{bmatrix} \quad B_{22} = \begin{bmatrix} 3 & 4 \\ 1 & 2 \end{bmatrix}$$

$$X_1 = A_{11} \times B_{11} = \begin{bmatrix} 13 & 14 \\ 57 & 58 \end{bmatrix} \quad X_2 = A_{12} \times B_{21} = \begin{bmatrix} 46 & 54 \\ 98 & 114 \end{bmatrix}$$

$$X_3 = A_{11} \times B_{12} = \begin{bmatrix} 9 & 15 \\ 41 & 55 \end{bmatrix} \quad X_4 = A_{12} \times B_{22} = \begin{bmatrix} 13 & 20 \\ 29 & 44 \end{bmatrix}$$

$$X_5 = A_{21} \times B_{11} = \begin{bmatrix} 13 & 14 \\ 57 & 58 \end{bmatrix} \quad X_6 = A_{22} \times B_{21} = \begin{bmatrix} 46 & 54 \\ 98 & 114 \end{bmatrix}$$

$$X_7 = A_{22} \times B_{12} = \begin{bmatrix} 9 & 15 \\ 41 & 55 \end{bmatrix} \quad X_8 = A_{22} \times B_{22} = \begin{bmatrix} 13 & 20 \\ 29 & 44 \end{bmatrix}$$

$$C_1 = X_1 \times X_2 = \begin{bmatrix} 59 & 68 \\ 155 & 172 \end{bmatrix} \quad C_2 = X_3 \times X_4 = \begin{bmatrix} 22 & 35 \\ 70 & 99 \end{bmatrix}$$

$$C_3 = X_5 \times X_6 = \begin{bmatrix} 59 & 68 \\ 155 & 172 \end{bmatrix} \quad C_4 = X_7 \times X_8 = \begin{bmatrix} 22 & 35 \\ 70 & 99 \end{bmatrix}$$

$$C = \begin{bmatrix} 59 & 68 & 22 & 35 \\ 155 & 172 & 70 & 99 \\ 59 & 68 & 22 & 35 \\ 155 & 172 & 70 & 99 \end{bmatrix}$$

123.

$$A = \begin{matrix} & A_{11} & A_{12} \\ & \begin{bmatrix} 0 & 1 & 2 & 3 \\ 1 & 2 & 3 & 0 \\ 2 & 3 & 0 & 1 \\ 3 & 0 & 1 & 2 \end{bmatrix}_{4\times4} \\ & A_{21} & A_{22} \end{matrix}$$

$$B = \begin{matrix} & B_{11} & B_{12} \\ & \begin{bmatrix} 3 & 2 & 1 & 0 \\ 0 & 2 & 3 & 1 \\ 2 & 2 & 1 & 0 \\ 1 & 1 & 0 & 3 \end{bmatrix}_{4\times4} \\ & B_{21} & B_{22} \end{matrix}$$

$$P_1 = A_{11}(B_{12} - B_{22}) = \begin{bmatrix} 0 & 1 \\ 1 & 2 \end{bmatrix} \left(\begin{bmatrix} 1 & 0 \\ 3 & 1 \end{bmatrix} - \begin{bmatrix} 1 & 0 \\ 0 & 3 \end{bmatrix} \right) = \begin{bmatrix} 3 & -2 \\ 6 & -4 \end{bmatrix}$$

$$P_2 = (A_{11} + A_{12})B_{22} = \left(\begin{bmatrix} 0 & 1 \\ 1 & 2 \end{bmatrix} + \begin{bmatrix} 2 & 3 \\ 3 & 0 \end{bmatrix}\right)\left(\begin{bmatrix} 1 & 0 \\ 0 & 3 \end{bmatrix}\right) = \begin{bmatrix} 2 & 12 \\ 4 & 6 \end{bmatrix}$$

$$P_3 = (A_{21} + A_{122})B_{11} = \left(\begin{bmatrix} 2 & 3 \\ 3 & 0 \end{bmatrix} + \begin{bmatrix} 0 & 1 \\ 1 & 2 \end{bmatrix}\right)\left(\begin{bmatrix} 3 & 2 \\ 0 & 2 \end{bmatrix}\right) = \begin{bmatrix} 6 & 12 \\ 12 & 12 \end{bmatrix}$$

$$P_4 = A_{22}(B_{21} - B_{11}) = \begin{bmatrix} 0 & 1 \\ 1 & 2 \end{bmatrix}\left(\begin{bmatrix} 2 & 2 \\ 1 & 1 \end{bmatrix} - \begin{bmatrix} 3 & 2 \\ 0 & 2 \end{bmatrix}\right) = \begin{bmatrix} 1 & -1 \\ 1 & -2 \end{bmatrix}$$

$$P_5 = (A_{11} + A_{22})(B_{11} + B_{22}) = \left(\begin{bmatrix} 0 & 1 \\ 1 & 2 \end{bmatrix} + \begin{bmatrix} 0 & 1 \\ 1 & 2 \end{bmatrix}\right)\left(\begin{bmatrix} 3 & 2 \\ 0 & 2 \end{bmatrix} + \begin{bmatrix} 1 & 0 \\ 0 & 3 \end{bmatrix}\right)$$
$$= \begin{bmatrix} 0 & 10 \\ 8 & 24 \end{bmatrix}$$

$$P_6 = (A_{12} - A_{22})(B_{21} + B_{22}) = \left(\begin{bmatrix} 2 & 3 \\ 3 & 0 \end{bmatrix} - \begin{bmatrix} 1 & 0 \\ 1 & 2 \end{bmatrix}\right)\left(\begin{bmatrix} 2 & 2 \\ 1 & 1 \end{bmatrix} + \begin{bmatrix} 1 & 0 \\ 0 & 3 \end{bmatrix}\right)$$
$$= \begin{bmatrix} 6 & 14 \\ 4 & -4 \end{bmatrix}$$

$$P_7 = (A_{21} - A_{11})(B_{11} + B_{12}) = \left(\begin{bmatrix} 2 & 3 \\ 3 & 0 \end{bmatrix} - \begin{bmatrix} 0 & 1 \\ 1 & 2 \end{bmatrix}\right)\left(\begin{bmatrix} 3 & 2 \\ 0 & 2 \end{bmatrix} + \begin{bmatrix} 1 & 0 \\ 3 & 1 \end{bmatrix}\right)$$
$$= \begin{bmatrix} 13 & 10 \\ 2 & -2 \end{bmatrix}$$

$$C_{11} = P_5 + P_4 - P_2 + P_6 = \begin{bmatrix} 0 & 10 \\ 8 & 24 \end{bmatrix} + \begin{bmatrix} 1 & -1 \\ 1 & -2 \end{bmatrix} - \begin{bmatrix} 2 & 12 \\ 4 & 6 \end{bmatrix} + \begin{bmatrix} 6 & 14 \\ 4 & -4 \end{bmatrix} = \begin{bmatrix} 5 & 11 \\ 9 & 12 \end{bmatrix}$$

$$C_{12} = P_1 + P_2 = \begin{bmatrix} 3 & -2 \\ 6 & -4 \end{bmatrix} + \begin{bmatrix} 2 & 12 \\ 4 & 6 \end{bmatrix} = \begin{bmatrix} 5 & 10 \\ 10 & 2 \end{bmatrix}$$

$$C_{21} = P_3 + P_4 = \begin{bmatrix} 4 & 10 \\ 2 & 8 \end{bmatrix} + \begin{bmatrix} 1 & -1 \\ 1 & -2 \end{bmatrix} = \begin{bmatrix} 5 & 9 \\ 3 & 6 \end{bmatrix}$$

$$C_{22} = P_1 + P_5 - P_3 + P_7 = \begin{bmatrix} 3 & -2 \\ 6 & -4 \end{bmatrix} + \begin{bmatrix} 0 & 10 \\ 8 & 24 \end{bmatrix} - \begin{bmatrix} 6 & 12 \\ 12 & 12 \end{bmatrix} + \begin{bmatrix} 13 & 10 \\ 2 & -2 \end{bmatrix} = \begin{bmatrix} 10 & 6 \\ 3 & 6 \end{bmatrix}$$

$$C = \begin{bmatrix} 5 & 11 & 5 & 10 \\ 9 & 12 & 10 & 2 \\ 5 & 9 & 10 & 6 \\ 3 & 6 & 3 & 6 \end{bmatrix} = A \cdot B$$

141.

1. Let the missing number, m, be n/2.
2. Count the numbers smaller than m.
3. If the count is lesser than m
4. Then, we know that the missing number is smaller than m.
5. Else, the missing number is greater than or equal to m.
6. Continue doing this until you find the missing number.

142.

```
Function power(a, n)
        if (n = 0) return(1)
        x = power(a, ⌈n/2⌉)
        if (n is even) then return (x²)
            else return (a × x²)
```

147.
This algorithm divides all points in two sets and recursively calls for two sets. Then we have to find strip (time complexity of this part is $O(n)$). Next we should sort this strip (time complexity of this part is $O(n\log n)$). And finally find the closest points in our strip(time complexity of this part is $O(n)$). As a result:

$$T(n) = 2T\left(\frac{n}{2}\right) + O(n) + O(n\log n) + O(n) = 2T\left(\frac{n}{2}\right) + O(n\log n) = T(n \times \log n \times \log n)$$

148.

Consider the master theorem:

$$T(n) = aT\left(\frac{n}{b}\right) + f(n)$$

In this algorithm, we divide array of points into 2 subarray, each subarray has half of points, so we have $a = 2, b = 2$
for $f(n)$, we use $O(n)$ time to find pairs and $O(n\log n)$ to sort that pairs, and find the closest points in $O(n)$, so, we have $f(n) = O(n) + O(n) + O(n\log n) = O(n\log n)$.
Recursive relation of algorithm:

$$T(n) = 2T\left(\frac{n}{2}\right) + O(n\log n)$$

Time complexity of algorithm: $T(n) = O(n(\log n)^2)$.

Chapter 11
Dynamic Programming

Abstract Dynamic programming is another way to design algorithms. Dynamic programming is similar to the divide and conquer techniques in that the problem is divided into smaller sub-problems. But in this method, we first solve the smaller sub-problems, save the results, and then retrieve it whenever one of them is needed, instead of recalculating. Dynamic programming is often used for optimization problems. In these types of problems, there can be many possible answers. We want to find the answer that has the optimal value (minimum or maximum). This chapter provides 62 exercises for addressing different aspects of the dynamic programming method. To this end, this chapter provides exercises on mathematics numbers, all-pairs shortest paths, matrix chain multiplication, the knapsack problem, optimal binary search tree, longest common subsequence (LCS), string matching, and traveling salesman problem (TSP).

11.1 Lecture Notes

Dynamic programming is another way to design algorithms. Dynamic programming is similar to the divide and conquer techniques in that the problem is divided into smaller subproblems. But in this method, we first solve the smaller subproblems, save the results, and then retrieve it whenever one of them is needed, instead of recalculating.

Dynamic programming is often used for optimization problems. In these types of problems, there can be many possible answers. We want to find the answer that has the optimal value (minimum or maximum). In all optimization problems solvable by the dynamic programming method, the principle of optimality must be established.

Definition: The principle of optimality holds in a problem, if an optimal solution for an instance of the problem always contains the optimal solution for all sub instances.

For example, if the shortest path from city V_1 to V_2 passes through V_3, then in order for this path to be optimal, the routes from V_1 to V_3 and V_3 to V_1 must also be independently optimized, in which case the principle of optimality will be established for this problem.

© The Author(s), under exclusive license to Springer Nature Switzerland AG 2022   401
H. Izadkhah, *Problems on Algorithms*,
https://doi.org/10.1007/978-3-031-17043-0_11

In this technique, the problem is solved from a bottom-up approach. That is, first the problem is solved in a simpler or smaller way, and then we store these solutions in memory, and in later steps, we use these answers and try to find the answer to the problem using the initial answers. The key step in designing a dynamic-programming algorithm is to develop a dynamic programming recurrence relation that relates the solution to a larger problem to that of smaller sub-problems. Then create a table or array according to the corresponding recursive formula in which the revious answers are stored. Consider the following example.

Example 11.1 The Fibonacci sequence $f_0, f_1, \ldots$ is recursively defined as follows:

1. base case. $f_0 = 0$ and $f_1 = 1$
2. recursive case. for n $\geq$ 2, $f_n = f_{n-1} + f_{n-2}$.

The following recursive algorithm for computing the nth Fibonacci number has exponential complexity with respect to n.

```
int fib(int n)
{
    if(n == 0)
        return 0;
    if(n==1)
        return 1;
    return fib(n-1) + fib(n-2);
}
```

Figure 11.1 depicts the recursion tree for n = 5. The function $fib(2)$ is called three times and $fib(3)$ is called two times. For large values of n, many common sub-problems arise. With regard to the above mentioned the recursive function for Fibonacci computes the same subproblems again and again.

In dynamic programming algorithm for $fib(int\ n)$, instead of making blind recursive calls, we use fib_0 and fib_1 values to compute fib_2. After storing fib_2, we then compute fib_3 using fib_1 and fib_2. This process continues until fib_n is calculated. We call this approach a bottom-up approach because the focus is on calculating the solution to the smaller sub-problems, then using those solutions to calculate the solution to the larger sub-problems, and finally solving the main problem.

We can avoid the repeated work done in above code by storing the Fibonacci numbers calculated so far. The following algorithm for computing the nth Fibonacci number has $O(n)$ complexity. It is clear that the complexity of time has greatly decreased.

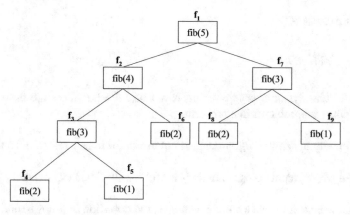

Fig. 11.1 Recursion tree for *fib(5)*

```
fib[0] = 0;
fib[1] = 1;
for(i = 2; i ≤ n; i++)
        fib[i] = fib[i - 1] + fib[i - 2];
return fib[n];
```

This algorithm shows the characteristics of a dynamic programming algorithm.

Example 11.2 Computing a Binomial Coefficient—A binomial coefficient $C(n, k)$ can be defined as the coefficient of x^k in the expansion of $(1 + x)^n$. Computing binomial coefficients is non optimization problem but can be solved using dynamic programming. The recursive relation is defined by the prior power. The binomial coefficient $C(n, k)$ is recursively defined as follows:

1. base case. $C(n, 0) = C(n, n) = 1$
2. recursive case. for $n > k > 0$, $C(n, k) = C(n - 1, k - 1) + C(n - 1, k)$.

Like the Fibonacci function, this function computes the same subproblems again and again. For example, for n = 5 and k = 2, the function C(3, 1) is called two times. For large values of n, there will be many common subproblems.

Dynamic algorithm constructs a $n \times k$ table. The table is then filled out iteratively, row by row using the recursive relation as follows.

```
Function BINOMIAL(n, k)
        for i = 0 to n do // fill out the table row wise
            for i = 0 to min(i, k) do
                if j==0 or j==i then C[i, j] = 1
                else C[i, j] = C[i-1, j-1] + C[i-1, j] // recursive relation
        return C[n, k]
```

11.2 Exercises

11.2.1 Preliminary

1. Explain why dynamic programming is not useful when there are no common (overlapping) subproblems in a problem.

2. Explain why dynamic programming is not useful for binary search in an array.

3. Explain why dynamic programming is useful for nth Fibonacci numbers.

4. Explain what is difference among recursive, memoization (top-down) and tabulation (bottom-up) solutions.

5. We first define *optimal substructure* property. If the optimal solution of a problem can be obtained by using the optimal solutions of its sub-problems, the problem has the optimal substructure property.
Explain why dynamic programming is useful for problems with optimal substructure property.

6. Suppose in a graph we want to find the shortest path between all pairs of nodes. Check whether the principle of optimality is true in this case.

7. Suppose in a graph we want to find the longest path between all pairs of nodes. Check whether the principle of optimality is true in this case.

11.2.2 Mathematics Numbers

8. The Fibonacci numbers are the numbers in the following integer sequence.

$$0, 1, 1, 2, 3, 5, 8, 13, 21, 34, 55, 89, 144, ...$$

The sequence F_n of Fibonacci numbers is defined by the following recurrence relation

$$F_n = F_{n-1} + F_{n-2} \ where \ F_0 = 0 \ and \ F_1 = 1$$

(a) Devise a divide and conquer to find nth Fibonacci number.
(b) Devise a dynamic programming to find nth Fibonacci number.
(c) Find the running time of both algorithms.

9. Catalan numbers are a sequence of natural numbers that occurs in many interesting counting problems such as following.

- Counting the number of expressions containing n pairs of parentheses which are correctly matched. For n = 3, possible expressions are ((())), ()(()), ()()(), (())(), (()()).
- Counting the number of possible binary search trees with n keys.
- Counting the number of full binary trees (A rooted binary tree is full if every vertex has either two children or no children) with $n+1$ leaves.
- Given a number n, finding the number of ways you can draw n chords in a circle with $2 \times n$ points such that no 2 chords intersect.

(a) Devise a divide and conquer to find nth Catalan number.
(b) Devise a dynamic programming to find nth Catalan number.
(c) Use binomial coefficient to find nth Catalan number.
(d) Find the running time of the designed algorithms.

10. Bell numbers indicates the number of ways to partition a set. Devise a dynamic programming to solve this problem.

11. Devise a dynamic programming algorithm that takes two parameters n and k and returns the value of binomial coefficient C(n, k).

12. Permutation refers to the process of arranging all the elements of a given set to form a sequence. There are $n!$ permutations on a set of n elements.
   The P(n, k), the permutation coefficient, is used to find the number of ways to obtain an ordered subset having k elements from a set of n elements. Devise a dynamic programming algorithm to find P(n, k).

13. *Lobb Number.* Using dynamic programming, devise an algorithm for this numbers.

14. *Eulerian Number.* Using dynamic programming, devise an algorithm for this numbers.

15. *Delannoy Number.* Using dynamic programming, devise an algorithm for this numbers.

16. *Entringer Number.* Using dynamic programming, devise an algorithm for this numbers.

17. *Rencontres Number.* Using dynamic programming, devise an algorithm for this numbers.

18. *Jacobsthal and Jacobsthal-Lucas numbers*. Using dynamic programming, devise an algorithm for this numbers.

19. *Super Ugly Number (number whose prime factors are in given set)*. Using dynamic programming, devise an algorithm for this numbers.

11.2.3 All-Pairs Shortest Paths

20. Floyd Warshall's algorithm is used for solving

   a. All pair shortest path problems
   b. Single Source shortest path problems
   c. Network flow problems
   d. Sorting problems

21. *All-pairs shortest-paths problem*. Let a weighted directed graph $G = (V, E)$ is given. Find for each pair of vertices $v, w \in V$, the cost of the shortest (i.e., the least-cost) path from v to w in G. The standard dynamic programming algorithm for the all pairs shortest paths problem (Floyd-Warshall's algorithm) is as follows. Suppose $V = \{1, 2, ..., n\}$.

$$
\begin{aligned}
&\text{Function FLOYD- WARSHALL(C, n)} \\
&\quad \text{for } i := 1 \text{ to n do} \\
&\quad\quad \text{for } j := 1 \text{ to n do} \\
&\quad\quad\quad \text{if } (i, j) \in E \\
&\quad\quad\quad\quad \text{then } D[i, j] := \text{cost of edge } (i, j) \\
&\quad\quad\quad\quad \text{else } D[i, j] := \infty \\
&\quad\quad D[i, i] := 0 \\
&\quad \text{for } k := 1 \text{ to n do} \\
&\quad\quad \text{for } i := 1 \text{ to n do} \\
&\quad\quad\quad \text{for } j := 1 \text{ to n do} \\
&\quad\quad\quad\quad \text{if } D[i, k] + D[k, j] < D[i, j] \text{ then } D[i, j] := D[i, k] + D[k, j] \\
&\quad \text{return}(D)
\end{aligned}
$$

(a) Fill the cost table D in Floyd-Warshall's algorithm on the graph shown in Fig. 11.2.
(b) Fill the cost table D in Floyd-Warshall's algorithm for the following matrix.

$$
\begin{bmatrix}
0 & 5 & \infty & \infty \\
50 & 0 & 15 & 5 \\
30 & \infty & 0 & 15 \\
15 & \infty & 5 & 0
\end{bmatrix}
$$

Fig. 11.2 A sample graph

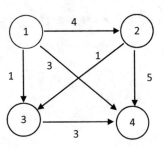

22. The matrix D specifies only the cost of the shortest path. However, it is not possible to create an optimal path due to the stored information. To store the information needed to build the optimal path, we use an auxiliary matrix P and change the Floyd-Warshall's algorithm as follows.

```
Function FLOYD- WARSHALL(C, n)
    for i := 1 to n do
        for j := 1 to n do
            P[i, j] = 0
    for i := 1 to n do
        for j := 1 to n do
            if (i, j) ∈ E
                then D[i, j] := cost of edge (i, j)
                else D[i, j] := ∞
    D[i, i] := 0
    for k := 1 to n do
        for i := 1 to n do
            for j := 1 to n do
                if D[i, k] + D[k, j] < D[i, j] then
                    P[i, j] = k
                    D[i, j] := D[i, k] + D[k, j]
    return(D)
```

After completing the algorithm, the following recursive algorithm can be used to print the shortest path between the two vertices.

```
Function PATH(q, r)
    if(P[i, j] ≠ 0) {
        path(q, P[q, r])
        print("v", P[q, r])
        path(P[q, r], r)}
```

Fig. 11.3 A sample graph

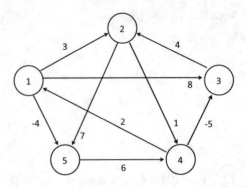

Fig. 11.4 A sample graph

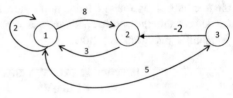

Fig. 11.5 A sample graph

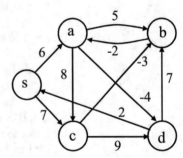

Using matrix P, print the path between all pairs of vertices in the graph shown in Fig. 11.3.

23. Using the Floyd-Warshall's algorithm, find the cost of shortest path between the nodes shown in Fig. 11.4.

24. Consider Fig. 11.5. Can the Floyd-Warshall's algorithm be applied to this graph to find the shortest path between the nodes?

25. Consider Fig. 11.6. Can the Floyd-Warshall's algorithm be applied to this graph to find the shortest path between the nodes?

26. Floyd Warshall algorithm can only be applied on directed graphs. True/False?

Fig. 11.6 A sample graph

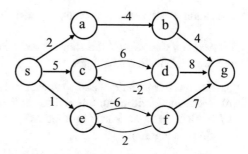

27. We apply Floyd-Warshall's algorithm to a weighted graph with no negative cycles. True/False?

28. (True/False?) In the Floyd-Warshall's algorithm

– the graph should not contain negative cycles.
– the graph can have positive and negative weight edges.

29. Analyze the running time of Floyd-Warshall's algorithm.

30. Let G is a directed graph. The aim is to determine whether the graph has negative cycle or not. A negative cycle is a cycle in which the total sum of the cycle is negative. Give an algorithm to detect negative cycle using Floyd-Warshall's algorithm.

31. A weighted directed graph consisting of V vertices and E edges are given. Write a function to print a cycle path whose sum of weight is negative. If no such path exists, print '-1'.

32. For transitive closure, construct a graph $G' = (V, E')$ with edge $(i, j) \in E'$ iff there is a directed path from i to j in G. Explain how transitive closure can be constructed by Floyd-Warshall's algorithm.

   Modify the Floyd-Warshall's algorithm to answer the questions 33–35 on directed graphs. Analyze your algorithms.

33. Find the number of paths between each pair of vertices.

34. Find the number of even-length paths and of odd-length paths between each pair of vertices.

35. Find the number of shortest paths between each pair of vertices.

36. Let G is a given graph containing n vertices where vertices are labeled with distinct integers from 1 to n. The label of each path is obtained by taking the labels of vertices on the path in order. Determine the label and length of the lexicographically

first shortest path between each pair of vertices.

37. Another way to find the shortest paths between all pairs of vertices is to use matrix multiplication technique.

(a) Explain this technique and write its recursive relation.
(b) Write an algorithm for this and get its time complexity.
(c) Explain whether matrix multiplication can detect the existence of a negative circle in a graph?

38. Using matrix multiplication technique, find the cost of shortest path between the nodes shown in Fig. 11.3.

39. The Floyd-Warshall's algorithm can be used to find the shortest cycle in a graph, called its *girth*. To this end, this algorithm computes d_{ii} for $1 \leq i \leq n$, which is the length of the shortest way to get from vertex i to i; in other words, the shortest cycle through i. Give an algorithm to find girth of a graph.

11.2.4 Matrix Chain Multiplication

40. A sequence of matrices is given, the aim being to find the most efficient way to multiply these matrices together. We have many options for multiplication of a chain of matrices because multiplication of the matrix is associative. In other words, no matter how we parenthesize the product, the result will be the same. For example, if we have four matrices A, B, C and D, we will have:
(ABC)D = (AB)(CD) = A(BCD) = ...

(a) Use Tabulation
(b) Use Memoization

41. What is the relationship between x, y, z and w so that multiplying the three matrices $A_{x \times y}$, $B_{y \times z}$ and $C_{z \times w}$ as (AB)C is better than A(BC), that is, it has fewer multiplication operations.

42. Consider the matrices A, B, and C which are 10×20, 20×30 and 30×40 matrices respectively. What is the minimum number of multiplications required to multiply the three matrices?

  a. 18000
  b. 12000
  c. 24000
  d. 32000

Given n matrices, M_1, M_2..., M_n, where for $1 \leq i \leq n$, M_i is a $r_{i-1} \times r_i$ matrix, parenthesize the product $M_1.M_2...M_n$ so as to minimize the total cost, assuming that the cost of multiplying an $r_{i-1} \times r_i$ matrix by a $r_i \times r_{i+1}$ matrix using the naive algorithm is $r_{i-1}r_i r_{i+1}$. Here is the dynamic programming algorithm for the matrix product problem:

```
Function MATRIX(n)
    for i := 1 to n do m[i, i] := 0
    for d := 1 to n-1 do
        for i := 1 to n-d do
            j := i + d
            m[i, j] := min_{i≤k<j}(m[i, k] + m[k + 1, j] + r_{i-1}r_k r_j)
    return(m[1, n])
```

43. Fill in the cost table m for $A_{5\times2}$, $B_{2\times3}$, $C_{3\times4}$, and $D_{4\times6}$ matrices.

44. Fill in the cost table m for $A_{5\times10}$, $B_{10\times30}$, $C_{30\times60}$, and $D_{60\times40}$ matrices.

45. Fill in the cost table m for $A_{10\times20}$, $B_{20\times50}$, $C_{50\times1}$, and $D_{1\times100}$ matrices.

46. Consider the matrices A, B, C, and D which are 20×15, 15×30, 30×5 and 5×40 matrices respectively. What is the minimum number of multiplications required to multiply the three matrices?

   a. 6050
   b. 7500
   c. 7750
   d. 12000

47. The time complexity of finding all the possible ways of multiplying a set of n matrices is given by $(n-1)$th Catalan number which is exponential. True/False?

48. Consider the following four matrices with the given dimensions:

$$A_{10\times2}, B_{2\times25}, C_{25\times3}, D_{3\times4}$$

Which of the following has the minimum number of multiplications required to multiply the four matrices.

   a. $(A\times B)\times(C\times D)$
   b. $(A\times(B\times C))\times D$
   c. $A\times((B\times C)\times D)$
   d. $((A\times B)\times C)\times D$

11.2.5 The Knapsack Problem

The 0–1 knapsack problem is as follows: given n items of weight $w_1, w_2, ..., w_n$, and a knapsack weight W, find a subset of the items that has total weight exactly W. The standard dynamic programming algorithm for the knapsack problem is as follows. Let t[i, j] be true if there is a subset of the first i items that has total weight exactly j.

$$\text{Function KNAPSACK}(w_1, w_2, ..., w_n, W)$$
$$t[0, 0] := \text{true}$$
$$\text{for } j := 1 \text{ to } W \text{ do } t[0, j] := \text{false}$$
$$\text{for } i := 1 \text{ to } n \text{ do}$$
$$\text{for } j := 0 \text{ to } W \text{ do}$$
$$t[i, j] := t[i - 1, j]$$
$$\text{if } j - w_i \geq 0 \text{ then } t[i, j] := t[i, j] \bigvee t[i - 1, j - w_i]$$
$$\text{return}(t[n, W])$$

49. Fill in table t for a knapsack of weight 19 with items of the following weights: 15, 5, 16, 7, 1, 15, 6, 3.

50. Fill in table t for a knapsack of weight 10 with items of the following weights: 5, 4, 3, 6.

11.2.6   Optimal Binary Search Tree

Binary search trees are used to organize a set of keys for fast access: the tree holds the keys in-order so that comparison with the query at any node either results in a match, or directs us to continue the search in left or right subtree. Given a sorted array key $k_1, k_2, ..., k_n$ of search keys and an array freq $p_1, p_2, ..., p_n$ showing their access probabilities. Optimal binary search tree is a tree of all keys such that the total cost of all the searches is as small as possible.

51. Given the following table of data ($k_1 < k_2 < k_3$) what are the minimum expected number of comparisons required for an optimal BST that can be constructed with the given data.

Keys	k_1	k_2	k_3
Frequency count	0.7	0.2	0.1

52. Given the following table of data what are the minimum expected number of comparisons required for an optimal BST that can be constructed with the given data.

Keys	0	1	2	3	4	5	6
Frequency count	22	18	20	5	25	2	8

53. Given the following table of data. Give the corresponding optimal BST.

Word	Probability
A	0.22
B	0.18
C	0.20
D	0.05
E	0.25
F	0.02
G	0.08

11.2.7 Longest Common Subsequence (LCS)

The longest common subsequence (LCS) problem is finding the longest subsequence present in the given two sequences. A subsequence is a sequence that appears in the same relative order, but are not required to occupy consecutive positions within the original sequences. Let X and Y are two strings of size m and n, respectively. The following algorithm is written for this task.

LCS- LENGTH(X, Y)
1. m = length(X)
2. n = length(Y)
3. for i = 1 to m   C[i, 0] = 0
4. for j = 1 to n   C[0, j] = 0
5. for i = 1 to m
6.   for j = 1 to n
7.   if (Xi == Yj)
8.   C[i,j] = C[i-1,j-1] + 1
9.   else C[i,j] = max(C[i-1,j], C[i,j-1])
10. return C

54. Fill the table C. What is the longest common subsequence of X and Y?

$$X = ABCB, \quad Y = BDCAB$$

55. Fill the table C. What is the longest common subsequence of X and Y?

$$X = BDCABA, \quad Y = ABCBDAB$$

56. The array C obtained from the *LCS-Length* algorithm shows only the length of the largest common sequence X and Y and does not specify the characters that make up the largest common sequence. We modify the *LCS-Length* algorithm to give the

characters of the longest path as well. To do this, we use the auxiliary array b.

LCS- LENGTH
1. m = length(X)
2. n = length(Y)
3. for i = 1 to m   C[i, 0] = 0
4. for j = 1 to n   C[0, j] = 0
5. for i = 1 to m
6.   for j = 1 to n
7.   if (Xi == Yj)
8.   C[i,j] = C[i-1,j-1] + 1
9.   b[i,j] = "↖"
10.   else if C[i-1,j] ≥ C[i,j-1]
11.   C[i,j] = C[i-1,j]
12.   b[i,j] = "↑"
13.   else
14.   C[i,j] = C[i,j-1]
15.   b[i,j] = "←"
10. return C

(a) Find the longest common subsequence of X and Y?

$$X = ABCB, \quad Y = BDCAB$$

(b) Find the longest common subsequence of X and Y?

$$X = BDCABA, \quad Y = ABCBDAB$$

11.2.8   String Matching

String matching is present in almost all text-processing applications. Every text editor contains a mechanism to perform string searching and matching.

57. Let t denotes a text string of length n and a pattern string p of length m.

(a) Assume the search patterns and/or texts are short. Give a simple and linear-time algorithm to find the patterns in the text.
(b) Build a finite state automaton that recognizes the patterns and returns to the appropriate start state on any character mismatch.

58. Assume the text is fixed and you are building a program to repeatedly search a particular pattern. Explain how to use the suffix tree and suffix array data structures to speed up search queries.

59. Assume the pattern is fixed and you are building a program to repeatedly search pattern in different texts. Give an algorithm to perform this.

60. *Edit distance*—Approximate string matching is a major problem in word processors because we live in a world prone to error. Spell checkers must be able to identify the closest match for any string not found in a dictionary.

Given two strings X and Y and insertions, deletions, and substitutions (replacing any character with any other) operations. The problem is to find minimum number of edits (operations) required to convert X into Y. There are three solutions for this problem as follows:

1. **Brute Force Approach**: This is a simple recursive solution where we will solve the problem via the solution of its sub-problems.
2. **Memoization Approach**: Here, we will calculate the overlapping problems and will save it for future use.
3. **Tabulation Approach**: This is the most optimized solution which uses iteration to calculate the result based on the bottom-up approach, a dynamic programming approach. Let two strings $x[1...m]$ and $y[1...n]$ are two strings. In this approach we need to somehow express $D(i, j)$ in terms of smaller subproblems $D(i-1, j)$, $D(i, j-1)$, $D(i-1, j-1)$. These subproblems indicate some prefix of the first string, $x[1...i]$, and some prefix of the second, $y[1...j]$. We have no idea which of them is the right one, so we need to try them all and pick the best:

$$E(i, j) = min\{1 + E(i-1, j), 1 + E(i, j-1), diff(i, j) + E(i-1, j-1)\}$$

where $diff(i, j)$ is defined to be 0 if $x[i] = y[j]$ and 1 otherwise. Our final objective, then, is to compute $D(m, n)$.

Give an algorithm for each of these approaches.

11.2.9 Traveling Salesman Problem (TSP)

A city and the distance between each pair of cities are given. The problem is finding the shortest possible tour to visit all the cities exactly once and without seeing a city twice, and finally, you have to go back to the starting city. It is important to note the difference between a Hamilton cycle and a traveling salesman. The Hamilton cycle is a way to find out if there is a cycle that meets each city exactly once.

61. Consider the naive solution for TSP as follows:

1. Consider city 1 as the starting and ending point.
2. Generate all (n-1)! permutations of cities.
3. Calculate cost of every permutation and keep track of minimum cost permutation.
4. Return the permutation with minimum cost.

Give time complexity of this algorithm.

62. In the set of vertices $\{1, 2, 3, 4, ..., n\}$, consider vertex 1 as the start and end point. For every other vertex i (other than 1), find the minimum cost path from 1 as the starting point, i as the ending point and all vertices appearing exactly once. Consider the following recursive relation to compute the sub-problems. Let us define $C(S, i)$ be the cost of the minimum cost path visiting each vertex in set S exactly once, starting at 1 and ending at i.

If size of S is 2, then S must be 1, i,
$C(S, i) = dist(1, i)$
Else if size of S is greater than 2.
$C(S, i) = min \{C(S - i, j) + dis(j, i)\}$ where j belongs to S, $j! = i$ and $j! = 1$.

11.3   Solutions

8.
a. Divide and conquer

```
int fibo(int n) {
        if (n == 0)
              return n;
        else if (n == 1)
              return n;
        else
                 return fibo(n - 1) + fibo(n - 2);}
```

b. Dynamic programming

```
int fibo (int n) {
        int arr[n];
        arr[0] = 0; arr[1] = 1;
        if (n == 0) return n;
        else if (n == 1) return n;
          else {
          for (int i = 2; i <= n; i++)
              arr[i] = arr[i - 1] + arr[i - 2];
              return n; }}
```

c.
Divide and conquer: $O(2^{n/2})$
Dynamic programming: $O(n)$

9.
a.
```
if (n <= 1)
    return 1
res = 0
for (int i = 0; i < n; i++)
    res += catalan(i) * catalan(n-i-1)
return res
```

$$T(n) = \begin{cases} if \ n < 1 & 1 \\ else & \sum_{i=0}^{n-1} T(i) * T(n-i-1) \end{cases}$$

It's equivalent to nth catalan number.

b.
```
catalan[0] = catalan[1] = 1
for (int i = 2; i <= n; i++) {
        catalan[i] = 0
        for (int j = 0; j < i; j++)
            catalan[i] += catalan[j] * catalan[i - j - 1]
}
return catalan[n]
```

c.
C: a = $O(3^n)$ b = $O(n^2)$ c = $O(n)$

14. Eulerian Number. Using dynamic programming, devise an algorithm for this numbers

```
int eulerian(int n,int m)
{
if(m≥n || n==0)
    return 0;
if(m==0)
    return 1;
return (n-m)*eulerian(n-1, m-1)+(m+1) * eulerian(n-1,m);
}
```

18.

```
f[0..n] = -1
f[1] = 1 f[0] = 0
Jacobsthal(n, f):
if f[n][k] != -1 return f[n][k]
        return Jacobsthal(n - 1, f) + 2 * Jacobsthal(n - 2, f)
```

21. a.

$$D_0 = \begin{bmatrix} 0 & 4 & 1 & 3 \\ \infty & 0 & 1 & 5 \\ \infty & \infty & 0 & 3 \\ \infty & \infty & \infty & 0 \end{bmatrix}$$

$$D_1[2][3] = Min\,(D_0[2][3],\, D_0[2][1] + D_0[1][3]) = (1, 1 + \infty) = 1$$

$$D_1[2][4] = Min\,(D_0[2][4],\, D_0[2][1] + D_0[1][4]) = (5, \infty + 3) = 5$$

$$D_1[3][2] = Min\,(D_0[3][2],\, D_0[3][1] + D_0[1][2]) = (\infty, \infty + 4) = \infty$$

$$D_1[3][4] = Min\,(D_0[3][4],\, D_0[3][1] + D_0[1][4]) = (3, \infty + 3) = 3$$

$$D_1[4][2] = Min\,(D_0[4][2],\, D_0[4][1] + D_0[1][2]) = (\infty, \infty + 3) = \infty$$

$$D_1[3][4] = Min\,(D_0[3][4],\, D_0[3][1] + D_0[1][4]) = (\infty, \infty + 1) = \infty$$

$$D_1 = \begin{bmatrix} 0 & 4 & 1 & 3 \\ \infty & 0 & 1 & 5 \\ \infty & \infty & 0 & 3 \\ \infty & \infty & \infty & 0 \end{bmatrix}$$

$$D_2[1][3] = Min\,(D_1[1][3],\, D_1[1][2] + D_1[2][3]) = (1, 4 + 1) = 1$$

$$D_2[1][4] = Min\,(D_1[1][4],\, D_1[1][2] + D_1[2][4]) = (3, 4 + 5) = 3$$

$$D_2[3][1] = Min\,(D_1[3][1],\, D_1[3][2] + D_1[2][1]) = (\infty, \infty + \infty) = \infty$$

$$D_2[3][4] = Min\,(D_1[3][4],\, D_1[3][2] + D_1[2][4]) = (3, \infty + 5) = 3$$

$$D_2[4][1] = Min\,(D_1[4][1],\, D_1[4][2] + D_1[2][1]) = (\infty, \infty + \infty) = \infty$$

$$D_2[4][3] = Min\,(D_1[4][3],\, D_1[4][2] + D_1[2][3]) = (\infty, \infty + 1) = \infty$$

$$D_2 = \begin{bmatrix} 0 & 4 & 1 & 3 \\ \infty & 0 & 1 & 5 \\ \infty & \infty & 0 & 3 \\ \infty & \infty & \infty & 0 \end{bmatrix}$$

$$D_3[1][2] = Min(D_2[1][2], D_2[1][3] + D_2[3][2]) = (4, 1 + \infty) = 4$$

$$D_3[1][4] = Min(D_2[1][4], D_2[1][3] + D_2[3][4]) = (3, 1 + 3) = 3$$

$$D_3[2][1] = Min(D_2[2][1], D_2[2][3] + D_2[3][1]) = (\infty, 1 + \infty) = \infty$$

$$D_3[2][4] = Min(D_2[2][4], D_2[2][3] + D_2[3][4]) = (5, 1 + 3) = 4$$

$$D_3[4][1] = Min(D_2[4][1], D_2[4][3] + D_2[3][1]) = (\infty, \infty + \infty) = \infty$$

$$D_3[4][2] = Min(D_2[4][2], D_2[4][3] + D_2[3][2]) = (\infty, \infty + \infty) = \infty$$

$$D_3 = \begin{bmatrix} 0 & 4 & 1 & 3 \\ \infty & 0 & 1 & 4 \\ \infty & \infty & 0 & 3 \\ \infty & \infty & \infty & 0 \end{bmatrix}$$

$$D_4[1][2] = Min(D_3[1][2], D_3[1][4] + D_3[4][2]) = (4, 3 + \infty) = 4$$

$$D_4[1][3] = Min(D_3[1][3], D_3[1][4] + D_3[4][3]) = (1, 3 + \infty) = 1$$

$$D_4[2][1] = Min(D_3[2][1], D_3[2][4] + D_3[4][1]) = (\infty, 4 + \infty) = \infty$$

$$D_4[2][3] = Min(D_3[2][3], D_3[2][4] + D_3[4][3]) = (1, 4 + \infty) = 1$$

$$D_4[3][1] = Min(D_3[3][1], D_3[3][4] + D_3[4][1]) = (\infty, 3 + \infty) = \infty$$

$$D_4[3][2] = Min(D_3[3][2], D_3[3][4] + D_3[4][2]) = (\infty, 3 + \infty) = \infty$$

$$D_4 = \begin{bmatrix} 0 & 4 & 1 & 3 \\ \infty & 0 & 1 & 4 \\ \infty & \infty & 0 & 3 \\ \infty & \infty & \infty & 0 \end{bmatrix}$$

21. b.

$$D_0 \begin{bmatrix} 0 & 5 & \infty & \infty \\ 50 & 0 & 15 & 5 \\ 30 & \infty & 0 & 15 \\ 15 & \infty & 5 & 0 \end{bmatrix} \rightarrow D_1 \begin{bmatrix} 0 & 5 & \infty & \infty \\ 50 & 0 & 15 & 5 \\ 30 & 35 & 0 & 15 \\ 15 & 20 & 5 & 0 \end{bmatrix} \rightarrow D_2 \begin{bmatrix} 0 & 5 & 20 & 10 \\ 50 & 0 & 15 & 5 \\ 30 & 35 & 0 & 15 \\ 15 & 20 & 5 & 0 \end{bmatrix}$$

$$D_3 \begin{bmatrix} 0 & 5 & 20 & 10 \\ 45 & 0 & 15 & 5 \\ 30 & 35 & 0 & 15 \\ 15 & 20 & 5 & 0 \end{bmatrix} \rightarrow D_4 \begin{bmatrix} 0 & 5 & 15 & 10 \\ 20 & 0 & 10 & 5 \\ 30 & 35 & 0 & 15 \\ 15 & 20 & 5 & 0 \end{bmatrix}$$

22.

$$P_0 = \begin{bmatrix} 0 & 1 & 1 & 0 & 1 \\ 0 & 0 & 0 & 2 & 2 \\ 0 & 3 & 0 & 3 & 0 \\ 4 & 0 & 0 & 0 & 0 \\ 0 & 0 & 0 & 5 & 0 \end{bmatrix}$$

$$P_1 = \begin{bmatrix} 0 & 1 & 1 & 0 & 1 \\ 0 & 0 & 0 & 2 & 2 \\ 0 & 3 & 0 & 3 & 0 \\ 4 & 1 & 1 & 0 & 1 \\ 0 & 0 & 0 & 5 & 0 \end{bmatrix}$$

$$P_2 = \begin{bmatrix} 0 & 1 & 1 & 2 & 1 \\ 0 & 0 & 0 & 2 & 2 \\ 0 & 3 & 0 & 3 & 2 \\ 4 & 1 & 1 & 0 & 1 \\ 0 & 0 & 0 & 5 & 0 \end{bmatrix}$$

$$P_3 = \begin{bmatrix} 0 & 1 & 1 & 3 & 1 \\ 0 & 0 & 0 & 2 & 2 \\ 0 & 3 & 0 & 3 & 2 \\ 4 & 1 & 1 & 0 & 1 \\ 0 & 0 & 0 & 5 & 0 \end{bmatrix}$$

$$P_4 = \begin{bmatrix} 0 & 1 & 1 & 3 & 1 \\ 4 & 0 & 4 & 2 & 2 \\ 4 & 4 & 0 & 3 & 2 \\ 4 & 1 & 1 & 0 & 1 \\ 4 & 4 & 4 & 5 & 0 \end{bmatrix}$$

$$P_5 = \begin{bmatrix} 0 & 1 & 1 & 5 & 1 \\ 4 & 0 & 4 & 2 & 4 \\ 4 & 4 & 0 & 3 & 4 \\ 4 & 1 & 1 & 0 & 1 \\ 4 & 4 & 4 & 5 & 0 \end{bmatrix}$$

26. False

27. True

28. True

33.

To find paths of different lengths, the matrix must be multiplied sequentially.

$$D = \begin{bmatrix} 0 & 1 & 1 & 0 & 0 \\ 0 & 0 & 0 & 1 & 1 \\ 0 & 1 & 0 & 0 & 0 \\ 2 & 0 & 1 & 0 & 0 \\ 0 & 0 & 0 & 1 & 0 \end{bmatrix}$$

$$D2 = D * D = \begin{bmatrix} 0 & 1 & 0 & 1 & 1 \\ 1 & 0 & 1 & 1 & 0 \\ 0 & 0 & 0 & 1 & 1 \\ 0 & 2 & 1 & 0 & 0 \\ 1 & 0 & 1 & 0 & 0 \end{bmatrix}$$

$$D3 = \begin{bmatrix} 1 & 0 & 1 & 2 & 1 \\ 1 & 2 & 2 & 0 & 0 \\ 1 & 0 & 1 & 1 & 0 \\ 0 & 1 & 0 & 2 & 2 \\ 0 & 2 & 1 & 0 & 0 \end{bmatrix}$$

$$D4 = \begin{bmatrix} 2 & 2 & 3 & 2 & 1 \\ 0 & 3 & 1 & 4 & 4 \\ 1 & 2 & 2 & 1 & 1 \\ 2 & 0 & 2 & 3 & 1 \\ 0 & 1 & 0 & 3 & 3 \end{bmatrix}$$

$$D1 + D2 + D3 + D4 = \begin{bmatrix} 3 & 4 & 5 & 5 & 3 \\ 2 & 5 & 4 & 6 & 5 \\ 3 & 3 & 3 & 3 & 3 \\ 4 & 3 & 4 & 5 & 3 \\ 1 & 3 & 2 & 4 & 3 \end{bmatrix}$$

34.

$$\text{Number of even paths: } D2 + D4 = \begin{bmatrix} 2 & 3 & 3 & 3 & 2 \\ 1 & 3 & 3 & 5 & 4 \\ 1 & 2 & 2 & 2 & 2 \\ 2 & 2 & 3 & 3 & 1 \\ 1 & 1 & 1 & 3 & 3 \end{bmatrix}$$

$$\text{Number of odd path: } D1 + D3 = \begin{bmatrix} 1 & 1 & 1 & 2 & 1 \\ 1 & 2 & 2 & 1 & 1 \\ 1 & 1 & 1 & 1 & 0 \\ 1 & 1 & 1 & 2 & 2 \\ 0 & 2 & 1 & 1 & 0 \end{bmatrix}$$

35. Determine the number of shortest paths between each pair of vertices.

We initialize the solution matrix same as the input graph matrix as a first step. Then we update the solution matrix by considering all vertices as an intermediate vertex. The idea is to one by one pick all vertices and updates all shortest paths which include the picked vertex as an intermediate vertex in the shortest path. When we pick vertex number k as an intermediate vertex, we already have considered vertices {0, 1, 2, ..., k-1} as intermediate vertices. For every pair (i, j) of the source and destination vertices respectively, there are two possible cases.

(1) k is not an intermediate vertex in shortest path from i to j. We keep the value of dist[i][j] as it is.

(2) k is an intermediate vertex in shortest path from i to j. We update the value of dist[i][j] as dist[i][k] + dist[k][j] if dist[i][j] > dist[i][k] + dist[k][j]

41.

$$\left(A_{x \times y} B_{y \times z} \right) C_{z \times w} \rightarrow (x \times y \times z) + x \times z \times w$$

$$A_{x \times y} \left(B_{y \times z} C_{z \times w} \right) \rightarrow x \times y \times w + (y \times z \times w)$$

$$\Rightarrow xyz + xzw \leq xyw + yzw \rightarrow \frac{dvided\ by}{xyzw} \quad \frac{1}{w} + \frac{1}{y} < \frac{1}{z} + \frac{1}{x}$$

42.

$$10 \times 20 \times 30 + 10 \times 30 \times 40 = 18000$$

45.

$$M_{ij} = \begin{cases} 0 & i = j \\ \min_{i \le k \le j-1} (M[i][k] + M[k+1][j] + d_{i-1}d_kd_j) & i < j \end{cases}$$

$$d_0 = 10, \ d_1 = 20, \ d_2 = 50, \ d_3 = 1, \ d_4 = 100$$

$$M_{11} = M_{22} = M_{33} = M_{44} = 0$$

$$\Rightarrow M_{12} = \min_{1 \le k \le 1} (M_{11} + M_{22} + d_0d_1d_2) = 0 + 0 + 10 \times 20 \times 50 = 10000$$

$$M_{23} = \min_{2 \le k \le 2} (M_{22} + M_{33} + d_1d_2d_3) = 0 + 0 + 20 \times 50 \times 1 = 1000$$

$$M_{34} = \min_{3 \le k \le 3} (M_{33} + M_{44} + d_2d_3d_4) = 0 + 0 + 50 \times 1 \times 100 = 5000$$

$$M_{13} = \min_{1 \le k \le 2} (M_{11} + M_{23} + d_0d_1d_3, \ M_{12} + M_{33} + d_0d_2d_3)$$

$$= \min(0 + 100 + 10 \times 20 \times 1, 10000 + 0 + 10 \times 50 \times 1) = 300$$

$$M_{24} = \min_{2 \le k \le 3} (M_{22} + M_{34} + d_1d_2d_4, \ M_{23} + M_{44} + d_1d_3d_4)$$

$$= \min(0 + 5000 + 20 \times 50 \times 100, 1000 + 0 + 20 \times 1 \times 100) = 3000$$

$$M_{14} = \min_{1 \le k \le 3} (M_{11} + M_{24} + d_0d_1d_4, \ M_{12} + M_{34} + d_0d_2d_4, M_{13} + M_{44} + d_0d_3d_4)$$

$$= \min(0 + 3000 + 10 \times 20 \times 100, 10000 + 5000 + 10 \times 50 \times 100, 300 + 0 + 10 \times 1 \times 100) = 1300$$

	1	2	3	4
1	0	10000	300	1300
2		0	1000	3000
3			0	5000
4				0

48.

(a) $(A \times B) \times (C \times D) \rightarrow 10 \times 2 \times 25 + 25 \times 3 \times 4 + 10 \times 25 \times 4 = 1800$

(b) $(A \times (B \times C)) \times D \rightarrow 2 \times 25 \times 3 + 10 \times 2 \times 3 + 10 \times 3 \times 4 = 330$

	Yj	B	D	C	A	B
Xi	0	0	0	0	0	0
A	0	0	0	0	1	1
B	0	1	1	1	1	2
C	0	1	1	2	2	2
B	0	1	1	2	2	3

Fig. 11.7   Answer to question 56(a)

		A	B	C	B	D	A	B
	0	0	0	0	0	0	0	0
B	0	0	1	1	1	1	1	1
D	0	0	1	1	1	2	2	2
C	0	0	1	2	2	2	2	2
A	0	1	1	2	2	2	3	3
B	0	1	2	2	3	3	3	4
A	0	1	2	2	3	3	4	4

Fig. 11.8   Answer to question 56(b)

$$(c)\ A \times ((B \times C) \times D) \rightarrow 2 \times 25 \times 3 + 2 \times 3 \times 4 + 10 \times 2 \times 4 = 254$$

$$(d)\ ((A \times B) \times C) \times D \rightarrow 10 \times 2 \times 25 + 10 \times 25 \times 3 + 10 \times 3 \times 4 = 1370$$

56. (a) See Fig. 11.7

56. (b) See Fig. 11.8

Chapter 12
Greedy Algorithms

Abstract Greedy algorithms are simple algorithms used in optimization problems. This algorithm makes the optimal choice in each step so that it can find the optimal way to solve the whole problem. This chapter provides 169 exercises for addressing different aspects of greedy algorithms. To this end, this chapter provides exercises on activity selection problem, minimum spanning tree, Huffman coding, Dijkstra's shortest path algorithm, job sequencing problem, knapsack problem, travelling salesman problem, and several other applications.

12.1 Lecture Notes

Greedy algorithms are a simple algorithm used in optimization problems. This algorithm makes the optimal choice in each step so that it can find the optimal way to solve the whole problem. It should be noted that:

- Greedy algorithms at each stage of problem solving, regardless of previous or subsequent choices, select the element that seems best.
- These algorithms do not guarantee the optimal answer because they choose the answer regardless of the previous or next steps.

Greedy algorithms is an iterative procedure in which each iteration has three steps:

1. Selection: The element to be added to the collection is selected. This is a greedy choice, that is, the element that is best chosen at that stage, regardless of previous or subsequent choices.
2. Feasibility check: After selecting an item greedily, the algorithm should consider whether it is possible to add it to the previous set of answers. Sometimes the addition of an element violates one of the basic conditions of the problem that must be considered. If adding this element does not violate any conditions, the element will be added; Otherwise it is left out and based on the first step another element is selected to be added. If there is no other option to select, the execution of the algorithm ends.
3. Solution check: At each step, after completing step 2 and adding a new element to the answer set, we must check whether we have reached a desired answer or

© The Author(s), under exclusive license to Springer Nature Switzerland AG 2022   425
H. Izadkhah, *Problems on Algorithms*,
https://doi.org/10.1007/978-3-031-17043-0_12

not. If we have not arrived, we will go to the first step and continue the cycle in the next steps.

A greedy algorithm works if a problem has two characteristics:

- Greedy choice property: By making an optimal local solution, an optimal global solution can be achieved. In other words, the "optimal" solution can be achieved with "greedy" choices.
- Optimal substructure. Optimal solutions contain optimal sub-solutions, i.e., solutions to sub problems of an optimal solution are optimal.

Among the popular applications of the greedy method are the problems that are described below.

1. The problem of fractional knapsack: In the knapsack problem, the goal is to fill a knapsack with valuable items that have different weights. This knapsack should be filled in such a way that its weight does not exceed the allowable limit and the value of the equipment inside it is maximum. In the case of fractional knapsack, unlike a 0-1 knapsack problem, it is possible to separate a fraction from a device—such as fabric—and add it to the equipment inside the knapsack.
2. Minimal spanning tree: Prim and Kruskal methods are two famous methods of finding minimum spanning tree from a weighted graph that use greedy method. The minimum spanning tree is a spanning tree of a graph whose total weight of edges is less than or equal to the sum of the weight of the edges of other trees spanning the graph.
3. Single-source shortest path: When we want to calculate the shortest path from a certain origin to all the other vertices of the graph, an algorithm such as Dijkstra's shortest path algorithm helps us using the greedy method.
4. Encoding and compressing information: Hoffman code is one of the methods of compressing information by trying to reduce the file size by re-encoding the characters in the information based on their usage. According to this method, a high-use character is replaced with a shorter code and a low-use character is replaced with a longer code.

12.2 Exercises

12.2.1 Basics

1. A greedy algorithm always makes a choice that looks best now, regardless of future consequences. True/False?

2. A difference between greedy algorithms and dynamic programming is that dynamic programming chooses the optimal solution to the subproblem, but in greedy algorithms, the subproblems are not necessarily optimal. True/False?

3. A greedy algorithm makes every choice that seems best at the moment and then solves the sub-problems that arise after making the choice. The choice made by a greedy algorithm may depend on the choices made so far, but it may not depend on any future choices. True/False?

4. Dynamic programming solves the sub-problems bottom up. Unlike dynamic programming, a greedy algorithm usually progresses in a top-down manner, making greedy choices one after the other, reducing each instance of the problem to a smaller instance. True/False?

5. Greedy algorithms do not always yield optimal solutions. True/False?

6. Sometimes greedy algorithms fail to find the globally optimal solution because they do not consider all the data. The choice made by a greedy algorithm may depend on choices it has made so far, but it is not aware of future choices it could make. True/False?

7. Does the selection sort algorithm use a greedy strategy to sort the data? (Yes/No?)

12.2.2 Activity Selection Problem

Let's consider that there are n activities with a start and finish times. The objective is to select maximum number of non-conflicting activities to perform within a given time frame, assuming that only one person or machine is available to execute.
8. Devise a greedy algorithm to maximize the count of activities that can be performed.

9. A greedy way to solve activity selection problem is to sort the given activities in ascending order according to their finishing time. True/False?

10. Find the maximize number of activities that can be executed.

Activity Name	A	B	C	D	E	F
Start Time (s)	5	1	3	0	5	8
Finish Time (f)	9	2	4	6	7	9

11. Find the maximize number of activities that can be executed.

Activity Name	A	B	C	D	E	F
Start Time (s)	1	3	0	5	8	5
Finish Time (f)	2	4	6	7	9	9

12. *n* seminars are supposed to use one room. Seminar *i* has a start time and an end time *f1*. We want to select the largest number of seminars that do not intersect in terms of time. The following greedy algorithms have been proposed to solve this problem:

– The "shortest seminar first" method, which examines the seminars in order and according to their length (from small to large), and any seminar that does not intersect with the seminars that have been selected so far is finally selected.
– The "earliest seminar first" method, which examines the seminars in order and in terms of their starting time (from small to large), and each seminar that does not intersect with the seminars that have been selected so far is finally selected.
– The "minimum number of intersection" method, which examines the seminars in the order of their number of intersections (from small to large), and any seminar that has not intersected with the seminars that have been selected so far is finally selected.

How many of these algorithms are incorrect:

a. 0
b. 1
c. 2
d. 3

13. Analyze the time complexity of activity selection problem.

14. Check if the following code can find the optimal number of activities correctly.

```
ACTIVITY-SELECTION(Activity, start, finish)
    Sort Activity by finish times stored in finish
    Selected = {Activity[1]}
    n = Activity.length
    j = 1
    for i = 2 to n:
        if start[i] ≥ finish[j]:
            Selected = Selected ∪ Activity[i]
            j = i
    return Selected
```

15. Which technique is more efficient for activity selection problem?

(a) Greedy technique
(b) Dynamic programming technique

16. Dynamic Programming Approach takes $O(n^3)$. Greedy Approach takes $O(n)$ time when sorted and $O(n \log n)$ when unsorted. True/False

17. Does Greedy Activity Selector work? Answer: Yes! True/False?

12.2.3 Minimum Spanning Tree

A spanning tree for a connected and undirected graph is a subgraph that (1) is a tree and (2) connects all vertices. There can be several spanning trees for a graph. Spanning tree weight is the sum of the weights given to each edge of the spanning tree. For a weighted, connected and undirected graph, the minimum spanning tree (MST) or minimum weight spanning tree is a spanning tree with a weight less than or equal to the weight of any other spanning tree.

1. Starting with any node of the graph, add the frontier edge with the smallest weight. This greedy algorithm is called Prim's algorithm.
2. Adding edges in increasing weight, ignoring the edges whose addition creates a cycle. This greedy algorithm is called Kruskal's algorithm.
3. Starting with all edges, remove them in decreasing order of weight, skipping the edges whose removal creates a disconnected graph. This greedy algorithm is called Reverse-Delete algorithm.

18. True/False?

– A minimum spanning tree has $(V-1)$ edges where V is the number of vertices in the given graph.
– In the unweighted graphs, any spanning tree will be a minimum spanning tree.

19. Network design such as telephone, electrical, hydraulic, TV cable, computer, road are minimum spanning tree application. True/False?

20. The following steps are used to find MST using Prim's algorithm.

1. Let $V' = \emptyset$ and $E' = \emptyset$
2. Let v_1 be an arbitrary vertex of V and $V'_1 = \{v_1\}$
3. Let e be an edge of minimum weight joining a vertex V' and a vertex of $V - V'$ such that $E' \cup \{e\}$ be acyclic, then add $E' = E' \cup \{e\}$
4. If $V' = V$, then output E'; otherwise, goto step 3

21. Explain how Prim's algorithm works?

22. For an undirected graph $G = (V, E)$, a connected graph $S = (V, E')$ is a spanning tree if S be a tree, and $E' \subseteq E$. Prim's algorithm constructs a spanning tree using a priority queue Q of edges, as follows:

```
E' := ∅; V₁ := {1}; Q := empty
for each w ∈ V such that (1, w) ∈ E do
    insert(Q, (1, w))
while |V₁| < n do
    e := deletemin(Q)
    Suppose e = (u, v), where u ∈ V₁
    if v∉V₁ then
```

Fig. 12.1 A sample graph

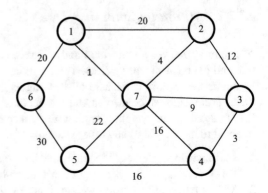

$$E' := E' \cup \{e\}$$
$$V_1 := V_1 \cup \{v\}$$
for each $w \in V$ such that $(v, w) \in E$ do
    insert(Q, (v, w))

23. Use Prim's algorithm to find the minimum spanning tree for the graph shown in Fig. 12.1.

24. The following steps are used to find MST using Kruskal's algorithm. Explain how we can find MST using Kruskal's algorithm.

1. Let $E' = \emptyset$
2. Let e be an edge of E – E' with minimum weight such that $E' \cup \{e\}$ is acyclic, then add $E' = E' \cup \{e\}$
3. If $|E| = n-1$, then output E'; otherwise, goto step 2

25. Explain how Prim's algorithm works?

26. Kruskal's algorithm constructs a spanning tree using the union-find algorithm. The union-find algorithm can also be used to find cycles which performs two following operations on a data structure:

– Find(x): Determine which subset element x is in. This can be used for determining if two elements are in the same subset.
– Union(x, y): Join the two subsets that x and y are in. Here first we have to check if the two subsets belong to same set. If no, then we cannot perform union.

For an undirected graph $G = (V, E)$, a connected graph $S = (V, E')$ is a spanning tree if S be a tree, and $E' \subseteq E$. Assume F is the set of vertex-sets in the forest.

$E' := \emptyset; F := \emptyset$
for each vertex $v \in V$ do $F := F \cup \{\{v\}\}$
Sort edges in E in ascending order of cost
while $|F| > 1$ do
    $(u, v) :=$ the next edge in sorted order
    if $find(u) \notin find(v)$ then
        union(u, v)
        $E' := E' \cup \{u, v\}$

27. Consider the graph shown in Fig. 12.1.

(a) Use Kruskal's algorithm to find the minimum spanning tree for this graph.
(b) How many minimum spanning trees does it have?

28. Kruskal's algorithm is known as a greedy algorithm, because it chooses at each step the lightest edge to add to the list. True/False?

29. Explain how we can find MST using Boruvka's algorithm. Analyze time complexity of Boruvka's algorithm.

30. Using the Kruskal's algorithm for a given graph, how many minimum spanning trees can be created?

31. Let $G = (V, E)$ be an undirected graph. Prove that if all its edge weights are distinct, then it has a unique minimum spanning tree.

32. Finding a spanning tree of maximum degree 2 is not possible in a polynomial time (it is an NP-complete problem). True/False?

33. Can Prim's and Kruskal's algorithm yield different minimum spanning trees?

34. Find minimum spanning tree for Fig. 12.2.
35. Find minimum spanning tree for Fig. 12.3.

Fig. 12.2 A sample graph

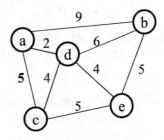

Fig. 12.3 A sample graph

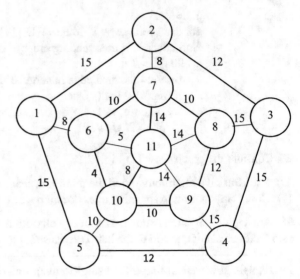

Fig. 12.4 A sample graph

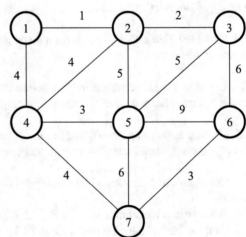

36. Consider Fig. 12.4.

– Use Prim's algorithm to find minimum spanning tree.
– Use Kruskal's algorithm to find minimum spanning tree.

37. Consider Fig. 12.5.

– Use Prim's algorithm to find minimum spanning tree.
– Use Kruskal's algorithm to find minimum spanning tree.

Fig. 12.5 A sample graph

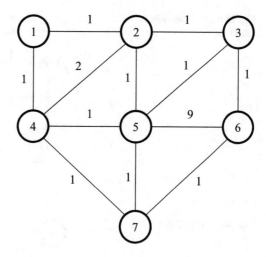

38. The table below shows the distances, in miles, along straight roads between six villages, A to F. The dash (–) indicates that there are no straight roads connecting the villages.

	A	B	C	D	E	F
A	–	6	3	–	–	–
B	6	–	5	6	–	14
C	3	5	–	8	4	10
D	–	6	8	–	3	8
E	–	–	4	3	–	–
F	–	14	10	8	–	–

– Use Prim's algorithm to find minimum spanning tree.
– Use Kruskal's algorithm to find minimum spanning tree.

39. An electrical engineer has designed a circuit with 9 terminals, to which a voltage equivalent to 5 volts must be connected. Suppose 5 volts is connected to one of the terminals (see Fig. 12.6). In order to use the least wiring in the circuit, what is the minimum number of wires needed? (Consider the distance between each row and column to be one centimeter.)

40. What is the order of edge selection in Kruskal's algorithm on the graph shown in Fig. 12.7?

a. (b, f), (b, d), (e, d), (c, d), (a, b)
b. (a, b), (b, f), (b, d), (f, d), (e, d)
c. (b, f), (b, d), (e, d), (e, c), (a, e)
d. None of the above

41. A company plans to create a LAN between 10 rooms of the company. Graph shown in Fig. 12.8 shows the rooms between which there is a possibility of direct

Fig. 12.6 A circuit with 9
terminals

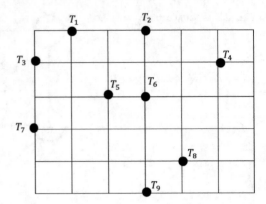

Fig. 12.7 A sample graph

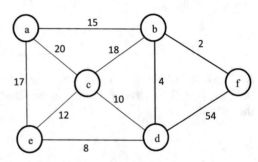

Fig. 12.8 The rooms of a
company

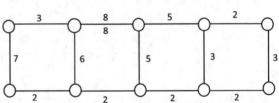

wiring. The weight of each edge is the cost of wiring between the two rooms. If
the goal is to do wiring at a minimum cost, which of the following algorithms is
appropriate to solve this problem?

a. Prim
b. Floyd
c. Kruskal
d. Dijkstra

42. For a complete graph containing n nodes, total number of spanning trees is $n^{(n-2)}$.
True/False?

43. For a complete graph containing n nodes with one edge removed, total number
of spanning trees is $(n-2)^{(n-3)}$. True/False?

44. For a graph, there may be several minimum spanning trees of the same weight. True/False?

45. The Prim's algorithm always finds the minimum spanning tree. True/False?

46. The Kruskal's algorithm always finds the minimum spanning tree. True/False?

47. If not all edge weights are distinct, both the Prim's and Kruskal's algorithms may not always produce the same minimum spanning tree. True/False?

48. To apply Kruskal's algorithm, the given graph must be weighted, connected and undirected. True/False?

49. Cost of the constructed MST would always be same in both Prim and Kruskal algorithms. True/False?

50. In Prim's algorithm, it does not matter which of the two edges is selected at the beginning of the algorithm. True/False?

51. In Kruskal's algorithm, it does not matter which of the two edges is selected at the beginning of the algorithm. True/False?

52. True or False?

- In MST, there is no source and no destination, but it is the subset (tree) of the graph (G) which connects all the vertices of the graph G without any cycles and with the minimum possible total edge weight.
- In the shortest path there is a source and destination, and the aim is to find out the shortest path between them.

53. True or False?

- In MST, graph G should be connected, undirected, edge-weighted, labeled.
- In the shortest path, it is not necessary for the graph G to be connected, undirected, edge-weighted, labeled.

54. True or False?

- In MST, relaxation of edges is not performed.
- In the shortest path, the relaxation of edges is performed.

55. True or False?

- In MST—for the disconnected graphs, the minimum spanning tree can not be formed but many spanning-tree forests can be formed.
- In the shortest path—for the disconnected graphs, the distance between two vertices present in two different components is infinity.

56. True or False?

– Network design activities e.g., computer networks, telecommunication networks, and water supply networks are application of finding MST.
– Finding direction between physical locations like in Google maps is an application of finding the shortest path.

57. True or False?

– In Prim's algorithm—the tree that we are making or growing is always connected.
– In Kruskal's algorithm—the tree that we are making or growing is usually disconnected.

58. True or False?

– Prim's algorithm starts finding MST from a random vertex and then continues by adding the next vertex with smallest weight edge connected to the existing tree.
– Kruskal's algorithm starts finding MST from the smallest weight edge and then continues by adding the next smallest weight edge to the existing tree/forest.

59. True or False?

– Prim's algorithm is faster for dense graphs.
– Kruskal's algorithm is faster for sparse graphs.

60. Justify why Kruskal's algorithm is preferred when-

(1) the graph is sparse i.e, $E = O(V)$; or
(2) the edges are already sorted or can be sorted in linear time.

61. Justify why Prim's algorithm is preferred when the graph is dense, i.e., $E = O(V^2)$, existing large number of edges in the graph.
62. Show the worst case time complexity of Kruskal's Algorithm is $O(E \log V)$ or $O(E \log E)$.

63. In Kruskal's algorithm, if the edges are already sorted, there is no need to create a min heap. Therefore, deletion from min heap time is saved. In this case, time complexity of Kruskal's Algorithm is $O(E + V)$. True/False?

64. Assume that Prim's algorithm is implemented by adjacency list representation using binary heap and Kruskal's algorithm is implemented using *union by rank*. Which of the following is true?

a. Worst case time complexity of both algorithms is the same.
b. Worst case time complexity of Kruskal's algorithm is better than Prim's algorithm.
c. Worst case time complexity of Prim's algorithm is better than Kruskal's algorithm.

65. Assume you are given a graph contains only k different edge weights. How to modify Prim's algorithm so that it runs on this graph in time of $O(n \log k)$.

66. Does the following algorithm find MST correctly? If not, explain why it is wrong and give a graph on which it fails.

$$E' := \emptyset;\ V_1 := \{1\};\ Q := empty$$
for each $w \in V$ such that $(1, w) \in E$ do
    insert(Q, (1, w))
while $|V_1| < < n$ do
    e := deletemin(Q)
    Suppose $e = (u, v)$, where $u \in V_1$
    $E' := E' \cup \{e\}$
    $V_1 := V_1 \cup \{v\}$
    for each $w \in V$ such that $(v, w) \in E$ do
        if $w \notin V_1$ then insert(Q, (v, w))

67. Consider a simple, connected, and undirected graph $G = (V, E)$ whose weight on all edges is 1. We want to find the minimum spanning tree of this graph. What is the time complexity of the fastest algorithm to do this?

a. $\theta(|v| + |E|)$
b. $\theta(|E| \log |E|)$
c. $\theta(|v| \log |E|)$
d. $\theta(|v| \times |E|)$

68. The "power" of a spanning tree in graph G is equal to the weight of its heaviest edge. In the given graph G, we want to choose the one with the least power from all the spanning trees. Which of the Prim or Kruskal algorithms produce always the least powerful spanning tree?

a. Prime
b. Kruskal
c. Both
d. none of the above

69. In the case of the minimum spanning tree on an undirected and weighted graph, which of the following is the best answer? (Suppose that the Prim algorithm uses a binary heap and the Kruskal algorithm uses disjoint sets as data structure.)

a. Prim and Kruskal algorithms may calculate different optimal trees.
b. If the number of graph edges is very large, the Prime algorithm is inevitably faster than the Kruskal algorithm.
c. If the edges are negative but the graph is cycle-free, both algorithms give incorrect answers.
d. In the Prime algorithm, the spanning tree is the same of the BFS.

70. Consider a graph $G = (V, E)$ and an edge $e = (u, v)$. We want to find the minimum spanning tree of G, which undoubtedly contains e. Which of the following is always true?

a. We delete e, (without removing u and v), we get the minimum spanning tree of the resulting graph and then we add e.
b. Combine u and v and obtain the minimum spanning tree of the resulting graph and then add e.
c. Remove all edges connected to u and v. Obtain the minimum spanning tree of the resulting graph and then add e.
d. There is no polynomial solution to this problem.

71. The weighted graph G is given by n vertices and e edges. A special edge $q = (u, v)$ is also marked. We want to find the minimum spanning tree that contains the edge q. If the best algorithm for finding the minimum spanning tree in this graph is the algorithm X from time $O(T_X(n, e))$, which option is correct for the best execution time of an algorithm for the required job?

Suppose we do not know the structure of the algorithm X and we always have to give this algorithm a precise graph (for example with the adjacency list).

a. $O(n + T_X(n, e))$
b. $O((n + e) + T_X(n, e))$
c. $O(n \times T_X(n, e))$
d. $O((n + e) \times T_X(n, e))$

72. A simple, weighted, undirected graph with n vertices and e edges (with different and positive weights) is given. For a spanning tree T of G whose edges are denoted by E(T), calculate the value of "ugliness" or U(T) according to the following formula:

$$U(T) = \sum_{0 \in E(T)} w(e) - \max w(e)$$

where w(e) is the weight of the edge e. Simply put, the ugliness of a spanning tree is equal to the sum of the weights of all its edges, except for its heaviest edge.

We want to find the most beautiful spanning tree (the tree with the least amount of ugliness) in G.

If the best algorithm for finding the minimum spanning tree in this graph is the unknown algorithm X from time $\theta(T_X(n, e))$, what is the best option for the best execution time of an algorithm for the requested job?

Suppose we do not know the structure of the algorithm X and we always have to give this algorithm a precise graph (with the adjacency list).

a. $\theta(T_X(n, e))$
b. $\theta((n + e) + T_X(n, e))$
c. $\theta(n \times T_X(n, e))$
d. $O((n + e) \times T_X(n, e))$

73. Suppose the graph G = (V, E) is an undirected graph and M is a minimum spanning tree for G, which statement is correct?

a. The path M is the shortest path between each pair of vertices V1 and V2.
b. The path M is not the shortest path between each pair of vertices V1 and V2.
c. All paths in M are the shortest paths.
d. There is no path between all vertices in M.

74. Let G = (V, E) be an undirected graph. The following statements may or may not be correct. In each case, either prove it (if it is correct) or give a counterexample (if it isn't correct).

(a) If graph G has more than $|V| - 1$ edges and there is a unique heaviest edge in the graph, then this edge cannot be part of a minimum spanning tree.
(b) If G has a cycle with a unique heaviest edge e, then e cannot be part of any MST.
(c) Let e be any edge of minimum weight in G. Then e must be part of some MST.
(d) If the lightest edge in a graph is unique, then it must be part of every MST.
(e) If e is part of some MST of G, then it must be a lightest edge across some cut of G.
(f) If G has a cycle with a unique lightest edge e, then e must be part of every MST.
(g) The shortest-path tree computed by Dijkstra's algorithm is necessarily an MST.
(h) The shortest path between two nodes is necessarily part of some MST.
(e) Prim's algorithm performs correctly when there are negative edges.

75. We have an undirected and connected graph with n nodes and n–2 edges. Which of the following algorithms is more suitable for generating a spanning tree with the least cost on the graph?

a. Prim
b. Kruskal
c. Sollin
d. none of the above

76. Which choice is correct for Kruskal's and Prime's algorithms to create the minimum spanning tree?

a. The execution time of both algorithms on the same graphs is equal.
b. Both algorithms produce the same spanning tree on the same graphs.
c. The total length of the sides of the spanning tree is the same in both algorithms.
d. Both algorithms produce a spanning tree by growing and merging a forest of trees.

77. Regarding Prim's algorithm, complete the following table.

Type of graph	Type of implementation and used data structure	Time complexity
dense graph ($|E| = \theta(V^2)$)	adjacent list + binary heap	
dense graph ($|E| = \theta(V^2)$)	adjacent list + Fibonacci heap	
sparse graph ($|E| = \theta(V)$)	adjacent list + binary heap	
dense graph ($|E| = \theta(V)$)	adjacent list + Fibonacci heap	
dense and sparse graph	adjacent matrix	

78. Regarding Kruskal's algorithm, complete the following table.

Type of graph	Time complexity
dense graph ($|E| = \theta(V^2)$)	
sparse graph ($|E| = \theta(V)$)	

79. The following algorithm is given to determine the spanning tree.

$T = \emptyset$
for each edge e, taken in arbitrary order
    $T = T \cup \{e\}$
    if T has a cycle c
        let e' be a maximum-weight edge on c
        $T = T - \{e\}$
return T

In this algorithm, the heaviest edge is removed in each cycle to finally obtain a connected graph without a circle (i.e., a tree).

(a) Apply this algorithm on graph shown in Fig. 12.9.
(a) Can the spanning tree from this algorithm be a minimum spanning tree?

A directed spanning tree of a directed graph is a spanning tree in which the directions on the edges point from parent to child.

80. Suppose Prim's algorithm is modified to take the lightest edge directed out of the tree.

(a) Does it find a directed spanning tree of a directed graph?
(a) If so, will it find a directed spanning tree of minimum cost?

81. Suppose Kruskal's algorithm is used to construct a directed spanning tree.

(a) Does it find a directed spanning tree of a directed graph?
(a) If so, will it find a directed spanning tree of minimum cost?

Fig. 12.9 A sample graph

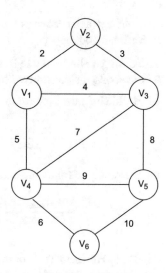

82. Which of the following algorithms does not use greedy strategy?

a. Dijkstra's algorithm
b. Kruskal's algorithm
c. Huffman Coding
d. Bellmen–Ford algorithm

83. Design an algorithm for the max-cost spanning tree.

84. The fastest implementations of Prim's and Kruskal's algorithms use Fibonacci heaps. However, *pairing heaps* are heap-ordered multiway tree structures that can be used in Prim's and Kruskal's algorithms with the same bounds but with less overhead. Discuss both algorithms.

85. Provide an algorithm that creates a spanning tree whose maximum degree is almost minimized.

86. Show how to find the maximum spanning tree of a graph, that is, the spanning tree of largest total weight.

12.2.4 Huffman Coding

Huffman encoding is a data compression algorithm. In this coding, considering the frequency of the input characters, codes with variable lengths are assigned to the input characters so that the most frequent character gets the smallest code and the least frequent character receives the largest code. There are mainly two major activities in Huffman coding

1. building a Huffman tree from input characters,
2. traversing the constructed tree and assigning codes to characters.

87. Build Huffman Tree for the following data.

character	A	B	C	D	E	F
frequency	5	9	12	13	16	45

88. Consider the following data.

character	A	B	C	D	E	F	G
frequency	27	15	3	1	1	1	1

(a) Build Huffman Tree for the data.
(b) Determine the codes associated with each character.

89. Consider the following data.

character	A	B	C	D	E	F	G
frequency	7	4	4	3	5	16	3

(a) Build Huffman tree for the data.
(b) Determine the codes associated with each character.
(c) Compress the following strings by considering the code obtained for each character.

  1. FADED
  2. ABED
  3. FEED

90. Encode the following message using Huffman coding.

ABAABAACAABAACACAACBAACBAACA

91. If we encode the string *abcabbaccaabdffe* using the Huffman coding method, how many bits will the code length be?

a. 38
b. 36
c. 34
d. none of the above

Fig. 12.10 A codded tree
using Huffman code

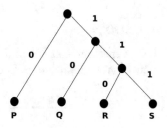

92. A networking company uses a compression technique used is Huffman coding to
encode the following message before transmitting over the network. Each character
in input message takes 1 byte. How many bits will be saved in the message?

character	A	B	C	D	E	F
frequency	5	9	12	13	16	45

93. Consider Fig. 12.10. Decode the following codes.

(a) 011011110
(b) 11011011100010

94. How many bits may be required for encoding the message 'mississippi'?

95. Given a char array ch[] and frequency of each characters as freq[]. The task is to
find Huffman Codes for every character in ch[] using Priority Queue.

96. Which of the following algorithms is the best method for Huffman coding?

a. exhaustive search
b. greedy algorithm
c. brute force algorithm
d. divide and conquer algorithm

97. The length of code assigned to each character in Huffman coding depends on the
number of characters (frequency of characters). The more the frequency of character,
the shorter the code length. True/False

98. In Huffman encoding, data is stored at the leaves of a tree. True/False?

99. Consider the following data. Using the Huffman coding,

character	A	B	C	D	E
frequency	0.17	0.11	0.24	0.33	0.15

100 00 01101 is decoded as

a. BACE
b. CADE

c. BAD

d. CADD

100. A message consists of the characters from the set $X = \{P, Q, R, S, T\}$. The table of probabilities for each of the characters is shown below:

character	P	Q	R	S	T
frequency	0.22	0.34	0.17	0.19	0.08

If a message of 100 characters over X is encoded using Huffman coding, then the expected length of the encoded message in bits is ....

101. Encode the following message using Huffman coding.

BCCABBDDABCCBBAEDDCC

102. Analyze time complexity of the Huffman coding.

103. If the given array is sorted (by non-decreasing order of frequency), we can generate Huffman codes in O(n) time. True/False?

104. What can be the maximum code length of an element in n elements with Huffmann compression method?

a. $\frac{n}{2}$

b. log n

c. n–2

d. n–1

105. If in Huffman's coding method, the frequency of the first data is greater than the sum of the n–1 frequencies of the other data and the frequency of the n–1 of the mentioned element is equal, then what is the maximum length of the generated codes?

a. $\lfloor \log n \rfloor + 1$

b. $\lceil \log n \rceil + 1$

c. $\lfloor \log(n - 1) \rfloor + 1$

d. $\lceil \log(n - 1) \rceil + 1$

106. In Huffman's compression algorithm, if a linear search is used instead of a heap to find two characters with the least frequency, what will be the execution time? (n is the number of characters to be encoded.)

a. $\theta(n \log n)$

b. $\theta(n^2)$

c. $\theta(n)$

d. $\theta(n^2 \log n)$

107. A text file consists of 256 types of 8-bit letters. Suppose the maximum number of repetitions of a letter in this file is M and the lowest number of repetitions of a letter is m. With this definition, assume that $M < 2 \times m$. If we call the compressed file size by Huffman's algorithm S_H and the original file size S_O, which option is always correct?

a. $S_H > S_O$
b. $S_H < S_O$
c. $S_H = S_O$
d. none of the above

108. "In Huffman's algorithm, if there is a character with a frequency greater than F, there must be a code of length 1."

What is the minimum value of F for the correctness of this statement in the general case?

a. $\frac{1}{5}$
b. $\frac{1}{4}$
c. $\frac{1}{3}$
d. $\frac{2}{5}$

109. "In Huffman's algorithm, if there are no characters with a frequency of F or more, we will never have a code of length 1."

What is the maximum value of F for the correctness of this statement in the general case?

a. $\frac{1}{5}$
b. $\frac{1}{4}$
c. $\frac{1}{3}$
d. $\frac{2}{5}$

110. In the Huffman tree construction algorithm, if the frequency of n input data is $F_1, F_2, \ldots, F_n$, where F_i is the i-th Fibonacci number, then what is the height of the tree?

a. $\log n$
b. $\frac{n}{2}$
c. $n-1$
d. none of the above

111. In Huffman's algorithm, if linear search is used instead of a heap to find the two characters with the least frequency, what will be the execution time? (n is the number of characters to be encoded.)

a. $\theta(n^2)$
b. $\theta(n \log n)$
c. $\theta(n^2)$
d. $\theta(n^2 \log n)$

12.2.5 *Dijkstra's Shortest Path Algorithm*

Given a graph and a source vertex in the graph, Dijkstra's algorithm finds the shortest paths from the source to all vertices in the given graph (the single-source shortest-paths problem).

112. The following algorithm is Dijkstra's algorithm for solving the single-source shortest-paths problem. Let C[i, j] denotes the cost of the edge from vertex *i* to vertex *j*. Vertex *s* is the source.

```
S := {s}
for each v ∈ V do
    D[v] := C[s, v]
    if (s, v) ∈ E then P[v] := s else P[v] := 0
for i := 1 to n – 1 do
    choose w ∈ V – S with smallest D[w]
    S := S ∪ {w}
    for each vertex v ∈ V do
        if D[w] + C[w, v] < D[v] then
            D[v] := D[w] + C[w, v]; P[v] := w
```

Fill the contents of arrays D and P applied on the graphs shown in Fig. 12.11, which the source vertex is 1. Use this information to print the shortest paths from vertex 1 to each of the other vertices.

113. To run Dijkstra's algorithm to find the single-source shortest-paths on unweighted graphs at a linear time, ... data structure is used.

a. Queue
b. Stack
c. Heap
d. B-Tree

Fig. 12.11 A sample graph

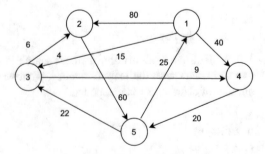

Fig. 12.12 A sample graph

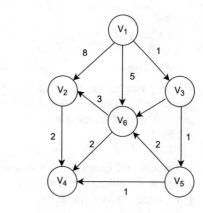

Fig. 12.13 A sample graph

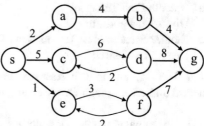

114. In an unweighted, undirected and connected graph, the shortest path from a node *S* to any other node is computed most efficiently in terms of time complexity by

a. Dijkstra's algorithm.
b. Warshall's algorithm
c. Performing a DFS.
d. Performing a BFS.

115. Suppose we run Dijkstra's algorithm on the weighted and directed graph shown in Fig. 12.12 to find the shortest path from node V_1 to every other node. In what order do the nodes get included into the set of vertices for which the shortest path distances are finalized?

a. $V_1, V_3, V_5, V_6, V_2, V_4$
b. $V_1, V_3, V_5, V_4, V_6, V_2$
c. $V_1, V_3, V_5, V_4, V_2, V_6$
d. $V_1, V_3, V_2, V_5, V_4, V_6$

116. Consider Fig. 12.13, using Dijkstra's algorithm, compute the shortest path from node S to every other node.

117. ....is the time complexity of Bellman–Ford single-source shortest path algorithm on a complete graph containing *n* vertices.

a. $\Theta(n^2)$
b. $\Theta(n^2 Log\ n)$
c. $\Theta(n^3)$
d. $\Theta(n^3 Log\ n)$

118. In a weighted graph, a shortest path algorithm is used to calculate the shortest path from the source 's' to a destination 't'. Is the following statement true? "If we increase weight of every edge by 1, the shortest path always remains the same." Yes/No?

119. A directed graph is given in which the weight of each edge is the same. Using which of the following methods can we efficiently find the shortest route from a given source to a destination?

a. Breadth First Traversal
b. Dijkstra's algorithm
c. Neither Breadth First Traversal nor Dijkstra's algorithm can be used
d. Depth First Search

120. The single-source shortest path created by the Dijkstra's algorithm on a connected undirected graph form a spanning tree. True/False?

121. Is the spanning tree constructed by Dijkstra's algorithm on a connected undirected graph a minimum spanning tree?

122. Dijkstra's algorithm is most widely used method in finding shortest possible distance between two geographical locations such as in Google Maps. True/False?

123. Dijkstra's algorithm is widely used in routing of data in networking and telecommunication domains for minimizing the delay occurred for transmission. True/False?

124. We can use Dijkstra's algorithm in robotics, transportation, embedded systems, factory or production plants to detect faults. True/False?

125. Analyze the time complexity of Dijkstra's algorithm.

126. Using binary heap for implementing priority queue instead of list, we can reduce the time complexity of Dijkstra's algorithm. True/False?

127. Dijkstra's algorithm can also be used for directed graphs as well. True/False?

128. Dijkstra's algorithm does not work for graphs containing negative edges. For such cases, algorithms such as Bellman–Ford can be used. True/False?

129. Dijkstra's algorithm is similar to the breadth-first search (BFS), except that a priority queue is used instead of first-in-first-out queue. True/False?

Fig. 12.14 A sample graph

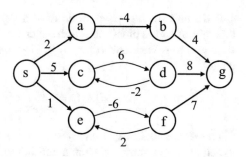

Fig. 12.15 A sample graph

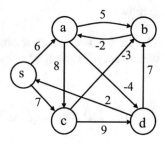

130. What is Dijkstra's algorithm space complexity?

131. Dijkstra's algorithm works only for connected graphs. True/False?

132. Dijkstra's algorithm works only for those graphs that do not contain any negative weight edge. True/False?

133. The actual Dijkstra's algorithm does not output the shortest paths. True/False?

134. It only provides the value or cost of the shortest paths. True/False?

135. By making minor modifications in the actual Dijkstra's algorithm, the shortest paths can be easily obtained. True/False?

136. Dijkstra's algorithm works for directed as well as undirected graphs. True/False?

137. Consider Fig. 12.14. Can the Bellman–Ford algorithm be applied to this graph to find the shortest path from node S to every other node?

138. Consider Fig. 12.15. Can the Bellman–Ford algorithm be applied to this graph to find the shortest path from node S to every other node?

139. To find the single-source longest path in a graph, can we use the Dijkstra's algorithm by changing minimum to maximum? If so, prove your algorithm correct. If not, give a counterexample.

140. To find shortest paths in directed acyclic graphs (DAGs), we can perform a topological sort to order the vertices such that all edges go from left to right starting from source s. We know $d(s, s) = 0$. The following can be used to find the shortest pathes.

$$d(s, j) = \min_{(x,i)\in E} d(s, i) + w(i, j)$$

Since we already know the shortest path $d(s, i)$ for all vertices to the left of j, using dynamic programming, give an algorithm to find shortest paths in directed acyclic graphs.

141. In the previous question, replace *min* with *max* to find the longest path in a directed acyclic graph.

12.2.6   Job Sequencing Problem

142. A set of jobs is given in which each job has a deadline and if a job is completed before the deadline, it will be given a profit. There may also be a possible deadline for each task. How to maximize total profit if only one task can be scheduled at a time.

143. Devise a greedy algorithm for job sequencing problem.

144. According to the table below, find the number of jobs done and the maximum total profit.

JobID	Deadline	Profit
a	4	20
b	1	10
c	1	40
d	1	30

145. Maximize total profit for the following jobs.

JobID	Deadline	Profit
a	2	100
b	1	19
c	2	27
d	1	25
e	3	15

146. Analyze time complexity of Job Sequencing Problem.

12.2.7 Knapsack Problem

n items and weights of each item are given. We need to put these items in a knapsack of capacity W.

In the 0-1 knapsack problem, we are not allowed to break items, or we put the whole item in the knapsack, or we do not put it at all.

The knapsack problem involves deciding which subset of items to consider from a set of items in order to optimize a value: perhaps the worth of the items, the size of the items, or the ratio of worth to size.

In this problem, it is assumed that we can take an item or ignore it (we can not take a fraction of an item). It is also assumed that there is only one item per item. The knapsack has a fixed size and we want to optimize the value of the items we get.

147. Consider the following 0-1 knapsack problem where P_i represents the value (price) and w_i represents the weight of object i. The total weight of the knapsack is 50. Which objects can be selected by considering the following greedy solutions?

Item	w	P
1	40	100
2	35	50
3	18	45
4	4	20
5	10	10
6	2	5

(a) Greedy through value (P). That is, in each step, select the object with the highest value.

(b) Greedy through $\frac{P_i}{w_i}$. That is, in each step, select the object with the highest $\frac{P_i}{w_i}$.

148. Our knapsack can hold up to 25 units of weight. Here is a list of items and their price.

Item	Weight	Price
Laptop	22	12
PlayStation	10	9
Textbook	9	9
Basketball	7	6

Which items do we choose to optimize for price?

Unlike a 0-1 knapsack, in a fractional knapsack, we can break items to maximize the total value of the knapsack. An efficient solution to this problem is to use greedy approach. The main idea of the greedy approach is to calculate the *value/weight ratio* for each item and sort the items based on this ratio. Then take the item with the highest ratio and add it to the knapsack until it is not possible to add the whole item, in which case an item is added as much as possible.

149. The capacity of the knapsack is 30 (W = 30). Here is the list of items and their worths.

Item	Weight	Price
I_1	18	25
I_2	15	24
I_3	10	15

Consider the following greedy approaches. In each approach, which fractional items do we choose to optimize the price?

(a) Because the value of objects is important, it is better to choose the objects with the highest value in order.
(b) Because the weight of objects has a negative effect, it is better to select the objects from the lowest weight in order.
(c) It is better to select objects in descending order $\frac{p_i}{w_i}$, that is, to select the object with the highest $\frac{p_i}{w_i}$.

150. Our knapsack can hold up to 25 units of weight. Here is a list of items and their price.

Item	Weight	Price
I_1	22	12
I_2	10	9
I_3	9	9
I_4	7	6

Which fractional items do we choose to optimize the price?

151. Check if the following algorithm performs correct for fractional knapsack.

```
GREEDY-FRACTIONAL-KNAPSACK(w[1..n], p[1..n], W)
    for i = 1 to n do
        x[i] = 0
        weight = 0
        for i = 1 to n
            if weight + w[i] ≤ W then
                x[i] = 1
                weight = weight + w[i]
            else
                x[i] = (W − weight)/w[i]
                weight = W
                break
    return x
```

152. The knapsack capacity is W = 60 and the list items are shown in the following table—

Item	A	B	C	D
Profit	280	100	120	120
Weight	40	10	20	24

Which fractional items do we choose to optimize for price?

12.2.8 Travelling Salesman Problem

The travelling salesman problem (TSP) asks the following question: "Given a list of cities and the distances between each pair of cities, what is the shortest possible cycle that visits each city exactly once and returns to the origin city?"

153. Consider the following naive solution for TSP.

1. Consider city i as the starting and ending point.
2. Generate all (n–1)! permutations of cities.
3. Calculate cost of every permutation and keep track of minimum cost permutation.
4. Return the permutation with minimum cost.

(a) What is the time complexity of this naive solution?
(b) Find TSP cycle for graph shown in Fig. 12.16 starting from city 1.

154. Devise a greedy algorithm for TSP. Do not worry about your solution being optimal and efficient.

155. There is no known algorithm to solve the TSP that is both optimal and efficient. True/False?

156. Consider the following greedy solution for TSP.

(1) Sort all of the edges in the graph.

Fig. 12.16 A sample graph

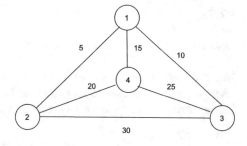

(2) Choose the edge with the lowest weight and add it to our tour if it does not violate any of the following conditions: there are no cycles in our tour with less than n edges or increase the degree of any node (city) to more than 2.
(3) Stop if we have n edges in the tour, else repeat step 2.

(a) Can this greedy algorithm find an acceptable tour?
(b) Suppose the graph has a Hamiltonian cycle. Will this algorithm find it?
(c) Suppose this greedy algorithm finds a Hamiltonian cycle. Is it necessarily a Hamiltonian cycle of minimum cost? Justify your answer.

157. Consider the following greedy algorithm for TSP. Start at node number 1, and move along the edge of minimum cost that leads to an unvisited node, repeat this until the algorithm reaches the vertex from which all edges lead to the visited nodes.

(a) Show that this greedy algorithm may fail to find a cycle.
(b) Suppose the graph has a Hamiltonian cycle. Will this algorithm find it?
(c) Suppose this greedy algorithm finds a Hamiltonian cycle. Is it necessarily a Hamiltonian cycle of minimum cost? Justify your answer.

12.2.9 Applications

158. *Graph coloring problem.* This problem finds minimum number of colors for coloring a given graph's vertices such that no two adjacent vertices share the same color. Devise a greedy algorithm for this problem. Do not worry about your solution being optimal.

159. *Manhattan distance.* There are N towns in a coordinate plane. Town i is located at coordinate (x_i, y_i). The distance between town i and town j is $|x_i - x_j| + |y_i - y_j|$. Devise an algorithm to compute the sum of the distance between each pair of towns.

160. Let array C contains coins and you can use each coin infinitely. Devise an algorithm to find the minimum number of coins that sum is equal to given value K.

Example:
C[] = {6, 3, 1, 4}
Ân: 23
Output: 6 6 6 4 1
Note 6 6 6 4 1 is minimum coins which sum is to 23.

161. *K Centers Problem.* Given n cities and distances between every pair of cities, devise a greedy algorithm to select k cities to place warehouses (or ATMs or Cloud Server) such that the maximum distance of a city to a warehouse is minimized.

162. n interval $A_i = [l_i, r_i]$ are given for $1 \le i \le n$. A set of these intervals is called "golden" if no two intervals in that set overlap or collide. We want to find the golden set S, which has the longest total among all gold sets. We know that the start and end points of intervals are all different. The following algorithm is proposed for this.

Sort the intervals based on a specific parameter (determined at the beginning of the algorithm), then scroll through the ordered intervals from beginning to end, and at each step if the desired interval does not overlap the previously selected intervals, select it.

This algorithm is optimal in terms of the answer.

(a) True, if we select the intervals in their order l_i.
(b) True, if we select the intervals in their order r_i.
(c) True, if we select the intervals in order of their length.
(d) None of the above parameters gives the maximum answer.

163. Given a schedule containing the arrival and departure time of trains in a station. The aim is to find the minimum number of platforms required for the railway station to avoid delay in any train's arrival (no train is kept waiting.). Devise a greedy algorithm for this problem.

Example:
Input: n = 6
arrival[] = {0800, 0840, 0850, 1000, 1400, 1700}
departure[] = {0810, 1100, 1020, 1030, 1800, 1900}
Output: 3
Explanation: At least three platforms are required to safely arrive and depart of all trains.

164. *Minimize the sum of product.* Let N and M are two arrays of equal size n. Devise an algorithm to find the minimum value of N[0] * M[0] + A[1] * M[1] + $\cdots$ + N[N–1] * M[N–1], where shuffling of elements of arrays N and M is allowed.

Example:
Input:
n = 3
N[] = {4, 1, 1}
M[] = {7, 6, 5}
Output: 1*7 + 1*6 + 4*5 = 7 + 6 + 20 = 33 is the minimum sum.

165. *Largest number with given sum.* The aim is to find the largest number with a given number of digits N and given sum of digits S.

Example:
Input:
N = 3, S = 15
Output:
960
Explanation: Sum of elements is 15. Largest possible three digit number is 960 with sum 15.

166. In the array A of N elements, give an algorithm to find the majority element in the array. In an array A of size N, an element is called majority if it appears more than $N/2$ times in the array. If each element appears only once in the array print there is no majority element.

167. *One-processor scheduling problem.* Assume there are n tasks in which each task i has a start time t_i, and a deadline d_i. A feasible schedule is a sequence of tasks so that when tasks are performed in sequence, each task is completed before the deadline. A greedy algorithm for this problem processes the tasks in order of deadline (the early deadlines before the late ones). Show that if there is a feasible schedule exists, the schedule generated by this greedy algorithm is feasible.

168. *Set Cover*—Let $S = \{S_1, \ldots, S_m\}$ denotes a collection of subsets of the universal set $U = \{1, \ldots, n\}$. Devise an algorithm to find the smallest subset T of S whose union equals the universal set—i.e.,

$$\cup_{i=1}^{|T|} T_i = U$$

169. *Set Packing*—Let $S = \{S_1, \ldots, S_m\}$ denotes a set of subsets of the universal set $U = \{1, \ldots, n\}$. Devise an algorithm to select (an ideally small) collection of mutually disjoint subsets from S whose union is the universal set.

Airline flight crew planning is an application of set of packaging. Each aircraft in the fleet must have a crew consisting of a pilot, co-pilot and navigator. There are limitations on the composition of possible crews, based on their training to fly different types of aircraft, personality conflicts, and work schedules. Given all the possible combinations of crew and aircraft, each of which is represented by a subset of items, we need an assignment such that each plane and each person is in exactly one chosen combination. However, one person cannot be on two different planes at the same time, and each plane needs a crew.

12.3 Solutions

10.

Activity Name	A	B	C	D	E	F
Start Time (s)	5	1	3	0	5	8
Finish Time (f)	9	2	4	6	7	9

First we sort ascending based on F[i].

Activity Name	1 (B)	2 (C)	3 (D)	4 (E)	5 (A)	6 (F)
S[i]	1	3	0	5	5	8
F[i]	2	4	6	7	9	9

$\Downarrow$

i	$S[i] \geq F[j]$	A	j
2	True	{1, 2}	2
3	False	–	–
4	True	{ 1, 2, 4 }	4
5	False	–	–
6	True	{1, 2, 4, 6}	6

$\implies \{B, C, E, F\}$

13.

First, we sort the activities in ascending order based on their end time, which is equal to: $\theta(n\log n)$

The activity selection algorithm is as follows:

```
ActivitySelection (S[], F[]){
A={ 1 };
j =1;
for (i=2 ; i<=n, i++)
{
   if (S [i] ≥ F [j])
   {
   A= A ∪{ i };
   j=i;
   }
}
return A;
}
```

This algorithm time complexity is $\theta(n)$. So, generally the activity selection algorithm time complexity will be $\theta(n\log n)$.

15. Greedy Approach

16. True

17. True

18. True

20.

1. Create a set S that keeps track of vertices already included in MST.
2. Assign a key value to all vertices in the input graph. Initialize all key values as INFINITE. Assign key value as 0 for the first vertex so that it is picked first.
3. While S does not include all vertices

   a. Pick a vertex u which is not there in S and has minimum key value.
   b. Include u to S.
   c. Update the key value of all adjacent vertices of u. To update the key values, iterate through all adjacent vertices. For every adjacent vertex v, if the weight of edge u-v is less than the previous key value of v, update the key value as the weight of u-v

O(E log V)

This time complexity can be reduced to O(E + V log V) using Fibonacci heap.

24.

(V is the number of vertices in the given graph.)

Step 1. Sort all the edges in non-decreasing order of their weight.

Step 2. Pick the smallest edge. Check if it forms a cycle with the spanning tree formed so far. If cycle is not formed, include this edge. Else, discard it.

Step 3. Repeat step 2 until there are (V-1) edges in the spanning tree.

Time Complexity: O(E log E)

28. True

30. In a graph, if the weights of all the edges are distinct, the minimum spanning tree is unique. But if several edges have the same weight, you can use the probabilities

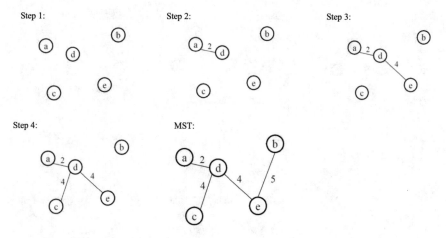

Fig. 12.17 Answer to question 34

to find the number of minimum spanning trees.

33. Yes, they can create MST but at the same cost.

34. Kruskal's algorithm (See Fig. 12.17):

$$ad = 2, \ de = 4, \ cd = 4, \ be = 5, \ ce = 5, \ ac = 5, \ db = 6, \ ab = 9$$

35. Kruskal's algorithm (See Fig. 12.18):

$1, 7 : 1, \ 3, 4 : 3, \ 2, 7 : 4, \ 3, 7 : 9, \ 2, 3 : 12, \ 4, 5 : 16, \ 4, 7 : 16, \ 1, 2 : 20, \ 1, 6 : 20, \ 5, 7 : 22, \ 5, 6 : 30$

36. Prim's algorithm (See Fig. 12.19).
Kruskal's algorithm (See Fig. 12.20):

$1, 2 : 1, \ 2, 3 : 2, \ 4, 5 : 3, \ 6, 7 : 3, \ 4, 7 : 4, \ 1, 4 : 4, \ 2, 4 : 4, \ 2, 5 : 5, \ 3, 5 : 5, \ 5, 7 : 6, \ 3, 6 : 6, \ 5, 6 : 9$

37. Prime (See Fig. 12.21).
38. Prime (See Fig. 12.22).

Kruskal (See Fig. 12.23).
39. See Fig. 12.24.

$$\sqrt{2} + 2 + \sqrt{5} + 1 + 2 + \sqrt{5} + \sqrt{10} + \sqrt{2}$$

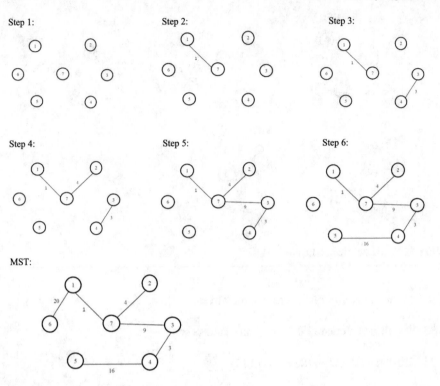

Fig. 12.18 Answer to question 35

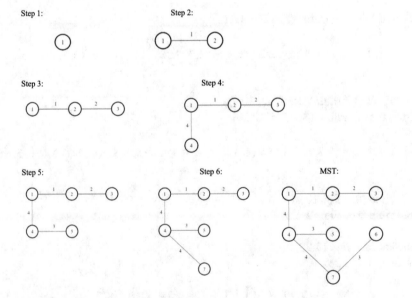

Fig. 12.19 Answer to question 36 (Prim's algorithm)

Step 1:

Step 2:

Step 3:

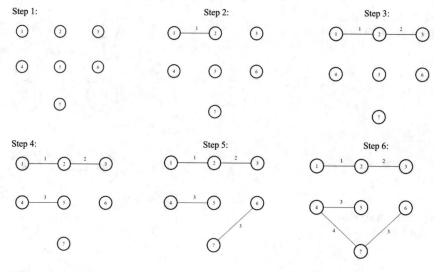

Step 4:

Step 5:

Step 6:

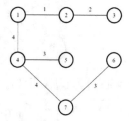

MST:

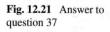

Fig. 12.20 Answer to question 36 (Kruskal's algorithm)

Fig. 12.21 Answer to question 37

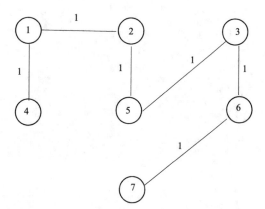

Fig. 12.22 Answer to
question 38 (Prime)

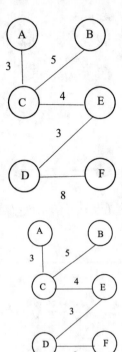

Fig. 12.23 Answer to
question 38 (Kruskal)

1	A, C	3	✓
2	D, E	3	✓
3	C, E	4	✓
4	B, C	5	✓
5	A, B	6	✗
6	B, D	6	✗
7	C, D	8	✗
8	D, F	8	✓
9	C, F	10	-
10	B, F	14	-

Fig. 12.24 Answer to question 39

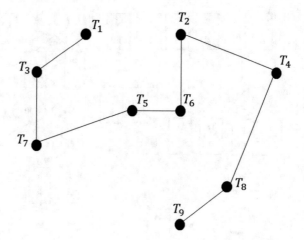

Fig. 12.25 Answer to question 40

1	b, f	2	✓
2	b, d	4	✓
3	e, d	8	✓
4	c, d	10	✓
5	e, d	12	✗
6	a, b	15	✓
7	a, e	17	-
8	c, b	18	-
9	c, a	20	-
10	d, f	54	-

40. a, see Fig. 12.25.

42. True

44. True

45. True

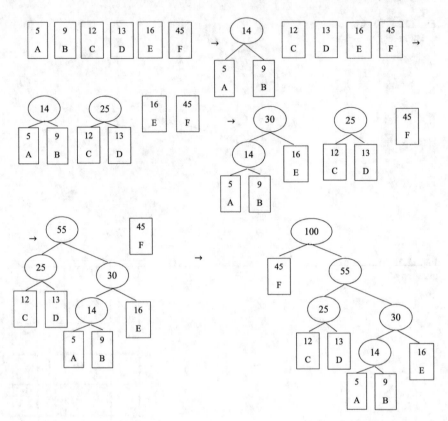

Fig. 12.26 Answer to question 87

49. True

57. (a) true, (b) false

58. (a) true, (b) true

59. (a) true, (b) true

87. See Fig. 12.26.

Character	A	B	C	D	E	F
frequency	5	9	12	13	16	45

Given that the edges on the right are assigned the number one and the edges on the left are assigned the number zero, the code for each letter will be as follows:

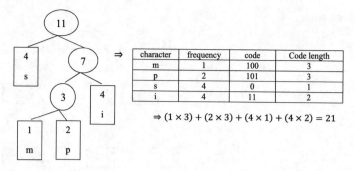

character	frequency	code	Code length
m	1	100	3
p	2	101	3
s	4	0	1
i	4	11	2

$$\Rightarrow (1 \times 3) + (2 \times 3) + (4 \times 1) + (4 \times 2) = 21$$

Fig. 12.27 Answer to question 94

A: 1100
B: 1101
C: 100
D: 101
E: 111
F: 0

97. Total number of characters in the message = 100. Each character takes 1 byte.
So total number of bits needed = 800.
After Huffman Coding, the characters can be represented with: f: 0
c: 100
d: 101
a: 1100
b: 1101
e: 111
Total number of bits needed = 224
Hence, number of bits saved = 800 − 224 = 576.

94. See Fig. 12.27.

Character	m	i	s	p
frequency	1	4	4	2

character	frequency	code	Code length
m	1	100	3
p	2	101	3
s	4	0	1
i	4	11	2

96. (b) greedy algorithm

Fig. 12.28 Answer to
question 100

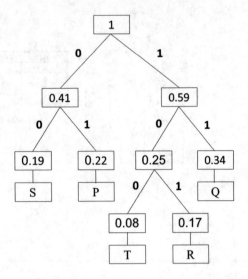

97. True

100. See Fig. 12.28.

According to the Huffman tree, the code of each letter will be as follows:

P: 01

Q: 11

R: 101

S: 00

T: 100

101. See Fig. 12.29.

character	A	B	C	D	E
frequency	3	6	6	4	1

A = 1101
D = 111
B = 10
E = 1100

Fig. 12.29 Answer to question 101

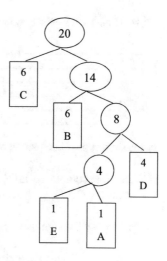

$C = 0$

$\Rightarrow$ 10 0 0 1101 10 10 111 111 1101 10 0 0 10 10 1101 1100 111 111 0 0

102. Time complexity is O(nlogn) where n is the number of unique characters.

117. (c) $\Theta (n^3)$

122. True

123. True

124. True

127. True

131. False

134. False

144.

1. Sort all jobs in decreasing order of profit.
2. Iterate on jobs in decreasing order of profit. For each job, do the following:

   a. Find a time slot i, such that slot is empty and i < deadline and i is greatest. Put the job in this slot and mark this slot filled.

    b. If no such i exists, then ignore the job.

Profit	Deadline	Job ID
40	1	C
30	1	D
20	4	A
10	1	B

So maximum profit sequence of jobs is: [C, A] $= 40 + 20 = 60$

146. The time complexity is O(n^2). The sorting process is $O(n\log n)$ and the addressing process is $O(n^2)$, so we have $O(n\log n) + O(n^2) = O(n^2)$.

147. a. Greedy through value (P). Items 1, 4, and 6 can be selected.

$$overalweigth = 46, \quad overalprofit = 125$$

    b.

Item	1	2	3	4	5	6
$\frac{Profit}{Weight}$	2.5	1.42	2.5	5	1	2.5

Items 4, 1, and 6 can be selected.

$$overalweigth = 46, \quad overalprofit = 125$$

150. Our knapsack can hold at most 25 units of space. Here is the list of items and their worths.

	Item	Size	Price
1	Laptop	22	12
2	PlayStation	10	9
3	Text book	9	9
4	Baskesball	7	6

Which fractional items do we choose to optimize for price?

$S[2] + S[3] = 19 \le 25$ *and* $P[2] + P[3] = 18 \implies$ PlayStation and Text book

152.

Item	A	B	C	D
Profit	280	100	120	120
Weight	40	10	20	24

$$W[A] + W[B] = 50 \le 60, \; P[A] + P[B] = 380$$

$$W[A] + W[C] = 60 \leq 60 , \, P[A] + P[C] = 400$$

$$W[B] + W[C] + W[D] = 54 \leq 60 , \, P[B] + P[C] + P[D] = 340$$

$$\Rightarrow A, C$$

or

Sorting

Item	B	A	C	D
$\frac{Profit}{Weight}$	10	7	6	$\frac{31}{6}$

Selecting

Item	Selected Weight	Profit
B	10	100
A	40	280
C	0	0
D	0	0
Sum	50	380

153. A TSP cycle in the graph is 1-2-4-3-1. The cost of the tour is $5 + 15 + 25 + 10$ which is 60.

Chapter 13
Graph

Abstract A graph is a nonlinear data structure that has a wide range of applications. Most of the concepts in computer science and the real world can be visualized and represented in terms of graph data structure. This chapter provides 70 exercises for addressing different applications of graph algorithms. To this end, this chapter provides exercises on graph traversal techniques, applications of DFS/BFS, graph cycle, topological sorting, shortest paths, connectivity, and maximum flow.

13.1 Lecture Notes

A graph is a nonlinear data structure that has a wide range of applications. Most of the concepts in computer science and real world can be visualized and represented in terms of graph data structure.

A graph is a visual representation of a set of objects that are related to each other. Each of these objects is called a "vertex" or "node". The vertices are also connected by "edges". We define a graph with (V, E) in which V is the set of vertices and E is the set of edges.

Graphs are used to solve many problems in mathematics and computer science. Many structures can be represented by graphs. For example, a graph can be used to show how websites relate to each other. In this way, we show each website with a vertex in the graph, and if there was a link to another website in this website, a directed edge connects this vertex to the vertex that displays the other website.

13.2 Exercises

13.2.1 Preliminary

1. The flavors of graphs are:

– Undirected/Directed

© The Author(s), under exclusive license to Springer Nature Switzerland AG 2022   471
H. Izadkhah, *Problems on Algorithms*,
https://doi.org/10.1007/978-3-031-17043-0_13

- Weighted/Unweighted
- Simple/Non-simple
- Sparse/Dense
- Cyclic/Acyclic
- Embedded/Topological
- Labeled/Unlabeled

(a) Define each of the above.
(b) The first step in any graph problem is to determine the flavors of graphs needed to model the problem. Which of the above flavors is required to model a social network e.g., a friendship graph?

2. There are two data structures for graphs to represent them: adjacency matrices and adjacency lists. Specify which of these data structures is more efficient in the following cases:

(a) Faster to test if (x, y) is in graph?
(b) Faster to find the degree of a vertex?
(c) Less memory on small graphs?
(d) Less memory on big graphs?
(e) Edge insertion or deletion?
(f) Faster to traverse the graph?
(g) Better for most problems?

3. Given n nodes of a forest (collection of trees), count the number of trees in a given forest.

4. *Transpose of a directed graph*. Transpose of a directed graph G is another directed graph on the same set of vertices in which all edges are inverted relative to the corresponding edges in G meaning that if G contains an edge (u, v) then the transpose of G contains an edge (v, u). Given graph G represented as adjacency list, find the transpose of the graph G.

5. Find minimum number of operation required to convert number x into y using the two following operations:

- Multiply number by 2.
- Subtract 1 from the number.

Constraint:
   $1 \leq x, y \leq 1000$

6. *Stepping Numbers*. Given two integers n and m, find all the stepping numbers in range $[n, m]$. If all adjacent digits have an absolute difference of 1, it is called a stepping number. For example, 541 is a stepping number while 641 is not.

13.2.2 Graph Traversal Techniques

The following code shows the traversing a graph by Breadth First Search (BFS). v represents the vertices of the graph.

```
void BREADTH_FIRST_SEARCH (Source_Vertex v)
{
visited [v] = true;
addq(queue, v);
while (not empty queue)
    {
    delq (queue, v);
    for all node w adjacent to v do
        {
        addq (queue, w);
        visited [w] = true;
        }
    }
}
```

7. Starting from node A, write BFS traversal for graph shown in Fig. 13.1.

8. Starting from node V1, write BFS traversal for graph shown in Fig. 13.2.

The following code shows the traversing a graph by Depth First Search (DFS). v represents the vertices of the graph.

```
void DEPTH_FIRST_SEARCH (Source_Vertex v)
{
print (Data (v));
visited [v] = ture;
for (each vertex w adjacent to v) do
```

Fig. 13.1 A sample graph

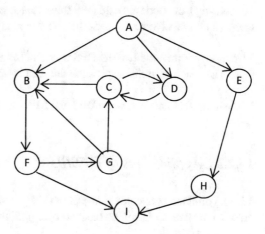

Fig. 13.2 A sample graph

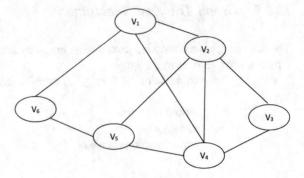

Fig. 13.3 A sample graph

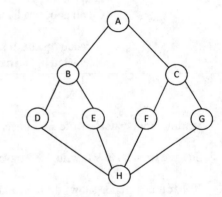

if (not visited [w])
    DEPTH_FIRST_SEARCH(w)
}

9. Starting from node A, write DFS traversal for graph shown in Fig. 13.3; whenever there's a choice of vertices, pick the one that is alphabetically first.

10. Starting from node a, write DFS traversal for graph shown in Fig. 13.4; whenever there's a choice of vertices, pick the one that is alphabetically first.

11. Starting from node 1, write DFS traversal for graph shown in Fig. 13.5; whenever there's a choice of vertices, pick the one that is numerically first.

12. Starting from node 1, write DFS and BFS traversals for graph shown in Fig. 13.6.

13.2.3 Applications of DFS/BFS

13. In a directed connected graph G = (V, E), a vertex v is called a mother vertex if all other vertices in G can be reached by a path from v.

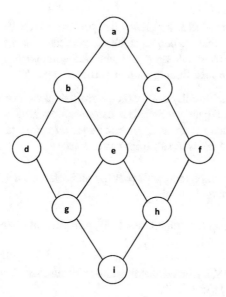

Fig. 13.4 A sample graph

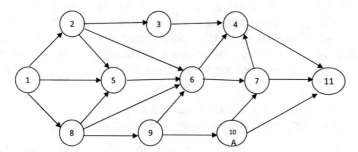

Fig. 13.5 A sample graph

Fig. 13.6 A sample graph

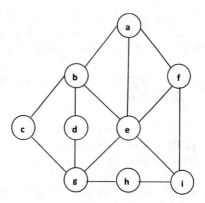

(a) In graph G represented as adjacency list, perform a DFS/BFS on all the vertices to determine whether we can reach all the vertices from a given vertex.

(b) Inspired from Kosaraju's strongly connected component algorithm, devise an efficient algorithm for finding mother vertices.

14. Construct the reach-ability matrix (i.e., transitive closure of a graph) for a given graph represented as adjacency list. To construct such matrix, using DFS determine if a vertex v is reachable from another vertex u for all vertex pairs (u, v) in the given graph meaning that there is a path from vertex u to v.

15. Given a directed graph, count all possible paths that exist between two vertices. These paths don't contain a cycle.

16. In a directed connected graph $G = (V, E)$, print all paths between a given pair of vertices (u, v) using BFS.

17. Given a graph $G(V, E)$, determine the minimum number of edges between a given pair of vertices (u, v) using DFS.

18. *Bidirectional Search.* Searching in a graph is a well-known problem and has many practical applications. Normal searching algorithms in a graph are BFS and DFS. These algorithms perform searching in one direction usually from source vertex toward the goal vertex. But we can do search form both direction simultaneously. Bidirectional search is a graph search algorithm which find smallest path form source to goal vertex. It runs two simultaneous search—

– Forward search form source/initial vertex toward goal vertex
– Backward search form goal/target vertex toward source vertex

   Using bidirectional search find smallest path form source to goal vertex for a given graph.

19. A connected component of an undirected graph is a maximal set of vertices so that there is a path between every pair of vertices. Using breadth-first search, give an algorithm to find the connected components of an undirected graph.

20. Modify BFS to apply on a disconnected graph or any vertex that is unreachable from all vertex.

21. Given a binary matrix where 0 represents water and 1 represents land, and a group of connected ones form an island, using breadth-first search (BFS), count the total islands.

22. We can say that G is strongly connected if:

(1) DFS(G, v) visits all vertices in the graph G, then there exists a path from v to every other vertex in G, and

(2) There exists a path from every other vertex in G to v.

Given a directed graph, check if it is strongly connected or not. A directed graph is said to be strongly connected if every vertex is reachable from every other vertex.

23. *Chess Knight Problem*. Given a chessboard, using breadth-first search (BFS), devise an algorithm to find the minimum number of steps (shortest distance) taken by a knight to reach a given destination from a given source.

24. The problem of coloring the vertex is to assign a label (or color) to each vertex of the graph so that no edge connects the two vertices of the same color. A graph is bipartite if it can be colored without conflicts while using only two colors. Using BFS, devise an algorithm to determine whether a graph is bipartite or not.

13.2.4 Graph Cycle

25. The following algorithm is used to detect cycles in a directed graph.

Algorithm: DETECT CYCLE (Source_Vertex S)
1. Mark S as visited.
2. Mark S as *in_path* vertex.
3. For all the *adjacent vertices* to S do
4.   If the *adjacent vertex* has been marked as *in_path* vertex, then
5.   Cycle found. return.
6.   If the *adjacent vertex* has not been visited, then
7.   DETECT CYCLE (*adjacent_vertex*)
8. Now that we are backtracking unmark the source_vertex in in_path as it might be revisited.

(a) Apply this algorithm to find cycle in graph shown in Fig. 13.7.
(b) Modify this algorithm to find all cycles in the graph.

26. Given a directed graph, using DFS, give an algorithm to check whether the directed graph contains a cycle or not. The algorithm should return true if the given graph contains at least one cycle, else return false.

27. Given a directed graph, using BFS, give an algorithm to check whether the directed graph contains a cycle or not. The algorithm should return true if the given graph contains at least one cycle, else return false.

28. The following algorithm is used to detect cycles in an undirected graph.

Fig. 13.7 A sample graph

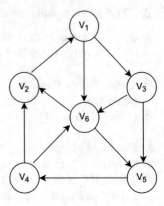

Fig. 13.8 A sample graph

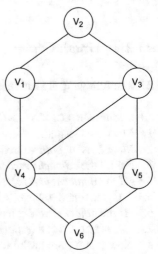

Algorithm: DETECT CYCLE (Source_Vertex *S*)
1. Mark *S* as visited.
2. For all the *adjacent vertices* to *S* do
3.   If the *adjacent_vertex* has not been visited, then
4.   Set parent[*adjacent_vertex*] = *S*.
5.   DETECT CYCLE (*adjacent_vertex*)
6.   Else if the parent[*S*] != *adjacent_vertex*, then
7. Cycle found. return.

(a) Apply this algorithm to find cycle in graph shown in Fig. 13.8.
(b) Modify this algorithm to find all cycles in the graph.

29. Given an undirected graph, using DFS, give an algorithm to check whether the graph contains a cycle or not. The algorithm should return true if the given graph contains at least one cycle, else return false.

30. Given an undirected graph, using BFS, give an algorithm to check whether the graph contains a cycle or not. The algorithm should return true if the given graph contains at least one cycle, else return false.

31. In a graph, the existence of a cycle in a graph can be detected using the degrees of the graph nodes. Give an algorithm to detect a cycle in the graph using degrees of the nodes and print all the nodes that are involved in any of the cycles. The algorithm should print -1, if there is no cycle in the graph.

32. In a directed weighted graph, using Bellman Ford algorithm, find any cycle of negative weight in it, if such a cycle exists. A negative cycle is one in which the overall sum of the cycle becomes negative.

33. A graph consists of directional and non-directional edges so that the directional edges do not form a circle. In this graph, how can directions be assigned to undirected edges so that the graph (with all directional edges) remains acyclic?

34. A undirected and connected graph and a number k are given. How to count the total number of cycles of length k in the graph. A cycle of length k simply means that the cycle consists of k vertices and k edges.

35. Given a weighted and undirected graph, determine if a cycle exist in this graph with odd weight sum.

36. Given a graph, the task is to find if it has a cycle of odd length or not.

13.2.5 *Topological Sorting*

37. A graph G is called a Directed Acyclic Graph (DAG) if it has at least one vertex with in-degree 0 and one vertex with out-degree 0. Topological sorting for DAG is a linear ordering of vertices so that for every directed edge $u->v$, vertex u comes before v in the ordering. Topological sorting for a graph is not possible if the graph is not a DAG.

Algorithm: TOPOLOGICAL SORT (Graph G, Source_Vertex S)
1. Mark the source vertex S as visited.
2. For every vertex v adjacent to the source vertex S.
3.   If the vertex v is not visited, then
4.   TOPOLOGICAL SORT (Graph G, Source_Vertex v)

Fig. 13.9 A sample DAG

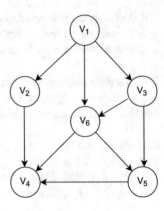

5. Push the source vertex S into the stack.

Printing the topological sorting order
1. While the stack is not empty.
2. Print the vertex v at the top of the stack.
3. Pop the vertex v at the top of the stack.

Apply this algorithm to find topological sort in graph shown in Fig. 13.9.

38. Consider the following Kahn's algorithm for topological sorting.

Algorithm Overview

(A) Output the nodes with indegree 0 in any order.
(B) Remove them and their outgoing edges. We now have a new set of nodes with indegree 0 which are the children of the earlier nodes.
(C) Output them in any order and remove them.
(D) This goes on.

Apply this algorithm to find topological sort in graph shown in Fig. 13.10.

39. Given a directed graph, using topological sorting, give an algorithm to check whether the directed graph contains a cycle or not. The algorithm should return true if the given graph contains at least one cycle, else return false.

40. Topological sort is simply a modification of DFS. Devise an algorithm for topological sorting using depth first search (DFS).

Fig. 13.10 A sample DAG

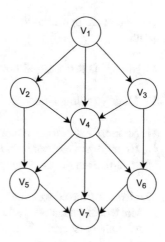

13.2.6 Shortest Paths

41. Using depth-first search (DFS), devise an algorithm to find the shortest paths in a graph without edge weights.

42. Modify BFS algorithm to find the shortest path in a graph without edge weights.

43. Using bidirectional search, devise an algorithm to find the shortest paths in an undirected graph without edge weights.

44. Suppose a weighted graph is given that all weights are positive. Using Dijkstra's shortest path algorithm find the shortest paths from the source to all vertices in the given graph. Don't try it on graphs that contain negative edge weights because termination is not guaranteed in this case.

45. Suppose a weighted graph with positive and negative edge weights is given. Using Bellman–Ford algorithm find the shortest paths from a given source to all vertices in the given graph. In contrast to Dijkstra's algorithm, it can deal with negative edge weights.

46. The Lee algorithm is one possible solution for maze routing problems based on breadth-first search. It always gives an optimal solution, if one exists. Given a maze in the form of the binary rectangular matrix, find the shortest path's length in a maze from a given source to a given destination using the Lee algorithm.

47. Suppose a directed graph $G = (V, E)$ is given in which the weight of each edge is 1 or 2. Find the shortest path from a given source vertex s to a specified destination vertex t. Expected time complexity is $O(V + E)$.

48. Devise an algorithm to find second shortest path between two nodes in a given graph.

49. Modify Dijkstra algorithm to find k shortest paths in a given graph.

50. Modify Bellman–Ford to find k shortest paths in a given graph.

51. Shortest Path Faster Algorithm (SPFA). Given a directed weighted graph $G = (V, E)$ and a source vertex s, find the shortest path from the source vertex s to all other vertices in the given graph.

52. Given a directed weighted graph G, using Shortest Path Faster Algorithm, check if there exists a negative cycle from the given source. If found to be true, then print "Yes". Otherwise, print "No".

53. Johnson's algorithm for all-pairs shortest paths. Use Johnson's algorithm to find shortest paths between every pair of vertices in a given weighted directed graph, where weights may be negative.

54. Given an unweighted graph, a source s, and a destination t, find the shortest path from source s to destination t in the graph.

55. Betweenness centrality. In network theory, betweenness centrality is a measure of centrality in a graph based on the shortest paths. This type of centralization is used to identify important and influential vertices in the graph. For example, identifying influential people in a social network. For each pair of vertices in a connected graph, there is at least one shortest path between the vertices, so that the number of edges the path passes (for unweighted graphs) or the total weight of the edges (for weighted graphs) is minimal. The betweenness centrality for each vertex is the number of these shortest paths that pass through that vertex. Explain how we can found the betweenness vertices for a given graph.

13.2.7   Connectivity

56. Suppose a directed graph with two vertices x and y is given. Check if there is a path from vertex x to y.

57. Given a directed graph, using depth first search (DFS), give an algorithm to determine if the graph is strongly connected. If there is a path between any two pairs of vertices, a directed graph is strongly connected.

58. Given a directed graph, using Kosaraju's algorithm, give an algorithm to determine whether s given graph is strongly connected or not.

59. Eulerian path is a path in graph that visits every edge exactly once. Eulerian circuit is an Eulerian path which starts and ends on the same vertex. Using Fleury's algorithm, find Eulerian path or circuit in a given graph.

60. A directed graph is strongly connected if there is a path between all pairs of vertices. A strongly connected component (SCC) on a directed graph is a maximal strongly connected subgraph. Using Tarjan's strongly connected components algorithm, find SCC in a directed graph.

61. Using DFS, devise an algorithm to find SCC in a directed graph.

62. Using Floyd-Warshall algorithm, devise an algorithm to find SCC in a directed graph.

63. Suppose the directed graph $G = (V, E)$ is represented by an adjacency matrix in which the number 1 indicates that there is an edge between the two vertices and the value 0 indicates that there is no edge between them. Count all possible walks from two given vertices u and v with exactly k edges.

64. Write a function that returns true if a given undirected graph is tree and false otherwise.

65. Let undirected graph $G = (V, E)$ and a vertex k are given. Devise an algorithm to count the number of non-reachable nodes from the vertex k using a DFS.

66. Given a graph, a source vertex in the graph and a number k. Devise an algorithm to determine if there is a simple path (without any cycle) starting from given source and ending at any other vertex such that the distance from source to that vertex is at least k length.

13.2.8 Maximum Flow

67. Maximum flow is defined as the maximum amount of flow that the graph or network allows to flow from the source node to the sink node. Suppose a graph showing a flow network in which each edge has a capacity, as well as two vertices source s and sink t are given. Using the Ford–Fulkerson's algorithm, find the maximum possible flow from s to t with following constraints:

1. Flow on an edge doesn't exceed the given capacity of the edge.
2. Incoming flow is equal to outgoing flow for every vertex except s and t.

68. Suppose a graph showing a flow network in which each edge has a capacity, as well as two vertices source s and sink t are given. Using the Dinic's algorithm, find the maximum possible flow from s to t.

69. Suppose a graph showing a flow network in which each edge has a capacity, as well as two vertices source s and sink t are given. Using the Edmonds–Karp's algorithm, find the maximum possible flow from s to t.

70. *Find minimum* s-t cut *in a flow network*—In a flow network, an s-t cut requires that the source s and the sink t be in different subsets and contain edges that go from s to t. The cut capacity of s-t is defined by the set of capacities of the vertices in the cut set. The problem described here is actually the problem of finding the minimum s-t capacity of a given network. Expected output is all edges of the minimum cut.

13.3 Solutions

2.
(a) adjacency matrices
(b) adjacency lists
(c) adjacency lists (m + n) versus $O(n^2)$
(d) adjacency matrices
(e) adjacency matrices O(1) versus O(d)
(f) adjacency lists $\theta(m + n)$ versus $\theta(n^2)$
(g) adjacency lists

3.
Step 1. Perform depth first search (DFS) traversal on every node.
Step 2. For every visited node, increase its count by one.
Step 3. If there are some unvisited nodes, again perform DFS on them.
Step 4. Count the number of times DFS is applied to the graph.

7. A-B-D-E-F-C-H-G-I

8. V1-V2-V6-V3-V4-V5

9. A-B-D-H-E-F-C-G

10. a-b-d-g-e-c-f-h-i

11. 1-2-3-4-11-5-6-7-8-9-10

15.

1. Create a recursive function that takes index of start node of the graph and the destination index node. Set variable count to zero aiming to store the count of pathes. Keep a record of the nodes visited in the current path.
2. Increase the variable count, if the current nodes is the destination.
3. Else for all the adjacent nodes, call the recursive function with the index of adjacent node and the destination.
4. Print the Count.

17. To do this, apply BFS to the vertex u. Simultaneously applying BFS, use an array called distances[n] and initialize it to zero for all vertices. Now, suppose that during BFS, the vertex x is removed from the queue and we add all the adjacent vertices (i) to the queue, at the same time we update the distance[i] = distance[x] + 1; Finally, distance[v] indicates the minimum number of edges between u and v.

19. We simply need to do the BFS or DFS starting from each unvisited vertex, and get all the connected components.

29. To detect the cycle in the undirected graph, we use the DFS as follows. For every visited vertex v, when we find each adjacent vertex u, so that u has already been visited, and u is not the parent of vertex v. Then, a cycle is then identified.

40.
Step 1: Create a stack.
Step 2: Perform DFS in a recursive way. Recursively call topological sorting for all its adjacent vertices, then push it to the stack (when all adjacent vertices are on stack). A vertex is pushed to stack only when all of its adjacent vertices are already in the stack.
Step 3: Print the contents of the stack.

41. No, DFS can not be used to find shortest path in an unweighted graph.

48. To find the second or kth shortest path between two nodes, K_shortest_path_ routing or Yen'_algorithm can be used. Another way is to use Dijkstra's algorithm. To this end, first find the shortest path using Dijkstra's algorithm. Second, remove each edge in the shortest path and then find the shortest path again.

56. Take a vertex as source in BFS (or DFS), follow the standard BFS (or DFS). If the second vertex is found in the traversal process, then return true else return false.

65. We perform DFS from a given source. All nodes that are not reachable after applying dfs will be belong to the disconnected component.

Chapter 14
Backtracking Algorithms

Abstract The backtracking algorithm is a problem-solving algorithm that tests all possible solutions and goes back wherever the solution was not appropriate (did not match the constraints of the problem) and corrects itself and finds a new way. Unlike brute force which tests all the ways and then we find the answer between them. This algorithm is usually used in problems that either look for the first possible answer (finds an answer to the problem) or look for all possible answers to the problem (find an answer, add it to the list of answers and other paths also check for other answers.). In this method, a search tree called the State Space Tree is used to find the solutions. In the state space tree, each branch is a variable and each level represents a solution. The backtracking method cannot be applied to all problems and is only used for certain types of problems. For example, it can be used to find a possible solution to a decision problem. This method is also very effective for optimization problems. In some cases, a backtracking algorithm is used to find a set of all possible solutions to the problem. On the other hand, this method is not considered an optimal method to solve a problem and is used when the solution required to solve a problem is not limited to time. This chapter provides 23 exercises for addressing different applications of the backtracking method.

14.1 Lecture Notes

The backtracking algorithm is a problem-solving algorithm that tests all possible solutions and goes back wherever the solution was not appropriate (did not match the constraints of the problem) and corrects itself and finds a new way. Unlike Brute Force which tests all the ways and then we find the answer between them.

This algorithm is usually used in problems that either look for the first possible answer (finds an answer to the problem) or look for all possible answers to the problem (find an answer, add it to the list of answers and other paths also checks for other answers.).

In this method, a search tree called the State Space Tree is used to find the solutions. In the state space tree, each branch is a variable and each level represents a solution.

© The Author(s), under exclusive license to Springer Nature Switzerland AG 2022
H. Izadkhah, *Problems on Algorithms*,
https://doi.org/10.1007/978-3-031-17043-0_14

The backtracking method cannot be applied to all problems and is only used for certain types of problems. For example, it can be used to find a possible solution to a decision problem. This method is also very effective for optimization problems. In some cases, a backtracking algorithm is used to find a set of all possible solutions to the problem. On the other hand, this method is not considered an optimal method to solve a problem and is used when the solution required to solve a problem is not limited to time.

14.2 Exercises

14.2.1 The Knight's Tour Problem

1. *The Knight's tour problem*—Given a $n \times n$ chessboard with the Knight placed on the first square of an empty chessboard (we can start from any initial position of the knight on the chessboard.). Moving according to the rules of chess knight must visit each square exactly once. Print the order of each cell in which they are visited (Use backtracking algorithm).

2. *Rat in a maze*—A maze is given as $n \times n$ binary matrix of cells, maz[][], where source cell is maze[0][0] and destination cell is maze[N–1][N–1]. A rat starts from the source and can only move forward and down directions and eventually has to reach the destination. In maz[][], 0 means the cell is a dead end and 1 means the cell can be used in the path from source to destination.

(a) Print the order of each cell visited. (Use backtracking algorithm)
(b) Count number of ways to reach destination in a Maze.
(c) Modify the maze problem so that a rat can move in all possible directions.
(d) Modify the maze problem to allow multiple steps or jump to a rat.
(e) Using stack, give an algorithm for the rat in a maze problem.

3. Given a $n \times m$ matrix where each element can either be 0 (meaning the cell is a dead) or 1 (meaning the cell can be used in the path from source to destination). Devise an algorithm to find the shortest path between a given source cell to a destination cell.

14.2.2 N-Queen Problem

4. *N-Queen Problem*—The aim is to place the n queen on an $n \times n$ chessboard so that no two queens threaten each other.

(a) Specify promising function for this problem.

(b) Devise a backtracking algorithm to solve this problem.

5. The tree representation of solution space is called state space tree. Construct the pruned state space tree for 4-Queen Problem.

6. The time complexity of N-queen problem is $O(N!)$. True/False?

7. Implement N-queen Problem for eight queen.

8. Which of the following provides a feasible solution to the 4-queen problem?

   a. (3, 1, 4, 2)
   b. (2, 3, 1, 4)
   c. (4, 3, 2, 1)
   d. (4, 2, 3, 1)

9. There are .... feasible solutions for the 8-queen problem.

   a. 120
   b. 96
   c. 92
   d. 85

14.2.3 The Sum-of-Subsets Problem

Subset Sum—In this problem, we want to find a subset of elements from a list of positive numbers whose sum adds up to a given number *Sum*.

10. Given the following set of positive numbers:

$$\{10, 7, 5, 18, 12, 20, 15\}$$

Find out the subsets whose sum are equal to 30.

11. *Binary state space tree*—The node at depth 0 is the starting node. The algorithm goes to the left from the root to include w_1 (writing w_1 on the edge where we include it), and goes to the right to exclude w_1 (writing 0 on the edge where we exclude w_1). Similarly, the algorithm goes to the left from a node at level 1 to include w_2, and goes to the right to exclude w_2, etc. Each subset is represented by a path from root to leaf. The nodes contain the sum of the weights included so far. See Fig. 14.1 for tree weights. The tree representation of solution space is called state space tree. Draw binary state space tree for the following set:

$$w_1 = 2, \ w_2 = 4, \ w_3 = 6, \ \text{and } Sum = 6$$

Fig. 14.1 Binary state space tree

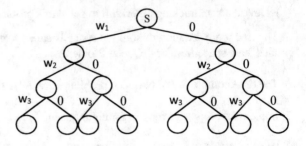

12. A node is said to be promising if a partial solution from that node is still possible to arrive at the solution. Whenever it is not possible to reach a solution from a node, that node is called a non-promising node (the node becomes infeasible). The non-promising nodes can be pruned from the state space tree. If we sort the weights in nondecreasing order before doing the search, for subset sum problem, we can define promising node as follows

```
bool promising (index i)
{
    return (weightSoFar + totalPossibleLeft ≥ Sum) &&
            (weightSoFar == Sum || weightSoFar + w[i + 1] ≤ Sum) };
```

We can also write the nonpromising node as follows:

```
bool nonpromising (index i)
{
    return (weightSoFar + w[i + 1] > Sum) ||
            (weightSoFar + totalPossibleLeft < Sum);
}
```

The totalPossibleLeft variable represents the sum of the weights remaining in the set and weightSoFar shows the sum of the weights included so far (level i). Whether the conditions of this function are met or not indicates whether a node is promising or not. Construct the pruned binary state space tree for the following:

$$w_1 = 2, \ w_2 = 4, \ w_3 = 6, \ and \ Sum = 6$$

13. Construct the pruned binary state space tree for the following:

$$w_1 = 3, \ w_2 = 4, \ w_3 = 5, \ w_4 = 6, \ and \ Sum = 13$$

14. For sum-of-subsets problem

(a) Devise a memoization technique for finding subset sum.
(b) Give a backtracking algorithm for subset sum.

15. Analyze the time complexity of subset sum problem.

16. Consider the following steps to solve subset sum problem.

(1) Start with an empty set
(2) Add the next element from the list to the set
(3) If the subset has the sum M, print that subset as the solution.
(4) If the subset is not feasible or if the algorithm has reached the end of the set, then backtrack through the subset until we find the most appropriate value.
(5) If the subset is feasible (i.e., $sum\ of\ seubset\ <\ M$), go to step 2.
(6) If the algorithm has visited all the elements without finding the appropriate subset, and if there is no possibility of backtracking, stop without a solution.

Answer the following questions:

(a) Apply this algorithm on set $\{8, 6, 7, 5, 3, 10, 9\}$ and sum = 15.
(b) Modify the above code so that all possible subsets with the sum k can be found.
(c) Apply the modified algorithm on set $\{8, 6, 7, 5, 3, 10, 9\}$ and sum = 15.

14.2.4 *M-Coloring Problem*

17. *m-Coloring Problem*—Given an undirected graph and an number m, the goal is to determine whether the graph can be colored (i.e., assigning the colors to all vertices) with a maximum of m so that no two adjacent vertices of the graph are colored with the same color. Give a backtracking algorithm for m coloring problem.

14.2.5 *Applications*

18. *Hamiltonian Cycle*—In graph theory, the Hamiltonian path is a path in an undirected graph that observes each vertex exactly once. The Hamiltonian circuit (or Hamiltonian cycle) is an cycle in the graph that observes each vertex exactly once and eventually returns to the starting vertex. Give a backtracking algorithm for finding Hamiltonian cycle.

19. *0–1 Knapsack Problem*—Given a Knapsack/Bag with W weight capacity and a list of N items with given v_i value (profit) and w_i weight. Put these items in the knapsack in order to maximise the value of all the placed items without exceeding the limit of the Knapsack. An item cannot be broken and you can select it completely (1) or not select it (0).

20. *Sudoku*—Sudoku is a puzzle in the form of a 9×9 2D array (a table) with a number of cells filled with numbers from 1 to 9. The goal is to fill in the blanks in the table so that, eventually, each of the numbers 1 through 9 appears exactly once in each row, each column, and each sub-square of size 3×3. Give a backtracking algorithm to solve this problem.

21. *Remove Invalid Parentheses*—An expression is given that contains open and closed parentheses and may contain a number of characters in the expression. We want to delete the minimum number of parentheses so that to make the input expression valid. If more than one valid output is possible, print all of these outputs. Give a backtracking algorithm to solve this problem.

22. Suppose three numbers sum A, prime B, and prime C are given. Give a backtracking algorithm to find all C prime numbers after prime B such that their sum is equal to A.

23. Let A is an array of *n* elements. Devise an algorithm to partition array A into *k* non-empty subsets such that the sum of elements in each subset is the same. Each element must be in only one subset, and all elements in this array must be exactly part of a subset.

Example:

Input: arr = [5, 4, 1, 8, 9], k = 3
Output: Yes
we can partition this array into three subsets with equal sum as [[5, 4], [1, 8], [9]].

14.3 Solutions

1.
If all squares are visited print the solution
else

(a) Add one of the next moves to solution vector and check if this move leads to a solution. (A Knight can perform maximum eight moves. We choose one of the eight moves in this step).
(b) If the move selected in the previous step does not lead to a solution, remove this move from the solution vector and try other alternative moves.
(c) If none of the alternatives work, return false (Returning false removes the previously added item in the recursion, and if false is returned with the initial recursion call, "there is no solution")

4. (a)
Two queens same row if column(i) == column(k)
Two queens same diagonal if column(i)–column(k) == i–k || column(i)—column(k)
== k–i

(b)

```
void queens (index i)
      {
      index j;
      if (promising(i))
      if (i == n)
                  cout << column[1] through column[n];
      else
                  for (j = 1; j ≤ n; j++ )
                        column[i +1] = j;
                        queens(i + 1);
                  }
   }

   bool promising (index i)
      {
      index k ;
      bool switch;
      k = 1;
      switch = true ;
      while (k < i && switch) {
         if (column[i] == column[k] || abs(column[i]—column[k] == i–k)
               switch = false;
      k++;
      }
      return switch;
```
14. (b)

```
void sum_ of_subsets (index i, int weightSoFar, int totalPossibleLeft)
      {
      if (promising(i))
                  if (weightSoFar == Sum)
                        cout << include[1] through include[i];
                  else
                        include[i+1] = "yes" ;
                        sum_ of_subsets (i +1, weightSoFar + w[i +1],
                        totalPossibleLeft–w[i + 1]);


      }
```

```
bool promising (index i)
{
return (weightSoFar + totalPossibleLeft ≥ Sum) &&
        (weightSoFar == Sum || weightSoFar + w[i + 1] ≤ Sum) };
```

17.

```
void m_coloring (index i)
{
int color;
if (promising (i))
        if (i == n)
                cout << vcolor[1] through vcolor [n];
        else
                for (color = 1; color ≤ m; color ++) {
                        vcolor[i + 1] = color ;
                        m_coloring(i + 1);
                }

}

bool promising (index i)
{
index j ;
bool switch ;
switch = true;
j = 1;
while (j < i && switch) {
                if (W[i][j] && vcolor[i] == vcolor[j])
                        switch = false;
j++;
}
return switch;
}
```
18.

```
void hamiltonian (index i)
{
index j;
if (promising (i)
        if (i == n–1)
                cout << vindex[0] through vindex[n-1];
        else
                for (j = 2; j ≤ n ; j++) {
                        vindex[i+1] = j;
                        hamiltonian (i +1);
```

```
                                      }
        }

bool promising (index i)
        {
        index j;
        bool switch;
        if(i == n–1 && !W[vindex[n-1]] [vindex[0]])
                    switch = false;
        else {
                    switch = true;
                    j = 1;
                    while (j < i && switch) {
                            if (vindex[i] == vindex[j])
                                    switch = false;
                    j++;
                    }
        }
        return switch;
        }
```

19.

```
void knapsack(index i, int profit, int weight)
        {
        if (weight ≤ W && profit > maxprofit) {
                    maxprofit = profit;
                    numbest = i;
                    bestset = include;
        }
        if (promising (i)) {
                    include[i + 1] = "yes";
                    knapsack(i + 1, profit + profit[i + 1], weight + weight[i +1]);
                    include[i +1] = "no";
                    knapsachk(i +1, profit, weight);
        }
        }

bool promising (index i)
        {
        index j , k;
        int totweight;
        float bound;
        if (weight ≥ W)
                    return false;
        {
```

```
j = i + 1;
bound = profit;
totweight = weight;
while (j ≤ n && totweight + w[j] ≤ W) {
        totweight = totweight + W[j];
        bound = bound + p[j];
        j++;
}
k = j;
if (k ≤ n)
        bound = bound + (W–totweight) * p [k]/w[k];
return bound > max profit;
}
}
```

Chapter 15
P, NP, NP-Complete, and NP-Hard Problems

Abstract The theory of computational complexity examines the difficulty of solving problems by computer (more precisely, algorithmically). In this theory, the complexity of problem definitions is classified into two sets; P which denotes "Polynomial" time and NP which indicates "Non-deterministic Polynomial" time. There are also NP-Hard and NP-Complete sets, which we use to express more complex problems. This chapter provides 87 exercises for addressing the theory of computational complexity.

15.1 Lecture Notes

The theory of computational complexity examines the difficulty of solving problems by computer (more precisely, algorithmically). In this theory, the complexity of problem definitions is classified into two sets; P which denotes "Polynomial" time and NP which indicates "Non-deterministic Polynomial" time. There are also NP-Hard and NP-Complete sets, which we use to express more complex problems. In the case of rating from easy to hard, we might label these as "easy" (P problems), "medium" (NP problems), "hard" (NP-Complete problems), and finally "hardest" (NP-Hard problems).

Let's quickly review some common Big-O notations:

- $O(1)$—constant-time
- $O(log_2(n))$—logarithmic-time
- $O(n)$—linear-time
- $O(n \times log n)$—n-log-n
- $O(n^2)$—quadratic-time
- $O(n^3)$—cubic
- $O(n^k)$—polynomial-time
- $O(k^n)$—exponential-time, e.g. $O(2^n)$
- $O(n!)$—factorial-time
- $O(n^n)$—super-exponential

H. Izadkhah, *Problems on Algorithms*,
https://doi.org/10.1007/978-3-031-17043-0_15

In terms of time complexity, computer scientists divide the problems into two classes:

- Polynomial time, problems whose time order is in one of the first seven time orderings listed above.
- Exponential time, problems whose time order is in one of the last three time orderings listed above.

Moreover, we can classify problems into two general classes:

- Tractable Problem: problems can be solved by a polynomial-time algorithm. The upper bound is polynomial.
- Intractable Problem: problems cannot be solved by a polynomial-time algorithm. The lower bound is exponential.

15.1.1 Polynomial Algorithms

The first set of problems, known P problems, are called tractable problems that can be solved in polynomial time (quickly solvable). The time complexity polynomial algorithms can formally defined as follows:

$$T(n) = O(C * n^k) \ \ where C > 0 \ \ and \ \ k > 0$$

where n is input size and C and k $(k < n)$ are constants. For example, the time complexity of the following problems is polynomial time:

- Sorting problems
- String operations
- Multiplication of integers
- Finding a minimum spanning tree in a graph
- Shortest Path Algorithms; Djikstra, Bellman-Ford, Floyd-Warshall
- Searching problems

15.1.2 NP Problems

Note that NP stands for "Nondeterministic Polynomial-time". The second set of problems are called intractable problems that are not solvable in polynomials but can be verified (quickly checkable) in polynomial time. Exponential complexity of computation is a feature of these problems. These issues are defined as follows.

$$T(n) = O(C_1 * k^{C_2 * n}) \ \ where \ \ C_1 > 0, C_2 > 0 \ \ and \ \ k > 0$$

where C_1, C_2 and k are constants and n is the input size. For example, we'll see complexities like $O(n^n)$, $O(2^n)$, $O(2^{0.000001 * n})$ in this set of problems. There are several algorithms that fit this description, for example salesman problem and graph isomorphism.

These problems have two important properties: their complexity is $O(k^n)$ for some k and their results can be verified (checked) in polynomial time. These two conditions are enough for us to say that a problem falls into the category of NP problem. It can be said that these problems should be decision problems—having a yes or no answer, although note that in practice, all functional problems can be turned into decision problems.

The output of optimization problems is an optimal solution and the output of decision problems is the answer 'yes' or 'no'. Each optimization problem can be considered corresponding to a decision problem. We can usually turn an optimization problem into a decision problem by limiting the optimal value. For example, in the problem of the shortest path of a graph without direction G, the value of k is defined as the limit, and the decision problem is expressed as follows: Is there a path between two vertices u and v whose edges are less than or equal to k?

An algorithm falls into the category of NP problems if the algorithm is not solvable in polynomial time, and the set of solutions for the decision version related to the main problem can be verified in polynomial time by a "Deterministic Turing Machine".

In principle, NP class problems do not have a polynomial time to solve, but a polynomial time to verify solutions meaning difficult to solve and easy to check a given answer. For example, consider a Sudoku game. Solving this problem does not have a polynomial time. But checking a sample solution have a polynomial time.

Reduction In computational theory and computational complexity theory, reduction is an algorithm for transferring one problem to another. Reducing from one problem to another may be used to show that the second problem is at least as difficult as the first.

For example, suppose there are two problems *P1* and *P2*, and we also know that problem *P1* is a class NP problems. If problem *P1* can be reduced in polynomial time to problem *P2*, we can say that *P1* is reducible reduced to *P2* meaning that *P1* is also a class NP problems.

Understanding this concept is important to understand the other two categories of problems (NP-Complete and NP-hard problems).

15.1.3 NP-Complete Problems

This set is very similar to the NP set. What makes this set apart from other NP problems is a useful distinction called completeness. If a problem is NP and all other NP problems are polynomial-time reducible to it, the problem is NP-complete.

NP-complete problems are the hardest problems in the NP set. A decision problem L is NP-complete if:

1. L is in NP, meaning that any given solution to the problem can be quickly verified (there is a polynomial algorithm to verify), but there is no known efficient solution.

2. Every problem in NP is reducible to L in polynomial time (i.e., a quick algorithm to solve this problem can be used to quickly solve other NP problems.)

There are several NP problems proven to be complete, for example:

− Traveling Salesman Problem
− Knapsack Problem
− Graph Coloring Problem

15.1.4   NP-Hard Problems

A problem is NP-Hard if it follows property two mentioned below, there is no need to follow property 1.

1. L is in NP, meaning that any given solution to the problem can be quickly verified (there is a polynomial algorithm to verify), but there is no known efficient solution.
2. Every problem in NP is reducible to L in polynomial time (i.e., a quick algorithm to solve this problem can be used to solve all other NP problems quickly.)

Two famous NP-hard problems are:

− Salesman Problem
− Maximum clique problem

We can conclude that:

- *P* problems are quick to solve,
- *NP* problems are quick to verify but slow to solve,
- *NP-Complete* problems are also quick to verify, slow to solve and can be reduced to any other NP-Complete problem,
- *NP-Hard* problems are slow to verify, slow to solve and can be reduced to any other NP problem.

15.2   Exercises

15.2.1   Basic

1. Trackable: Problems that can be solvable in a reasonable (i.e., polynomial) time. True/False?

2. Intrackable: Problems that when grow large in terms of input size, we can not solve them in reasonable time. True/False

3. Polynomial-time algorithms are considered to be efficient, while exponential-time algorithms are considered inefficient. True/False?

4. What constitutes reasonable time?

5. Polynomial time- on an input of size n the worst-case running time is $O(n^k)$ for some constant k. True/False?

6. Polynomial time: $O(n^2)$, $O(n^3)$, $O(1)$, $O(n \log n)$. True/False?

7. Not in polynomial time: $O(2^n)$, $O(n^n)$, $O(n!)$. True/False?

8. Turing's "Halting Problem" is not solvable by any computer, no matter how much time is given. True/False?

9. An optimization problem is one which asks, "What is the optimal solution to problem X?" True/False?

10. An optimization problem tries to find an optimal solution. True/False?

11. 0–1 Knapsack, fractional knapsack, and minimum spanning tree are optimization problems. True/False?

12. A decision problem tries to answer a yes/no question. True/False?

13. "Does a graph G have a minimum spanning tree of weight $\leq W$?" is an examples of decision problems. True/False?

14. Satisfiability (SAT) is the problem of deciding whether a given Boolean formula is satisfiable. True/False?

15. Many problems will have decision and optimization versions. True/False?

16. Write decision and optimization versions of traveling salesman problem.

17. How many of the following problems can be solved in polynomial time? Assume that each input operation is performed on a bit at time $O(1)$.

(a) Computing power of a number
(b) Calculating the largest divisor of the common of a fixed number with n
(c) Addition of two n-bit numbers

18. Consider the following functions:

F(x)

if x<1 then return 1   else return F(x-1) + G(x)

G(x)

if x<2 then return 1   else return F(x-1) + G(x/2)

Which of the following is the best representation for the growth of function F(n) in terms of n?

a. logarithmic
b. linear
c. quadratic
d. exponential

19. Which algorithm does not have exponential complexity?

a. Finding the Hamiltonian Cycle in a graph
b. Finding the common sub-string length between two strings
c. 0/1 knapsack problem
d. Fibonacci recursive algorithm

20. There are n interval $I_i = [x_i, y_i]$, where $1 \leq i \leq n$. Each interval represents the time of a class, and each class needs a room independently. We want the minimum number of rooms needed so that we can allocate rooms to all classes without time interference. When is the fastest algorithm for finding this minimum number of rooms?

a. $O(n^3)$
b. $O(n\log n)$
c. $O(n^2)$
d. There is no a polynomial solution.

21. G = (V, E) is a directed, connected and weighted graph, and a and b are two separate vertices. Suppose P1 is the problem of finding the shortest path between a and b and P2 is the problem of finding the longest simple path between a and b. Which of the following is true about P1 and P2?

a. P1 and P2 can be solved in polynomial time.
b. P1 and P2 can not be solved in polynomial time.
c. P1 can be solved in polynomial time but not P2.
d. P2 can be solved in polynomial time but not P1.

15.2.2   Classification of Problems: Class P and NP

22. P: the class of problems that have polynomial-time deterministic algorithms. True/False?

23. P: the class of problems that are solvable in $O(p(n))$, where $p(n)$ is a polynomial on n. True/False?

24. A deterministic algorithm is (essentially) one that always computes the correct answer. True/False?

25. Fractional knapsack, minimum spanning tree, and sorting are three examples of P class. True/False?

26. Write five examples of class P.

27. NP (NP stands for "Nondeterministic Polynomial-time"): the class of decision problems that are solvable in polynomial time on a nondeterministic machine (or with a nondeterministic algorithm). True/False?

28. NP—set of problems for which a solution can be verified in polynomial time. True/False?

29. NP—set of problems for which a solution can be verified quickly, but there is no efficient known solution. True/False?

30. Fractional knapsack, and minimum spanning tree are two examples of NP class. True/False?

31. Traveling Salesman, graph coloring, and satisfiability (SAT) are three examples of NP class. True/False?

32. What is difference between NP and co-NP?

33. True/False?

(a) NP is the class of decision problems for which there is a polynomial time algorithm that can verify "yes" instances.
(b) Co-NP is the class of decision problems for which there is a polynomial time algorithm that can verify "no" instances.

34. $P \subseteq co - NP$? True/False?

35. $P \subseteq co - NP \cap NP \implies co - NP \cap NP \neq \emptyset$? True/False?

36. If a problem can be solved in polynomial time, then which of the following is definitely true?

a. $A \in P$
b. $A \in NP$
c. $A \in co\text{-}NP$

   d. a and b

37. Which of the following relationships has not been proven?

   a. $P \subseteq NP$
   b. $P \subseteq co - NP$
   c. $NP \cap co - NP \neq \emptyset$
   d. $P = co - NP$

38. (True/False?) A problem is NP-complete if the problem is both

   • NP-hard, and
   • NP.

39. If a problem is NP and all other NP problems are polynomial-time reducible to it, the problem is NP-complete. True/False?

40. NP-complete problems are the hardest problems in the NP set. True/False?

41. (True/False?) A decision problem P is NP-complete if:

   1. P is in NP,
   2. every problem in NP is reducible to P in polynomial time.

42. A problem is NP-hard if all problems in NP are polynomial time reducible to it. True/False?

43. Consider the following properties for problem P:

   1. every problem in NP is reducible to P in polynomial time.
   2. The given solution for P can be verified quickly (polynomial time).

Problem P is NP-Hard if it follows property 1, doesn't need to follow property 2. True/False?

44. Let L is an NP-complete problem that L ∈ P. We have:

   a. P = NP
   b. P ≠ NP
   c. L ∈ NP-hard
   d. L ∉ NP-hard

45. A search problem is NP-complete if all other search problems reduce to it. True/False?

46. There are n persons who want to cross an old bridge in night. The walking speed of the person i is si and the length of the bridge is L. They have an old lantern that should be held when crossing the bridge. Because the bridge is old, a maximum of k people can cross it at the same time, which will be their slowest speed. Note that the lantern must always be carried and it is not possible to throw it from one side of

the bridge to the other. We want to help these people and with $n, k, L, si,$ we provide an algorithm that finds how these people travel so that all of them cross the bridge in the fastest time. Make sure that anyone who crosses the bridge, if he does not return immediately (to return the lantern), the other side of the bridge will not be suspended and will go home immediately! And normally, if the lantern is on the destination and some people are still on the origin, one of the people who went to the destination in the last series must return the lantern to the origin and be on the origin himself. What is the complexity of the fastest algorithm to find the optimal answer to this problem?

a. $O(n)$
b. $O(nlogn)$
c. $O(n^2)$
d. It is an NP-hard problem.

47. Person goes to the store and, due to his limited budget, wants to buy the most weight items. We assume that this person knows the weight and price of all the goods. Which of the following is true about this issue?

a. It can be solved in polynomial time and with greedy solutions.
b. It can be solved in polynomial time with a dynamic solution.
c. It is an NP-hard problem.
d. It is an NP-complete problem.

48. Which of the following is incorrect?

a. All P problems are solved by a nondeterministic algorithm in polynomial time.
b. All NP problems are solved by a nondeterministic algorithm in polynomial time.
c. All NP-hard problems are solved by a nondeterministic algorithm in polynomial time.
d. All NP-complete problems are solved by a nondeterministic algorithm in polynomial time.

49. Assume $P \neq NP$. Which of the following problems can be solved in polynomial time?

(A) Finding longest simple path in an undirected graph.
(B) Finding shortest simple path in an undirected graph.
(C) Finding all spanning trees in an undirected graph.

a. B
b. A and B
c. B and C
d. A, B, and C

50. Consider the following problem.

$$\sum_{1 \le i \le n} x_i w_i \le M$$

$$max \sum_{1 \le i \le n} x_i p_i$$

$$s.t \quad x_i = \begin{cases} 0 \\ 1 \end{cases}$$

Which of the following is not correct?

a. This problem is not an NP problem.
b. This problem is an NP-hard problem.
c. This problem is a P problem
d. There is a polynomial solution for this problem in terms of n and M.

51. How many of the following statements are true?

(A) $f_i(n) = O(n) \implies \sum_{i=1}^{n} f_i(n) = O(n^2)$
(B) $f(n) + o(f(n)) = \Theta(f(n))$
(C) $L \in NP$ and $L \le_P 3 - SAT \implies L$ in an NP-complete problem.

a. 0
b. 1
c. 2
d. 3

52. Two decision problems are given A and B. We know that A is an NP-complete problem, but B can be solved in $O(n^2 log^4 n)$. We also know $B \le_P A$ meaning that problem B is reducible to A in polynomial time. Which of the following is true?

a. $P = NP$, and any NP problem can be solved in $O(n^3)$.
b. $P = NP$, but some NP problems are solved in more than $O(n^3)$.
c. $P \ne NP$
d. We do not know that is $P = NP$.

53. Which of the following is true for 2-SAT problem?

(a) 2-SAT is in both P and NP.
(b) 2-SAT is an NP-complete problem.

(c) 2-SAT is an NP-hard problem.

54. For a given graph, consider the following problems:

Problem A: What is the size of the largest clique in the graph?

Problem B: What is the size of the smallest vertex cover in the graph?

Which of the following options correctly shows the relationship between these problems?

a. Problem A is reducible to B in polynomial time.
b. problem B is reducible to A in polynomial time.
c. both a and b.
d. Neither 1 nor 2

55. We know P is set of problems that can be solved in polynomial time and NP is set of problems for which a solution can be verified in polynomial time. If $P \subseteq NP$?

56. Open question: Does P = NP?

57. What does NP-hard mean?

58. A lot of times you can solve a problem by reducing it to a different problem. (True/False?)

59. I can reduce problem B to problem A if, given a solution to problem A, I can easily construct a solution to problem B. (In this case, "easily" means "in polynomial time."). (True/False?)

60. Hamiltonian Cycle (HC) is an NP-hard because every problem in NP is reducible to HC in polynomial time. (True/False?)

61. TSP is reducible to Hamiltonian Cycle. (True/False?)

62. All problems in NP is reducible to SAT. (True/False?)

63. If R reduces in polynomial time to Q, R is "no harder to solve" than Q. For example, lcm(m, n) = m × n / gcd(m, n), lcm(m, n) problem is reduced to gcd(m, n) problem. (True/False?)

64. If all problems R ∈ NP are polynomial-time reducible to Q, then Q is NP-Hard. (True/False?)

65. We say Q is NP-Complete if Q is NP-Hard and Q ∈ NP. (True/False?)

66. If R is polynomial-time reducible to Q, we denote this R $\leq_P$ Q. If R $\leq_P$ Q and R is NP-Hard, Q is also NP-Hard. (True/False?)

67. Circuit Satisfiability is an NP-complete problem. (True/False?)

68. If:

(a) Problem A is NP-complete.
(b) Problem B is in NP.
(c) A is polynomial time reducible to B.

Then: B is NP-complete. True/False?

69. How many of the following problems are NP-complete?

– Packing + covering problems: set cover,vertex cover, independent set.
– Constraint satisfaction problems: circuit sat,SAT, 3-SAT.
– Sequencing problems: Hamilton cycle, TSP.
– Partitioning problems: 3D-matching, 3-color.
– Numerical problems: subset sum, partition.

70. By adding other problems, complete the following table.

NP-complete problems	P problems
3SAT	2SAT, HORN SAT
TRAVELING SALESMAN PROBLEM	MINIMUM SPANNING TREE
LONGEST PATH	SHORTEST PATH
3D MATCHING	BIPARTITE MATCHING
KNAPSACK	UNARY KNAPSACK
INDEPENDENT SET	INDEPENDENT SET on trees
INTEGER LINEAR PROGRAMMING	LINEAR PROGRAMMING
RUDRATA PATH	EULER PATH
BALANCED CUT	MINIMUM CUT

71. Let G = (V, E) is an undirected graph. Determine which of the following problems are NP-complete and which are solvable in polynomial time.

(a) Finding a spanning tree such that its set of leaves includes the set L (L $\subseteq$ V).
(b) Finding a spanning tree such that its set of leaves is precisely the set L (L $\subseteq$ V).
(c) Finding a spanning tree such that its set of leaves is included in the set L (L $\subseteq$ V).
(d) Finding a spanning tree with k or fewer leaves.
(e) Finding a spanning tree with k or more leaves.
(f) Finding a spanning tree with exactly k leaves.

72. Proof CLIQUE is NP-complete.

73. For each of the following graph problems, prove that it is NP-complete by showing that it is a generalization of some NP-complete problems.

(a) Subgraph isomorphism
(b) Finding the longest simple path.
(c) Max SAT problem
(d) Finding vertices of a graph that there are at least b edges between them.
(e) SET COVER

74. Let G = (V, E) is an undirected graph. We define k-spanning tree problem as "a spanning tree of G in which each node has degree $\leq k$".
Show that for any $k \geq 2$:

(a) K-spanning tree is a search problem.
(b) K-spanning tree is NP-complete problem.

75. Show that if P = NP then the RSA cryptosystem can be broken in polynomial time.

15.2.3 Reduction

76. Reduce SAT to 3-SAT.

77. Reduce 3-SAT to independent-set.

78. Reduce 3-SAT to subset-sum.

79. Reduce 3-SAT to Three-dimensional matching.

80. Reduce Three-dimensional matching to ZERO-ONE EQUATIONS (ZOE).

81. Reduce ZOE to subset-sum.

82. Reduce ZOE to INTEGER LINEAR PROGRAMMING (ILP).

83. Reduce ZOE to RUDRATA CYCLE.

84. Reduce RUDRATA CYCLE to TSP.

85. Reduce independent-set to vertex cover.

86. Reduce independent-set to clique.

87. Reduce the 3CNF-satisfiability problem to the CLIQUE problem.

15.3   Solutions

1. True

2. True

3. True

5. True

12. True

14. True

16.

– optimization: find hamiltonian cycle of minimum weight
– decision: is there a hamiltonian cycle of weight $\leq k$

18. d

19. b

21. c

22. True

27. True

28. True

33. True

34. True

35. True

36. d

37. d

43. True

44. a

48. c

49. a

50. b; 0/1 knapsack

51. c

52. d

53. a

54. c

69. all of them

References

1. Cormen TH, Leiserson CE, Rivest RL, Stein C (2009) Introduction to algorithms. MIT Press, Cambridge
2. Bhasin H (2015) Algorithms: design and analysis. Oxford University Press, Oxford
3. Skiena SS (1998) The algorithm design manual, vol 2. Springer, New York
4. Dasgupta S, Papadimitriou CH, Vazirani UV (2008) Algorithms. McGraw-Hill Higher Education, New York
5. Ian Parberry (1995) Problems on algorithms. ACM SIGACT News 26(2):50–56
6. Gondran M, Minoux M, Vajda S (1984) Graphs and algorithms. Wiley, New York

© The Editor(s) (if applicable) The Author(s), under exclusive license to Springer Nature Switzerland AG 2022
H. Izadkhah, *Problems on Algorithms*,
https://doi.org/10.1007/978-3-031-17043-0